Ditch of Dreams

The Florida History and Culture Series

UNIVERSITY PRESS OF FLORIDA

Florida A&M University, Tallahassee
Florida Atlantic University, Boca Raton
Florida Gulf Coast University, Ft. Myers
Florida International University, Miami
Florida State University, Tallahassee
New College of Florida, Sarasota
University of Central Florida, Orlando
University of Florida, Gainesville
University of North Florida, Jacksonville
University of South Florida, Tampa
University of West Florida, Pensacola

DITCH OF DREAMS

*The Cross Florida Barge Canal
and the Struggle for Florida's Future*

Steven Noll and David Tegeder

Foreword by Raymond Arsenault
and Gary R. Mormino

University Press of Florida

Gainesville · Tallahassee · Tampa · Boca Raton
Pensacola · Orlando · Miami · Jacksonville · Ft. Myers · Sarasota

Copyright 2009 by Steven Noll and David Tegeder
Printed in the United States of America. This book is printed on
Glatfelter Natures Book, a paper certified under the standards of
the Forestry Stewardship Council (FSC). It is a recycled stock that
contains 30 percent post-consumer waste and is acid-free.

14 13 12 11 10 09   6 5 4 3 2 1

Library of Congress Cataloging-in-Publication Data
Noll, Steven.
Ditch of dreams : the Cross Florida Barge Canal and the struggle
for Florida's future / Steven Noll and David Tegeder ; foreword
by Raymond Arsenault and Gary R. Mormino.
p. cm. — (Florida history and culture series)
Includes bibliographical references and index.
ISBN 978-0-8130-3406-5 (alk. paper)
1. Canals—Florida—Design and construction—History.
2. Marjorie Harris Carr Cross Florida Greenway (Fla.)—
History. 3. Florida—Economic conditions. 4. Oklawaha
River (Fla.)—Environmental conditions. 5. Florida—
Environmental conditions. 6. Environmentalism—Florida—
History. 7. Carr, Marjorie Harris. 8. Florida—Politics
and government—1865-1950. 9. Florida—Politics and
government—1951- I. Tegeder, David. II. Title.
TC624.F6N65 2009
386'.4609759—dc22   2009019226

The University Press of Florida is the scholarly publishing agency
for the State University System of Florida, comprising Florida
A&M University, Florida Atlantic University, Florida Gulf Coast
University, Florida International University, Florida State Uni-
versity, New College of Florida, University of Central Florida,
University of Florida, University of North Florida, University of
South Florida, and University of West Florida.

University Press of Florida
15 Northwest 15th Street
Gainesville, FL 32611-2079
http://www.upf.com

For our families,
Beverly, Jody, and Amanda
Faith and Tai Min

Contents

Foreword

Ditch of Dreams: The Cross Florida Barge Canal and the Struggle for Florida's Future is the latest volume in a series devoted to the study of Florida history and culture. During the past half-century, the burgeoning population and increased national and international visibility of Florida have sparked a great deal of popular interest in the state's past, present, and future. As the favorite destination of countless tourists and as the new home for millions of retirees and transplants, modern Florida has become a demographic, political, and cultural bellwether. Florida has also emerged as a popular subject and setting for scholars and writers. The Florida History and Culture Series provides an attractive and accessible format for Florida-related books. From killer hurricanes to disputed elections, from tales of the Everglades to profiles of Sunbelt cities, the topics covered by the more than forty books published so far represent a broad spectrum of regional history and culture.

The University Press of Florida is committed to the creation of an eclectic but carefully crafted set of books that will provide the field of Florida studies with a new focus and that will encourage Florida researchers and writers to consider the broader implications and context of their work. The series includes standard academic monographs as well as works of synthesis, memoirs, and anthologies. And while the series features books of historical interest, authors researching Florida's environment, politics, literature, and popular or material culture are encouraged to submit their manuscripts as well. Each book offers a distinct personality and voice, but the ultimate goal of the series is to foster a sense of community and collaboration among Florida scholars.

In *Ditch of Dreams*, Steven Noll and David Tegeder examine one of the most enduring and consequential controversies in Florida history: the struggle over the feasibility—and ultimately the advisability—of constructing a canal across the Florida peninsula. The story of the Cross

Florida Barge Canal, as Noll and Tegeder readily acknowledge, is "the history of something that never happened." But it is a dramatic and important story nonetheless. The saga began in the early-nineteenth-century territorial period, when Florida boosters first envisioned a canal, and with each succeeding generation there were new plans and proposals, many developed in cooperation with the Army Corps of Engineers. Over the years the Corps of Engineers considered 28 potential routes across the peninsula, finally settling on a circuitous path from Jacksonville to Yankeetown in 1932.

With the advent of the New Deal a year later, a federally funded 200-mile long ship canal became a distinct possibility. The first phase of construction began in 1935 but was halted a few months later when an eclectic coalition of political conservatives and conservationists forced the Roosevelt administration to cancel the project. Seven years later, Congress authorized a more modest barge canal along the same route, but the actual allocation of federal construction funds did not come until 1964, when President Lyndon Johnson presided over the groundbreaking ceremony in the St. Johns river town of Palatka. During the next seven years, construction crews completed nearly one-third of the canal, despite the vehement protests of Marjorie Harris Carr and other environmental activists intent on saving the natural beauty of the Ocklawaha River and surrounding lands. Finally, in 1971, after a series of court challenges and considerable political controversy, President Richard Nixon terminated the barge canal project. While this surprising turn of events brought construction to a halt, it merely set the stage for a related and continuing controversy over a proposed restoration of the once-wild Ocklawaha, a river rendered unrecognizable by a massive dam and reservoir.

These are the basic facts of the Cross Florida Barge Canal saga. But there is so much more to learn from Noll and Tegeder's eye-opening book. Based on years of careful primary research, *Ditch of Dreams* represents a major contribution to the environmental and political history of Florida, the history of the Army Corps of Engineers, and the study of federal public works. Like such classics as John Barry's *Rising Tide: The Great Mississippi Flood of 1927 and How It Changed America* (1997) and Marc Reisner's *Cadillac Desert: The American West and Its Disappearing Water* (1986), *Ditch of Dreams* offers a compelling and persuasive analysis of the politics of public works. And it does so not only with a clear-eyed vision of the past but also with an appreciation for the environmental dilemmas posed by technological intervention and large-scale commercial development in the fragile ecosystem we call Florida.

Raymond Arsenault and Gary R. Mormino
Series Editors

Introduction

HISTORIANS OFTEN TALK about the importance of getting immersed in their work. Usually this expression is figurative in nature, but late one Sunday morning in the spring of 2002, we took the advice quite literally. On a canoe trip on the St. Johns and Ocklawaha rivers, we flipped our craft not once, but twice into the placid blue green waters of the twisting Ocklawaha. Our companions, accompanying us in their kayaks, quickly dubbed the spot "Wet Man's Bend," and added to our embarrassment by chronicling the experience in the *Orlando Sentinel*. "I tease [the historians]," wrote travel reporter Lisa Carden, "that they are taking their research of the river a little far." The journey, even with the unexpected swim, convinced us that our research of the Cross Florida Barge Canal was not just another investigation of the past, for the canal and its history are central to understanding modern-day Florida. Still, why is the history of something that never happened so important? Started twice, once in the 1930s and again in the 1960s, the canal was never completed. All that remains today are the faded dreams of supporters and the physical remains of two failed efforts to cross the peninsula. And yet, the abortive canal touched far more than the river we were canoeing, a quiet stream that would have been obliterated in the wake of a giant ditch filled with relentless barge traffic.[1]

The story of the canal—both as a ship canal in the 1930s and a more modest barge canal along the same path in the 1960s—reveals much about competing visions of progress and preservation in Florida. From the halls of Congress to the committee rooms of the Tallahassee statehouse, from the boardrooms of the Jacksonville Chamber of Commerce to the drawing boards of the Army Corps of Engineers, and even to the very banks of the Ocklawaha River, the waterway, in its many iterations and designs, has played a major part in the development of the Sunshine State since its territorial days. Its twin legacies, the Marjorie Harris Carr

Cross Florida Greenway and the George Kirkpatrick Dam, with its accompanying Rodman Reservoir, remain contentious issues well into the twenty-first century, decades after canal construction finally ceased in 1971.

The story of the canal reveals much about those competing visions of progress, economic growth, and preservation, as well as the use of political power to achieve those goals. It also chronicles an emerging environmental movement in the 1960s and how that movement halted the construction of a major federal public-works project. It illustrates the importance of citizen activism, showing how a rag-tag bunch of north Florida residents with little power and influence faced down a formidable alliance of business interests and state and federal officials—including the Army Corps of Engineers—and won. At the same time, it is the story of Marjorie Harris Carr, the "mere Micanopy housewife" who took it upon herself to mobilize these activists and oppose what she saw as the destruction of natural Florida; in the process, she became a national symbol of the power of the environmental movement. This is also the story of a similar group of citizens in the 1930s who challenged the authority of the federal government under Franklin Roosevelt's New Deal—and won as well. But the canal's importance goes beyond these environmental struggles to shed light on the changing nature of politics in twentieth-century Florida and the nation at large, as New Deal liberalism lost its power to shape the destinies of average citizens. Finally, it is a story of redemption and how an ugly scar across the state's midsection became a jewel-like linear park: the location once designated for the canal, from land purchased to fulfill the commercial and financial aspirations of business leaders, now provides a piece of natural Florida, serving as a buffer against the constant incursions of growth and development.

The Cross Florida Barge Canal's history is as rich and long as the history of Florida itself. Well before the Sunshine State became associated with citrus production, retirement communities, and Disney World, many Floridians dreamed of digging a ditch across the peninsula to connect the Atlantic Ocean to the Gulf of Mexico. To do so, they believed, would invite unprecedented growth and prosperity and place Florida at the very center of American commerce. It was an ambitious vision, one in keeping with the emergence of a sophisticated national economy that began with the establishment of New York's Erie Canal. Early on, canal boosters and their legislative allies understood that such a large-scale project would depend upon the largesse of the federal government. From the 1820s onward, any debate over canal construction therefore centered in Washington, D.C., making a cross-Florida canal a truly national issue. The federal connection was further strengthened by the institutional presence of the U.S. Army Corps of Engineers, which was intimately involved in the planning and development of the waterway. As decades passed, sheer bureaucratic inertia—reinforced by the persistent drumbeat of local boosterism—strengthened the Corps' commitment to the project regardless of economic costs, potential benefits, and environmental consequences. Thus, by the 1960s, the canal appeared to have a life of its own, even as economic and environmental issues seriously challenged its legitimacy.

Although it became a divisive issue by the 1960s, the canal's early history was marked by relative consensus. The late nineteenth and early twentieth centuries saw canal advocates pushing state and local officials to build the project. These boosters, usually local businessmen from cities such as Jacksonville and Ocala, envisioned the canal as an engine of economic growth for the state. Until the Great Depression, few doubted the necessity of the project; rather, the debate centered on its funding and location. From the 1840s onward, the Army Corps of Engineers conducted numerous surveys that resulted in the selection of twenty-eight potential canal routes ranging from southern Georgia to Lake Okeechobee. In 1932, the Corps finally chose a path, designated Route 13-B, as the most practical and cost-effective course for a waterway across the state.

Starting at the Atlantic Ocean, Route 13-B began in Jacksonville and followed the St. Johns River to Palatka, and then along the Ocklawaha River to a point near Silver Springs. It then cut westward across land below Ocala to Dunnellon and finally along the course of the Withlacoochee River until it entered the Gulf of Mexico near Yankeetown. More than just carving a route directly through the Central Florida Ridge, the nearly two-hundred-mile long, thirty-foot-deep Ship Canal included significant alterations to the St. Johns, Ocklawaha, and Withlacoochee rivers, as well as a dredged channel nearly twenty miles into the Gulf of Mexico. The design was of such a monumental scale that, when completed, it would significantly dwarf such iconic projects as the Panama and Suez canals.

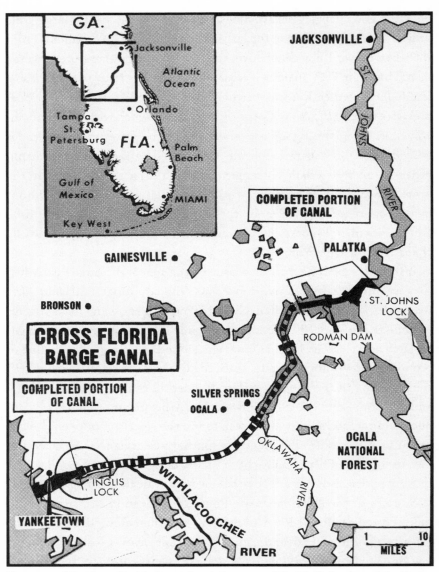

The route of the Cross Florida Barge Canal. Courtesy of Florida Defenders of the Environment (FDE), Gainesville.

The groundbreaking of the 1930s Ship Canal was brought about more by the politics of the Great Depression than by continued calls for economic growth and development. With the national unemployment rate reaching nearly 25 percent, the need for a work project trumped all other considerations, and construction began in September 1935 as part of Franklin Roosevelt's New Deal. But with that start, a host of new arguments appeared that questioned the efficacy of the project. A strange coalition of opponents, combining cranky fiscal conservatives, central Florida citrus and vegetable growers, port officials from Miami and Tampa, and conservation groups from around the state, organized to stop the canal. This primarily local movement soon found national allies like Michigan senator Arthur Vandenberg, who saw the Ship Canal as a symbol of Franklin Roosevelt's profligate spending and unprecedented federal power. Roosevelt's support for the project waned in the face of this intense local and national opposition. Within a year, the cranes and bulldozers fell silent as the federal government pulled the plug on the project. It was a rare defeat for one of FDR's major programs. By all rights, this decision should have spelled the end of the canal, but local boosters and congressional supporters would not let their dream die so quickly. By altering the design to make the project ostensibly less destructive to Florida's aquifer and tying it to broader concerns about national defense in a world on the edge of war in the late 1930s, these supporters brought the canal project back to life.

In 1942, as German U-boats ravaged merchant shipping off the Florida coast, Congress passed a bill authorizing the construction of another canal across the state following the same route as the abortive New Deal work project. This one, however, would be a barge canal, only twelve feet deep with attendant locks and dams. The canal would be an important link in the protected shipping of oil and gas from the fields of Texas to East Coast markets. Though it appeared as if canal boosters had won a great victory, Congress never appropriated the money to build the project, as more pressing war needs took higher priority. Once again, the canal was on hold. For the next twenty-two years, the struggle for canal appropriations continued. Finally, in 1964, with the money encumbered, President Lyndon Johnson traveled to Palatka to initiate the groundbreaking of the Cross Florida Barge Canal. The dream of canal supporters seemed well on its way to reality until it ran up against a new generation of environmental activists determined to preserve the Ocklawaha River. Led by Marjorie Carr, these critics forged a movement centered around scientific expertise, citizen activism, and the use of the legal system that challenged a political and institutional juggernaut intent on completing the canal. Through sheer persistence—as Carr later put it, "timing, knowledge of the facts, and staying in the fight until it is well and truly won"—they overcame enormous obstacles and in seven years convinced both the courts and President Richard Nixon to halt construction. Once again, the cranes and bulldozers stood silent—this time for good. Less than one-third complete, the Cross Florida Barge Canal soon became a monument to the lack of consensus over the meaning of development and progress. It also vividly underscored the growing power of a grassroots environmental movement determined to prevent the canal from destroying natural Florida, particularly the relatively untouched Ocklawaha River. Though construction halted in January 1971, controversy over the canal and its legacy did not stop. Since canal supporters had waited so long for the fulfillment of their dream, they were not about to yield easily to their adversaries. As John Pennekamp, the great editorialist of the *Miami Herald*, wrote in 1972, "The only thing certain about the future of the cross-state canal and the Oklawaha river is that the argument will continue unabated and none of the involved interests are giving up."[2]

This time the contentious debate swirled around the fate of existing canal structures on the Ocklawaha itself—especially the Rodman Dam, which the State of Florida officially designated as the George Kirkpatrick Dam in 1998. Though canal construction halted in 1971, the dam and its adjoining reservoir had already been completed and still remain in place today. For environmentalists, stopping the canal was only half the battle; from the beginning their efforts centered on preserving the Ocklawaha River. To this day, they will not claim complete victory until the dam is removed and the Ocklawaha is restored to its original free-flowing condition. This will provide, according to a prorestoration press release, "a rare opportunity to correct an environmental disaster" and will allow "the restored Ocklawaha [to become] a legacy we leave for future generations." On the other hand, the environmentalists' opponents no longer seek the completion of the canal. Instead, their goal is to retain the dam and Rodman Reservoir. In their eyes, the dam established a "viable and complex ecosystem that supports a wide variety of native plants and wildlife." The once-inaccessible river has now been replaced by an expansive reservoir providing significant recreational opportunities for the general public. Removal of the dam would drain the water and thus eliminate one of the premier bass-fishing spots in America. Since 1971, neither group has had the political clout to impose its will on the other and bring an end to the controversy. The debate concerning the dam's removal has become a perennial rite of spring in the Florida Legislature. As a 2007 op-ed piece in the *Daytona Beach News-Journal* noted:

> The fight over Rodman Reservoir transcends party loyalty. It crosses legislative alliances, defies political ideology and has defeated Florida governors as canny as Lawton Chiles and Jeb Bush. In a state with a short political attention span, this issue revolves around a dam that will be 40 years old next year. Over that time, the Vietnam War ended, the Soviet Union dissolved, apartheid ended in South Africa and the European Union was formed. But the issue of what to do about the Rodman Dam—excuse me, the George Kirkpatrick Dam—defies a political solution.[3]

The history of the Cross Florida Barge Canal, therefore, is not just another story about modern Florida—in many respects it is *the* story of modern Florida. As Floridians struggle with concerns about growth and preservation in a fragile and increasingly finite environment, the almost two-hundred-year legacy of this project looms as a cautionary tale over present-day public policy discussions concerning transportation and water use. At the start of the twenty-first century, newspapers and business journals have filled their pages with articles reminiscent of the issues surrounding canal construction. The widening and deepening of the Panama Canal, expected to be completed in 2014, has reopened discussions about the expansion of Florida's numerous port facilities and the state's accompanying transportation infrastructure. The prospect of giant Asian cargo ships—each capable of carrying more than five thousand mobile home–sized containers—reaching America's east coast holds the potential for tens of billions of dollars in trade. As a result, according to a recent article from *Florida Trend*, "all of the state's major ports are scrambling to cash in on an expected doubling of waterborne trade in and out of the state within the next twenty years." That "scramble" would no doubt include massive, as well as potentially wasteful and inefficient, expenditures for deepwater dredging, land acquisitions for warehousing, and further development of subsidiary rail and trucking facilities—all in an era of increasingly scarce financial and ecological resources. While such cities as Tampa, Miami, and Jacksonville jockey for position in a growing global marketplace, they are also competing over diminishing water resources. In the last half of the twentieth century, Florida's rapid population growth created overwhelming pressure on what seemed like a never-ending supply of water. Current public-policy debates increasingly focus on finding solutions to problems caused by 150 years of misallocation, reckless consumption, the continuing draining and dredging of rivers and wetlands, and the relentless pumping of the Floridan Aquifer. Proposed solutions have ranged from simple conservation measures to enormously expensive water projects like desalination plants and restoring the natural flow of the Everglades. One such proposal involves diverting flow from the Ocklawaha and St. Johns rivers to meet the demands of central and south Florida. Seen against the historical backdrop of the Cross Florida Barge Canal, Florida's current water war is but the latest skirmish in a never-ending struggle between different interests and visions. As Deirdre Connor of the *Florida Times-Union* put it in May 2008: "Here, deep in the heart of Florida and miles away from anything, the Ocklawaha River meets an abandoned dam. There's a history lesson, but the history is ebbing from memory. And the lesson couldn't be more relevant."[4]

The story of the canal also tells us much about the equally fragile nature of America's political system. Accusations of "boondoggles" and "pork-barrel spending" continue to resonate as citizens still wrestle with questions raised by large public-works projects. Why should federal taxpayers foot the bill for an endeavor that benefits a relatively small local population? Does that local constituency reflect the greater public good or simply the narrow interests of their pocketbooks? These questions are as applicable to the Florida Ship Canal of the 1930s as to the Alaskan Bridge to Nowhere of today. Moreover, the history of the Cross Florida Barge Canal raises questions as to whether politics really "works." Canal supporters and opponents used every aspect of state and national power—elected officials in the executive and legislative branches, the courts, even ostensible apolitical bureaucracies—to achieve their goals. The ambivalent results—whether it was the end of canal construction or the persistence of Rodman Dam—suggest that the American political system, especially in Florida, either functioned as a vibrant responsive form of democracy, or that it was controlled by special interests far removed from the lives of ordinary citizens.

The mixed legacy of the Cross Florida Barge Canal serves as a reminder that history provides no easy answers for the future. In 1971, environmentalists like Marjorie Carr celebrated President Richard Nixon's decision to halt canal construction to save the Ocklawaha River. For both them and their adversaries, the cross-Florida waterway was dead. Yet, only sixty miles south, another cross-Florida connection was just being completed as the canal received its death notice. I-4, the interstate highway that tied Florida's Gulf coast to the Atlantic, profoundly reshaped the state and provided the transportation infrastructure necessary for the kind of unprecedented economic growth and rapid development of which canal boosters could only dream. The highway also allowed access to another megaproject associated with 1971—Disney World. Congressman Charles Bennett, a Jacksonville Democrat who vigorously supported the canal, saw the ironies inherent in those events when he proclaimed that "Walt Disney's never-never land of Donald Duck and Mickey Mouse will damage Florida's ecology more than the now-stymied Cross-State Barge Canal." Predicting that the new theme park, especially in comparison to an enhanced waterway, would be nothing but "concrete buildings and asphalt," Bennett presciently pointed to the prevailing environmental problem facing the Sunshine State today: urban sprawl. With more than a little bitterness, he observed that Disney's creation near Orlando "will be helpful economically, but it will destroy more ecologically than the canal ever would."[5]

Bennett was no doubt on target about Disney World's impact, but he certainly failed to appreciate the significance of the canal debate and the environmental movement it spawned. In the end, the story of the canal is the story of the Ocklawaha River. The Florida balladeer Gamble Rogers recognized the river's importance in many of his songs. In 1992, after his untimely death, his wife remembered that "to him, that beautiful river was not only the heart of his beloved state, but its crowning glory as well." For Marjorie Carr and her allies, that most certainly was the case. They understood the importance of preserving one of Florida's last free-flowing natural rivers not only from the direct threat of the canal but also from the broader challenge of rapid development associated with a booming postwar economy. As early as 1965, Carr warned that "the Ocklawaha will become a symbol—whether of man's folly or man's wisdom remains to be seen." The anticanal movement may have saved the Ocklawaha but not before the Army Corps of Engineers had left its mark. The

COPYRIGHT-11-HARRIS.

VISTA ON OCKLAWAHA RIVER,
FLORIDA.

The natural beauty of the Ocklawaha enticed thousands of northern tourists in the late nineteenth century and provided the rationale to stop the Cross Florida Barge Canal in the twentieth century. Courtesy of State of Florida Photographic Archives, Tallahassee.

result is a linear park that not only bears Carr's name but also the scars of the Kirkpatrick Dam and Rodman Reservoir. The river remains a vestige of a natural Florida where, in the words of an *Orlando Sentinel* reporter, "it is important to understand this mystical place and, in the midst of its kaleidoscope of natural beauty, to contemplate our relationship with it." And yet, because the fate of the Ocklawaha still remains unresolved, many environmentalists feel they cannot rest. As Bill Partington, one of the primary leaders in the struggle against the canal, explained in 1990, "being an environmentalist in Florida means never having to say you're satisfied." While Partington was speaking directly about the river, he could have easily been referring to the larger environmental challenges facing the Sunshine State. Marjorie Carr, as usual, put it best only a year before her death. The effort to save the Ocklawaha is "not a north central Florida local issue. The Ocklawaha River is a glorious part of Florida's natural heritage. Floridians should be aware that if they can't save the Ocklawaha they have little chance of saving any of the remaining lovely wild places in Florida."[6]

Surveys and Steamboats

THE DREAM WAS THERE from the beginning. Starting with the Spanish in the sixteenth century, Europeans imagined a watery shortcut across the Florida peninsula, connecting the Gulf of Mexico and the Atlantic Ocean. Cutting hundreds of miles off the voyage through the treacherous Florida Straits, the passage would not only save time but also offer a sheltered waterway, safe from threats of violent storms, dangerous reefs and shoals, and the predations of pirates and unscrupulous treasure seekers. For Spanish and then English colonial administrators, the possibility of such a canal remained illusory at best. Though officials of both empires explored the peninsula in anticipation of a natural passage that would provide a shorter and safer route than the one around it, little came of their efforts. Like the elusive Northwest Passage, a watercourse across Florida seemed always around a river's bend. Though a natural path could never be found, the narrow width of the peninsula still held promise for a manmade waterway. Apocryphal stories still abound of Spanish desires to build a canal, many written by twentieth-century boosters of a new Cross Florida Barge Canal who tied their vision—in an effort to legitimize it—to that of earlier Spanish explorers. Little documentation from Spanish archives bears out these tales, however. With the advent of British control over Florida in 1763 came the first concrete proposals for a cross-peninsula waterway. Only two years after acquiring the colony, English governor James Grant suggested building an "east-west river-canal" across the province, to be completed in less than a year by slave labor. Nothing came of Grant's suggestion, and Florida remained relatively undeveloped. The 1783 Treaty of Paris ending the American Revolution stipulated Florida's return to Spain. This second era of Spanish control marked a time of defensive posturing designed to fend off the predatory advances of the United States, as administrators made little effort to develop the colony's economic potential. For both Spain and

England, therefore, the colony seemed an unimportant afterthought to their broader colonial enterprises. In addition, the cost of such a massive undertaking as building a cross-peninsula canal appeared prohibitively expensive, despite Grant's rosy assessment, and the technology necessary to build it remained decades from development. With the American acquisition of Florida on February 22, 1821, when Spain officially ceded the province under the terms of the 1819 Adams-Onis Treaty, a renewed vision of canal development took root.[1]

Almost immediately, Floridians clamored for a cross-territorial canal. Advocated in the midst of a "transportation revolution" that included the development of canals, steamboats, turnpikes, and railroads, their demands found a receptive audience already obsessed with canal construction. In many respects, the fever began in 1808 with U.S. treasury secretary Albert Gallatin's report on public roads and canals acting as a stimulus for a new nation's territorial and economic expansion. For Gallatin, America's inland waterways became critically important, especially in binding a nation that had more than doubled in size with the recent Louisiana Purchase. The federal government, argued Gallatin, had to step into the breach by coordinating engineering surveys that determined routes most beneficial to society at large and then launching large construction projects that would in turn stimulate further economic development for a rapidly growing nation.

The War of 1812 delayed the implementation of Gallatin's program. At the same time, it also strengthened the argument for federal works projects. Exposing problems in logistics and internal communications, the war established a need for greater development and coordination of the nation's inland transportation systems. National defense simply demanded internal improvements, which conveniently coincided with the larger commercial objectives that had so captured the imagination of the new republic. In 1817, just two years after the war's end, a congressional committee reported that transportation in general, and canals in particular, seemed "an essential ingredient to the general economy of the nation." By supporting a series of canals and roads, Washington would provide "due value to every production of land and labor." Looking southward, the committee recommended particularly the development of "a canal communication, if practicable, from the Altamaha, and its waters, to Mobile, and from thence to the Mississippi." This connection between the Atlantic and the Gulf would become the basis for a similar Florida canal, once that Spanish province became part of the United States. Despite the vigorous support for canal construction, Congress failed to appropriate, and President Monroe did not support, expenditures for these federal projects. Regardless of the setback, transportation enthusiasts held out hope that federal assistance would soon be forthcoming.[2]

The chance for a cross-Florida canal gained greater credence with the construction of a massive artificial river more than a thousand miles away. The completion of the Erie Canal in 1825 vividly demonstrated that canal transportation was both feasible and profitable. Seemingly overnight, canals become symbols of economic transformation as an impressive wave of manufacturing and commerce followed in their paths. Despite the promise of wealth canals seemed to offer, money remained a key obstacle for canal construction as Florida was a poor and sparsely settled territory. A commitment to small government and local control did not prevent Floridians from seeking federal help—especially since Florida was a territory and not a state at the time—in developing a canal across the peninsula. The rationale for such a waterway was one that canal supporters would rely on for the next 150 years: such a project would amplify Florida's economic development, improve commerce and transportation throughout the nation, and enhance national defense. Not simply designed to benefit a few Floridians, such a waterway would, in the words of an 1826 request to Congress, be a "work partaking in its character more of national than local interest," designed to "strengthen the bonds of the Union."[3]

Throughout the antebellum era, the fight for a canal took place in the broader context of a political struggle over the federal government's role in the development of "internal improvements," a nineteenth-century term for transportation infrastructure. Many, like John Quincy Adams, saw the national government as the key to providing funding for roads, canals, and later railroads, all of which would enhance the nation's economic development. Others, particularly Andrew Jackson, viewed support for such projects, especially those perceived as benefitting only a small group of individuals and not the nation at large, as problematic and even unconstitutional. Therefore canal boosters quickly learned to lobby Congress for sympathetic allies and couched their efforts in terms of benefits to the nation at large. These requests would change little over the course of the canal's history. Starting in the 1820s, Florida's Legislative Council routinely issued its requests to Congress in the form of memorials—formalized petitions asking that body to act on the concerns raised—and in terms of substance and tone they sounded remarkably similar to those produced by the Florida Canal Authority in the 1960s.

In December 1824, the Legislative Council endorsed such a memorial and sent it to Washington. The petition asked Congress to "consider as another object of primary importance, the opening of a canal across the Peninsula of East Florida from the river Suwaney to St. Johns or between such other points as on examination may prove to be more eligible. . . . It is believed that in no part of the United States could an object of so much public utility be promoted with less difficulty or expenditure." Two months later, Richard Keith Call, Florida's delegate to Congress and a future territorial governor, relied on the same argument to ask the House Committee on Roads and Canals for aid in canal construction. Taking Call's request under consideration, the committee reported that an Atlantic-Gulf canal was both necessary and feasible. Indeed, the cost and length of the connection would be significantly reduced with the recent "acquisition of Florida," since "the distance to be canalled would not exceed 120 miles." The committee concluded that "the cost of this undertaking, from the information received, would be about six millions of dollars." Though it issued a favorable report, it did not fund construction. Undeterred by the setback, Floridians continued their campaign to garner federal support.[4]

While the national government seemed reluctant to sponsor major internal improvement projects, it appeared more than willing to at least survey possible routes for potential roads and canals. This became especially important in the newly acquired territory of Florida, where much of the land was wild and uncharted. In 1824, Congress authorized the army to establish a board to initiate surveys of potential canals of "national importance . . . which could facilitat[e] commerce and intercourse among the states." Among the possibilities stood a canal "from the river St. Johns across Florida Neck, to the Gulf of Mexico. . . . Should it prove practicable, its beneficial effects would be great, comprehensible, and durable, the whole of the Atlantic and western states would deeply partake in its advantages. Besides the facility of intercourse which it would afford between these states, our trade with Mexico, Guatimala [sic], and the central parts of the continent, would not only be greatly facilitated, but rendered much more secure." Capitalizing on the government's interest, Florida's territorial officials took it upon themselves to keep the canal in the public eye. Legislative memorials to Congress in 1824 and 1825 reiterated concern for such a waterway. Congressional representative Call wrote to that body in February 1825, extolling the virtues of a "plan to connect the Mississippi with the Atlantic, by an internal communication extending along the northern margin of the Gulf of Mexico." This would best be accomplished, he concluded, "by uniting the waters of the St. John's with those of the river Suwannee, and will require a canal of not more than twenty miles in length." Later that year, the territorial Legislative Council passed a bill authorizing a three-man commission to examine "the expediency of opening a canal from the waters of the Gulf of Mexico to those of the Atlantic" and to determine to "extent of assistance that may be derived from the General Government in aid of the undertaking." It seemed the idea of a canal was taking shape.[5]

On March 3, 1826, Congress took the first step toward canal construction as it appropriated twenty thousand dollars for a preliminary survey of a canal route by the Army Corps of Engineers. Upon receiving the authorization, Chief Army Engineer General Alexander Macomb drew from a pool of exceptionally capable topographical engineers, many of them French expatriates who had studied at the prestigious Ecole Polytechnique and had served under Napoleon. Comprising an elite cadre of highly trained specialized officers, these engineers built dozens of coastal fortifications and mapped much of America's newly acquired territory in the antebellum period. Macomb chose Major Paul Perrault, one of the French émigrés and an officer in the American army since 1817, to lead the expedition. He then directed Perrault's team to evaluate two potential routes across the peninsula. The northerly passage went from the mouth of the St. Marys River on the Georgia-Florida border, through the Okefenokee Swamp to the St. Marks River, where it emptied into the Gulf of Mexico at Apalachicola Bay. The other route began at the St. Johns River and traversed the territory, entering the Gulf at Waccasassa Bay. In addition to choosing an appropriate passage, Macomb also charged the engineers to determine the feasibility of an oceangoing ship canal or whether a more modest barge and steamboat project would prove more viable and cost-effective. The army surveyors arrived in July 1826 to slog through Florida's swamps and marshes at the height of summer. Divided into two teams, the engineers encountered a variety of setbacks ranging from conflict with the local Seminole Indians to bureaucratic rivalries that impeded their progress. Results came so slowly that Perrault's superior, General Simon Bernard, another French refugee, visited Florida himself in April 1827. After a whirlwind tour of two months in the territory, Bernard returned to Washington and unofficially reported his findings directly to President John Quincy Adams. His experience in Florida led to a conclusion dramatically different from the positive pronouncements emanating from Washington. Bernard asserted in no uncertain terms that "a ship canal across the isthmus of Florida was impracticable and that the most that could be effected was a canal six feet deep for steamboats."[6]

Unaware of Bernard's negative appraisal, Floridians became increasingly impatient with the Corps' seeming lack of progress. An article in the September–December 1827 issue of the important *American Quarterly Review* whetted their appetites: "It was our intention to speak at length . . . concerning the canal . . . , and to show how visionary are the schemes. . . . This is now rendered unnecessary, as General Bernard is engaged in making an exploration of the country; his reports on this subject will doubtless be definitive." Unable to wait any longer, congressional delegate Joseph White wrote Secretary of War Peter Porter in July 1828, expressing outrage that "it is now two-and-a-half years since Congress pass[ed] a special act . . . in the two years and a half we have not even had the report of the first reconnaissance. . . . I feel exceedingly anxious to have that report presented. The reports that have been unofficially made [rumors that had spread about Bernard's meeting with Adams] has [sic] caused the subject to gradually fade from the public mind." Things did not improve for White and his fellow canal supporters when the Corps submitted its definitive report to Congress in February 1829. Comprised of summaries from Perrault and his various subordinates, as well as conclusions drawn by Bernard himself, the document affirmed Bernard's earlier assessment that a deepwater ship canal seemed unrealistic. Claiming "the ridge of the peninsula is at a mean elevation of 150 feet above the ocean and possesses no sources of water from which so large a canal could be supplied," the report asserted a lock canal for smaller-draft boats could provide a viable alternative. In spite of its rather negative assessment, the report did express a preference for the St. Johns route over the more northerly St. Marys–St. Marks route. The account also suggested that a canal, though impractical for oceangoing vessels, would provide an important link to a larger waterways system, but it provided no cost estimates for the ultimate completion of the project.[7]

This left the door open for further study, which would begin in 1830 as Congress appropriated another $10,400 to both complete the survey and provide a detailed accounting of the costs to build a canal. In its mandate, Congress authorized another team of army engineers to survey territorial Florida. Under the command of Major William Pousin, the three-man team left Washington for Florida in the fall of 1830. Pousin led the engineers until the summer of 1831, when he resigned from the army to return to his native France. Lieutenant John Pickell assumed command of the project and submitted the final report to Congress in March 1832. While not a veteran of the previous survey, Pickell seemed well versed in the duties and responsibilities of topographical engineers. His report provided a voluminous and detailed technical examination of the groundwater and soil of the land west of the St. Johns, the area specifically chosen for the canal in the previous survey. Pickell concluded that a lock canal was feasible along the route, as "waters may be easily commanded for the supply of any portion of the prism of the canal upon the summit." His report, however, said nothing regarding the cost and practicality of such an undertaking. Reflecting its inconclusiveness, the final report languished in Washington as Andrew Jackson forwarded it without comment to the House Committee on Roads and Canals on December 9, 1833.[8]

While the federal government's initiative toward canal construction seemed rather tepid, some Floridians began to consider other forms of transportation as the key to economic development. In February 1834, the territory's Legislative Council sent a memorial to Congress addressing these issues. Recognizing the "important project of effecting a communication between the Atlantic Ocean and the Gulf of Mexico across the Florida Peninsula," the council decried the fact that "the excavation of the Ship Channel so long, and so anxiously contemplated, it is believed will not probably be undertaken by the United States." Instead of expressing outrage, however, the council simply requested an alternative.

They sought another "project for creating such a communication, which they deem feasable, to wit: by the Construction of a Rail Road across the upper neck of the Pinsula [sic] from the town of Jacksonville on the St. Johns River, to the Gulf of Mexico at the disemboguement of the river St. Marks." However, a major economic downturn in the late 1830s, known as the Panic of 1837, threw the American economy into a tailspin for more than five years. With economic contraction, little money was available for large public-works projects of any kind. When the economy rebounded in the early 1840s, the canal craze had died and Americans seemed more willing to push for railroad construction. Worse, the Second Seminole War broke out in the 1830s as white Floridians sought the removal of Native Americans and their black allies to make way for increased white settlement. Far from surveying the topographical features of the territory in anticipation of a cross-peninsula canal, the United States Army became engaged in a protracted, costly, and bloody guerilla struggle against a determined and skillful enemy. Survival, rather than economic growth, seemed the order of the day. Canal construction would have to wait.[9]

As Florida achieved statehood in 1845, the clamor for a cross-state connection continued unabated. While some called for a waterway, others pushed for a railway to draw east and west closer together. Surveys continued through early statehood as army engineers pondered potential routes for both methods of transportation. As if to hedge their bets, state officials chartered four companies to also develop rail lines across the state. In August 1852, Congress appropriated twenty thousand dollars for yet another study of a cross-Florida canal. Led by Army Lieutenant Martin Luther Smith—a man so committed to the project that he suggested "a canal across Florida is too obvious to need more than a passing remark"—a group of engineers spent nearly three years determining the efficacy of a route from the St. Johns across the state to Tampa Bay. In a voluminous and highly technical summary, Smith found that "there are at least four lines across the peninsula which should be surveyed before any one can be decided upon." Calling for further surveys, he suggested encumbering an additional twenty thousand dollars to continue the work. He seemed especially excited about "a rather favorable line, starting from the harbor of Crystal River, and striking the St. Johns at the mouth of the Ocklawaha. How favorable the ridge dividing the Withlacoochee and Ocklawaha rivers may be for crossing, is not known; but before deciding upon any line as the best for crossing the peninsula, this one should be examined: want of funds prevented this from being done by myself." In many respects, the Smith report mirrored its predecessors by simply continuing the drumbeat of support for the project. What set the survey apart, however, was the first official mention of the eventual route for both the 1930s ship canal and the 1960s barge canal.[10]

Smith's report reaffirmed the dream of canal supporters. Particularly excited were the editors of *DeBow's Review*, the leading journal of southern commerce and industry. Hailing Florida as the "Key to the Gulf," the journal declared that "the ship canal across the St. Johns River . . . is looked upon as one of the most important . . . proposed improvements brought to our notice." The editors then set a rationale for the project so complete that it laid the foundation for many of the procanal claims of a century later. With an artificial waterway, shipping "from the great northern lakes" can "commence on a most remarkable voyage, descending the Mississippi [until it] reaches the western terminus of the Florida Canal, crosses the peninsula and again takes the inland navigation along the coast of Georgia" and then continue northward to New York. The canal would also provide "one of the greatest safeguards to our commerce in times of war." Finally, "it would allow for a savings of almost $2 million per year on insurance costs for ships traveling through the dangerous straits of Florida." The editors concluded by rhetorically asking, "why should this important project be overlooked?"[11]

The answer seemed obvious to this most southern of American publications. Caught up in the sectional politics of the 1850s, the editors of *DeBow's* asserted that the project's delay stemmed from "the natural result of retrogression created by that pestilential black hole of Northern fanaticism." For years they had noted how northerners consistently opposed projects designed to improve southern manufacturing and commerce. The canal "would be a southern one, and although in all respects a national object, yet the past legislation of Congress has shown that the South never has and probably never will obtain the aid of the government in any extensive works of public improvement. . . . Of the annual appropriations for rivers and harbors, but a pittance comes to the South." Not willing to easily concede defeat, they proposed an alternative solution. As early as 1852, *DeBow's* suggested, "it seems to us that the true interests of commerce require the immediate construction of a rail-road across the peninsula." Noting that the price of a ship canal could run as high as $100 million, the editors argued that a "first-class rail-road may be constructed to connect the Atlantic and Gulf at a cost of less than $3 millions of dollars fully furnished and equipped."[12]

In the end, the railroad provided the first real answer to Florida's economic demands. Incorporated in 1853 with an authorized initial capital investment of $1 million, the Florida Railway was organized with David Levy Yulee as its president. Yulee, antebellum Florida's most powerful politician and a tireless advocate for improved transportation, served as Florida's territorial delegate to Congress and then as the state's first senator. In 1854, the state's Internal Improvement Board recommended support for the railroad, designed to connect Fernandina on the Atlantic to Cedar Keys on the Gulf. The federal government also encouraged the development of the railway as it authorized a survey of the proposed cross-state rail link. The final report submitted to Congress in March 1855 that embraced this route as the most feasible was written by Lieutenant Martin L. Smith, the same topographical engineer who was so determined to see a waterway across Florida. For Smith, the key to Florida's economic future was in the connection between ocean and gulf and not necessarily in the means of achieving it. A New Yorker by birth, Smith represented a new breed of entrepreneur who optimistically tied his fortune to the Sunshine State. Listed as the Florida Railway's chief engineer, he soon rejected his northern roots and joined the Confederate engineering corps when Florida seceded from the Union in 1861.

David Yulee, a senator again by 1855, proved instrumental in securing the funding and political support for the railroad. Obtaining significant federal and state land grants and financial contributions, Yulee's corporation slowly constructed the track with contracted slave labor. By April 14, 1860, the *Tallahassee Floridian* reported that "only eighteen miles of track are to be laid and the long desired rail connection between the waters of the Atlantic and Gulf of Mexico will be established." Eleven months later, as the first locomotive, the "Abner McGehee," pulled into Cedar Keys, it appeared the fulfillment of a centuries-old dream had finally occurred. However, the shelling of Fort Sumter in April 1861 and the subsequent four years of the violent and destructive Civil War put that dream on hold again.[13]

While entrepreneurs like Yulee saw Florida's peninsula as a wasteland destined for conquest either by water or rail, others considered the state as a vast primordial wilderness, a place where nature ruled supreme. This was certainly true of the great inland rivers—the St. Johns, the Suwannee, the Withlacoochee, and especially the Ocklawaha. With no good harbors at their outlets, these streams never became major water routes for interstate trade and instead remained fairly remote until well into the late nineteenth century. In particular, the isolated area of west central Florida drained by the Withlacoochee stayed almost completely undeveloped until the late 1880s, with just a few small groups of self-sufficient families inhabiting the region. Only the St. Johns provided a significant amount of waterborne commerce, and that was only after significant dredging of both the river itself and the treacherous sandbars at its mouth allowed Jacksonville to become a major seaport. Thus the relatively untouched Ocklawaha, its name a corruption of the Seminole term for "muddy" or "crooked" river, became the idealized vision of wilderness Florida that would carry into the next century and mark its survival as a wild river (whatever that really meant) crucial to a nascent environmental movement.

In 1884, the travel writer George Barbour described the Ocklawaha in his guidebook to the state:

It is an extensive region of dense jungle, lying low and flat, undrainable, and impossible to improve for human use; and will always remain wild and unmolested, a paradise for all the strange reptiles, insects, birds, and fish that seek its innermost recesses. To the pleasure-seeking tourist and the sportsman it affords an inexhaustible field of interest, but to the invalid, health seeker, or practical settler it offers no attractions. As the steamer follows the vaguely defined course of the channel, there are frequent landings, localities where points of the mainland extend like a peninsula into this watery jungle, affording access and outlets to the more profitable and healthy regions lying inland all along the route.

The Ocklawaha, whose name is variously spelled in promotional literature and magazine articles of the late nineteenth century, enticed northern visitors and entrepreneurs alike in the years following the Civil War. Its exotic subtropical landscape, heavily forested and swampy banks, bizarre and atavistic wildlife, and otherworldly springs made it seem more like the jungles of South America or Africa than the United States. That a trip up the river by steamboat led to the remarkable fishbowl of Silver Springs only added to the appeal. From the end of the Civil War through the second decade of the twentieth century, the Ocklawaha River cast a mystical spell on its visitors.[14]

Yet, simultaneous to its discovery as a natural tourist attraction, the Ocklawaha proved just as appealing as a resource ripe for exploitation. The very environment that enchanted visitors also held the prospect of financial gain through the harvesting of cypress lumber. By the second decade of the twentieth century, the Ocklawaha's heyday as a tourist attraction had ended: the railroad had displaced the steamboat as the major method for accessing Florida, and the beach and grand resort hotel had replaced the exotic river as a prime vacation destination. Still, large-scale economic development continued on the Ocklawaha. Lumbering persisted until the 1940s, as cypress logs from the swamps and bottomlands of the river fed the mills in nearby Palatka. When the Cross Florida Barge Canal controversy raged in the 1960s, environmental activists such as Marjorie Carr relied on the literature of nineteenth-century sojourners to build a case for protecting this pristine "forest land, set with springs and lakes, and through which flows a beautiful wild river— the Oklawaha." Nineteenth-century tourists and Floridians, however, were not the proto-environmentalists envisioned by the Florida Defenders of the Environment (FDE) and other Barge Canal opponents. Like all humankind, their relationship to and incursion into this land shaped and reshaped the Ocklawaha Basin.[15]

The Ocklawaha River is one of Florida's major streams. Starting in a chain of lakes in Lake County, the river flows north and then east approximately seventy-five miles through east central Florida, eventually emptying into the St. Johns River across from the steamboat-landing town of Welaka. Draining a watershed area of about 2,800 square miles, the Ocklawaha is a narrow and relatively shallow river, with turns so "contorted and looped" that distance "doubles and redoubles itself for any navigator who takes a boat on it." Approximately halfway down its course, the Ocklawaha is joined by its major tributary, the Silver River. From its source at Silver Springs, the Silver empties over 500 million gallons a day into the Ocklawaha; and "[u]nlike the Ocklawaha, . . . [the]

waters of the Silver Run are perpetually clear." One of the oldest rivers in Florida, the Ocklawaha considerably predates the St. Johns, into which it flows. At one time, the river marked the coastline of prehistoric Florida, originating as the "runoff of the depression that remained behind after the recession of the sea level." Although Native Americans had used the river's resources for thousands of years, its lack of lofty banks, its dense swamplands, and its inclination to flood, as well as continued intrusions by Seminole Indians, made the river valley rather inaccessible to white settlement until after the Civil War.[16]

For much of the nineteenth century, Florida—as both territory and state—was "frontier" country. In addition to a cultural boundary, the frontier marked an ecological one. More than simply dividing civilization from wilderness, or settled areas from virgin land, the borderland separated European and Native American notions of land use. Certainly, the Ocklawaha Valley could be categorized as a primitive frontier region. And as Seminole control of the area ended in the mid-1850s with the destruction of the central Florida bands, white conceptions of the natural world took precedence, especially regarding the jewel of the valley's ecosystem—Silver Springs.

In 1855, Lady Amelia Murray, a minor member of British royalty, traveled by stage from Palatka to the springs. "If I had known that we should not arrive there till after midnight, with one man driving four horses through a pine barren which harbours wolves, bears, and panthers," she complained, "my courage would have failed me." That same year, Boston Brahmin and avocational historian George Bancroft also visited Silver Springs. Bancroft seemed less concerned about the vileness of the place and more fascinated with the beauty of the springs themselves. "The whole pool and every fountain and the large river that flows aft are thus transparently clear, the most perfectly pellucid that you can imagine," he wrote to his wife. "The river is from the fish gushing up, at the fountain head so broad and deep that steamboats may come up to a landing on the bank at the head of the fountain." Murray's and Bancroft's experiences at Silver Springs exemplified divergent Euro-American reactions to the natural world—particularly, fear and wonder. It would take Hubbard Hart, a transplanted Yankee from Vermont, to add another dimension to that response. Hart's conceptualization of the Ocklawaha as a commodity worthy of exploitation, albeit as a tourist attraction, would have

Hubbard Hart, late nineteenth-century Ocklawaha River steamboat entrepreneur. Courtesy of Special Collections, Putnam County Library, Palatka.

broad implications on the region's growth and ecology for decades to come. His vision would help shape the arguments both for and against the future development of the river.[17]

In 1855, the twenty-eight-year-old Hart moved to Florida to establish a stage mail line between Palatka and Tampa. The route passed through the tiny hamlet that had grown up at the Silver Springs boil, and Hart immediately recognized the potential for tourism at the springs. Since stage travel was, at best, difficult and time-consuming, Hart thought river access to the springs would provide the best method to transport tourists. By 1860, he had a acquired a steamboat, the *James Burt*, and soon booked tourists on two-day trips from Palatka on the St. Johns River to Silver Springs. His boats ran up the St. Johns to Welaka, then up the Ocklawaha to the Silver, culminating at the spring head itself. With business increasing, Hart soon purchased another boat, which he appropriately named *Silver Spring*, to accommodate the increasing trade. Yet, he was not simply a transporter of tourists. As an enterprising young entrepreneur, he also realized the broader economic potential of the Ocklawaha Valley. By 1861, Hart had developed a series of orange groves along the river, established a lumber trade on the Ocklawaha, and contracted with state authorities to clear the river of snags and other navigational hazards. He was in the process of adding a third boat to his fleet when the Civil War intervened.

The Civil War in central Florida was generally a riverine conflict, with Union gunboats sailing up and down the St. Johns, shelling Confederate gun batteries and attempting to prevent Confederate ships from running the blockade to supply southern soldiers and export cotton. While helping the Confederate war effort, blockade runners also took opportunities to quickly enrich themselves. As one of these wartime entrepreneurs, Hart used the Ocklawaha as a shipping point for Confederate supplies. His boats could not be pursued up the Ocklawaha by Union ships because its shallow waters and twisting narrows prevented deeper-draft vessels from navigating the river. Hart managed not only to be a cog in the Confederate supply line and a transporter of goods for export, but also to make money doing it. In December 1864, he contracted his steamboat *Silver Spring* to the Confederate government at a rate of two hundred dollars per day to carry supplies down the Ocklawaha to the Fort Brooke landing, from whence they could be transported by wagon to the train depot in Waldo on the Fernandina–Cedar Keys railroad. By war's end, the enterprise had netted Hart more than eleven thousand dollars. Since few boats traveled the river during the war years, however, its passage became more and more clogged with snags and obstructions. Recognizing the river's importance as a supply line, the Confederate government awarded Hart another $4,500 contract in February 1865 to "remove the obstructions cut into the Ocklawaha River from Fort Brook [sic] to the St. Johns River."[18]

Two months later the war was over, and Hart, after being cleared of charges of wartime smuggling (of which he was obviously guilty), quickly shifted allegiances to the federal side. By 1867, he was again clearing river obstructions, this time using black freedmen as his labor force, in anticipation of restarting his river tourist trade. Contracting with Florida's Reconstruction government, Hart again had the exclusive authority to "remove the obstructions to the navigation of the Ocklawaha River [in return for] donations of state owned land to enable him to do so. . . . Hart shall receive the amount thereof in lands at the present prices, provided that said expenditures do not exceed $20,000."[19]

Hart also tried to continue the lumber operations he had begun before the war. He applied for the state's "permission to cut cypress upon the Ocklawaha River and swamp" for ten cents a tree. The state denied Hart's request, but he continued to exploit the valley's timber resources even as he returned to the tourist trade. Perhaps to honor his Civil War service, the entrepreneur took to calling himself "Colonel Hart," and the name stuck. With northerners looking to central Florida and his tourist business beginning to return to profitability, Colonel Hart seemed poised to inaugurate the golden era of steamboats on the Ocklawaha, a time when boosterism turned the river itself into "the sweetest water-lane in the world."[20]

Hart's dream of using the river to transport tourists to Silver Springs depended upon reliable water transportation. Keeping the river clear of obstructions helped make that possible, but Hart also had to develop a new type of steamboat, one designed specifically to navigate the Ocklawaha's "very narrow and wonderfully crooked waters." By the early 1870s, he placed into service the first of these boats, each an "aquatic curiosity." Built with a specialized recessed stern paddlewheel to protect the mechanism from the river's snags and tight turns, these steamboats were much smaller than those that plied most American rivers, even the St. Johns. Their unique appearance, resembling "nothing in the world so much as a Pensacola Gopher with a preposterously exaggerated back," did not necessarily provide tourists with a feeling of safety and security, much less the luxury to which Gilded Age travelers were accustomed. In 1873, the New York newspaper editor and publisher William Cullen Bryant characterized Hart's steamer as "a little thing of its kind, rudely constructed, with slight attention to comfort and convenience." That same year, the Mandarin resident Harriet Beecher Stowe refused to journey down the river. "The aspect of this same boat on a hot night was not inspiring," she wrote in *Palmetto Leaves*, her book on life as a northern homesteader (or carpetbagger) in Florida. "We looked at this thing as it lay like a gigantic coffin in the twilight, and thought even the Silver Springs would not pay for being immured there, and turned away."[21]

In spite of this rather stinging appraisal of the Hart Line's travel arrangements, Stowe overcame her fears and later in 1873 took a Hart boat up the river through a landscape she deemed a "fairy land." She published her favorable impressions of the journey in the *Christian Union*,

Hart's *Osceola* steaming down the Ocklawaha, ca. 1880. Often alligators were added to steamboat pictures to emphasize the wildness of the scenery. Courtesy of State of Florida Photographic Archives, Tallahassee.

a small religious journal. Echoing the misgivings expressed in *Palmetto Leaves*, Stowe announced that she "shuddered at the idea of going on such a bush-wacking tour through the native swamps of the alligator in such a suspicious looking craft as that." She quickly became enamored of the river cruise, however, proclaiming that "it was a spectacle, weird, wondrous, magical—to be remembered as one of the things of a lifetime." She concluded rapturously that "we seemed [to be] floating through an immense cathedral, whose white marble columns met in vast arches overhead and were reflected in the glassy depths below."[22]

To those traveling up the Ocklawaha, nights on the river rivaled Silver Springs itself for a transcendent natural experience. "[W]e entered what appears to be an endless colonnade of beautifully-proportioned shafts," reported an anonymous author of an 1870 article, "suggesting the highest possible effects of Gothic architecture. . . . So absorbing were these wonderful effects of a brilliant light upon these . . . Florida swamps, that we had forgotten to look for the cause of the artificial glare, but, when we did, we found a faithful negro had suspended from cranes two cages, one on each side of the boat, into which we constantly placed unctuous pine-knots, that blazed and crackled." Harriet Beecher Stowe similarly

The *Metamora* steaming up the Ocklawaha. The enclosed paddle wheel was specially designed for Ocklawaha River steamboats to protect the machinery from the snags and twists of the river. Courtesy of Special Collections, Putnam County Library, Palatka.

enjoyed the evening show when "the soft vivid feathers of the cypress had a magical brilliancy as our light passed through the wooded aisles. The reflected fire-light gave the most peculiar light."[23]

Hart's entrepreneurial bent led him to dream of a heady future for Florida's waterways. He viewed the Ocklawaha as but a small part of an integrated system of rivers and connections that would open up Florida's interior to trade and commerce. By the 1870s, he had cleared the river of snags and obstructions south of where the Silver River entered, allowing boats to travel all the way down to Leesburg, located on Lake Griffin, near the Ocklawaha's source. This small town lay not far from Lake Panasofkee, which drained not into the Ocklawaha, but into the Withlacoochee and eventually the Gulf of Mexico. Hart toyed with the idea of a connection between Lakes Griffin and Panasofkee, essentially connecting the Atlantic and the Gulf. Hart even christened one of his smaller Ocklawaha steamboats the *Pansofkee* [sic] in the hope it would someday ply Florida's waters from ocean to gulf. In 1881, he reminded the board of Florida's Internal Improvement Fund that he had a contract "to cut a canal from Lake Harris [another lake on the upper Ocklawaha, near Lake Griffin] to Lake Panasoffkee [sic], and to improve the navigation of the Withlacoochee river." That same year, Hart wrote the board again, this time in his capacity as president of the Atlantic & Gulf Transit Canal Company. The company's plans seemed much more grandiose than Hart's earlier vision. Hart asked the board for assistance in building "one of the longest and widest [canals] ever constructed on this continent." Requesting a state grant for five hundred thousand acres, Hart pictured a 495-mile canal, stretching from the Perdido River on the Florida-Alabama border across the state to Fernandina and then paralleling the Atlantic south to the Halifax River near Daytona. With state support, Hart anticipated canal completion within five years of groundbreaking. Though Hart was an important businessman, Florida officials refused to allocate the funds or acreage for the project, citing its prohibitive costs and exorbitant projections. Lacking finances and government support, the company soon collapsed, and plans for a canal remained on hold.[24]

Hart's company was only one of several that emerged in an overheated Gilded Age economy that proposed construction of a cross-peninsula canal somewhere across the state. Among others was the similarly named Florida Atlantic and Gulf Ship-Canal Company, chartered in 1881 and incorporated by the state two years later. Controlled by a syndicate of nationally prominent businessmen and politicians organized to "survey, locate, construct, own and operate a ship canal or ship railway across the said Peninsula of Florida," the company sought further investors in a scheme they hoped would become "a greater success than the Suez or Panama Canals." Despite the wherewithal of New York banking interests and a board of directors that involved the governors of Tennessee and Virginia, as well as a Nevada senator, the project also went nowhere. Other companies similarly incorporated included those organized by such major Florida businessmen and promoters as Hamilton Disston and former lieutenant governor William Gleason. Never intended to stand alone as a cross-state canal, these projects were invariably tied to a variety of other drainage, dredging, land, and railroad enterprises designed to develop the state. Disston, a wealthy Philadelphia businessman who had moved to Florida in the late 1870s to become a major player in the effort to drain and develop the Everglades, envisioned a canal much farther south than most previous plans. He called for a connection of the Atlantic with the Caloosahatchee River through Lake Okeechobee as only a small part of a grander scheme to reclaim millions of acres of Everglades marshland. Gleason, another northerner drawn to the promise of easy money in an undeveloped state, also planned for a canal as part of a broader waterways network. For him, a more northerly cross-state canal would tie into his dream of an east coast deepwater channel from Jacksonville to Biscayne Bay. Like other smaller entrepreneurs, Disston and Gleason clamored for capital in New York and London's financial markets. They also worked backroom sweetheart deals with amenable Florida politicians who saw these projects as a way to encourage economic development while simultaneously lining their own pockets. Never completely wed to the idea of a cross-state passage, and with little of their own money tied to their schemes, Disston and Gleason did not suffer greatly when shrewd investors naturally gravitated toward railroads and land development in comparison to the relatively passé technology of canal construction. Thus Hart's vision of a cross-state canal would have to wait as the lack of progress on these proposals forced him to remain content with his Ocklawaha operations for the time being.[25]

While northern tourists used popular monthlies of the time such as *Harper's* and *Scribner's* to extol the virtues of Hart's boats on the Ocklawaha, it took a native southerner to provide the first major promotion of the river's natural wonders. Sidney Lanier, Georgia poet and Confederate veteran, arrived in Florida in 1875, commissioned by the Atlantic Coast Line Railway to write a guidebook to the state and, not coincidentally, to drum up business for the railroad. In 1876, Lanier published *Florida: Its Scenery, Climate, and History with an Account of Charleston, Savannah, Augusta, and Aiken and a Chapter for Consumptives; Being a Complete Handbook and Guide.* Conceptualized as a means to turn a quick profit, the guidebook was, in the words of the literary critic Lena Jackson, "essentially hack-work, quickly done." The promotional literature nevertheless revealed a talented metaphorical style that led Jackson to conclude that Lanier "put into it much poetry and much of himself," especially in his chapter on the Ocklawaha. "The stream," he wrote, "which in its broader stretches reflected the sky so perfectly that it seemed a riband [sic] of heaven bound in lovely doubling along the breast of the land, now beginning to narrow." He ended his chapter on the Ocklawaha with an ode to Silver Springs:

> The fundamental hues of the pool when at the rest were distributed into innumerable kaleidoscopic flashes and brilliancies, the

multitudes of fish became multitudes of animated gems, and the prismatic light seemed actually to waver and play through their translucent bodies, until the whole spring, in a great blaze of sunlight, shone like an enormous fluid jewel that without decreasing forever lapsed away upward in successive exhalations of dissolving and glittering colors.[26]

Ninety years later, the Florida Defenders of the Environment would revive Lanier's overly poetic prose in legal briefs and public statements opposing construction of the Cross Florida Barge Canal, which would have destroyed the Ocklawaha River and its environs. In a 1965 article entitled "The Oklawaha River Wilderness," Marjorie Carr relied on Lanier's words to describe the river as a "lane which runs for more than hundreds of miles of pure delight . . . a lane which is as if a typical woodstroll had taken shape and as if God had turned into water and trees the recollection of some meditative ramble through the lonely seclusion of His own soul." Though Lanier was no doubt sincere in his portrayal of the river and springs, his motivations seemed fundamentally at odds with the later environmentalists who fought to preserve the river. Lanier could certainly appreciate the river and its appeal, but he saw it as a commodity and had little difficulty reconciling signs of "progress" and commerce on the river with its vision of wilderness and beauty.[27]

While the river and springs enchanted visitors like Lanier, their beauty was appreciated through a nineteenth-century notion of environmental conduct that was anything but passive. Photos and drawings of late nineteenth-century river steamers often showed armed men sitting on the decks, ready to shoot alligators, birds, and occasionally such large shore mammals as deer, bears, and panthers. An 1870 article in the national *Appleton's Journal* reported that "a successful hunter, after much experience, seldom lets one of the reptiles [alligators] escape. If any philanthropist has ever objected to the slaughter, the circumstance is not remembered in the swamps and everglades of Florida." Three years later, Harriet Beecher Stowe traveled the river with a boatload of "a dozen or two of mighty hunters . . . who know how to hit what they fired at, but about an equal number of inexperienced hands foaming at the mouth with excitement, and quite as likely to hit any one of us as the alligators." In 1874, river tourist Martha Holmes recorded similar impressions of a trip up the Ocklawaha. "The gunners," she wrote, "are a loathsome set of fellows . . . crackwhacking [*sic*] at the animals not often to their damage, but making the alligators skoot [*sic*] before we can see them."[28]

Whether evidenced through Hart's capitalistic ventures, Lanier's literary prose, or the incongruous intermingling of alligator hunting with Victorian gentility, the Ocklawaha had become much more than a simple vacation destination by 1890. By then, cypress logging rivaled tourism as the primary economic activity in the region. With increased demand for rot-resistant cypress shingles following the Civil War, local woodsmen developed small-scale "shingle yards" along the water's edge, marking the beginnings of an industry that would profoundly alter the riverine environment. Northern tourists viewed these workers as just another part of a strange and alien landscape. In 1890, a traveler relished the sight of "two 'Florida crackers,'" who had "established a camp in a grove of the finest cypress-trees we ever saw. . . . [T]heir hut was a very model of the picturesque, and the smouldering fire, over which their dinner pot was cooking, sent up a wreath of blue smoke against the dark openings of the deep forest that gave a quiet charm."[29]

By the 1880s, the charm of such minor operations gave way to larger ones that included the recovery of cypress logs along the muddy banks or bottom of the Ocklawaha as well as the felling of large stands of trees at the river's edge. "Here and there we encounter great rafts of cypress logs which almost block the channel," reported a 1904 guidebook to the river. "The timber is owned in tracts of thousands of acres, and is cut for Northern markets. The trees are girdled some weeks before they are felled, for the wood is very heavy and when full of sap sinks in water like lead." By the early 1890s, national corporations dominated their smaller rivals. Chief among them was the company established by Henry and A. E. Wilson, two Michigan brothers involved in the lumber trade in their home state. Visiting the Ocklawaha as tourists, they came away from the river impressed less by its beauty than by its potential as a source of high-grade cypress wood. Returning home with dollar signs in their eyes, they raised enough capital to buy out Palatka's Tilghman Cypress Mill and began operations as the Wilson Cypress Company in 1893. The company logged the river's banks and valley until 1944, when economic conditions and a lack of suitable cypress forced an end to its operations.[30]

During the heyday of the Ocklawaha's timber industry, the river itself became the focal point of lumber operations as the swampy terrain

Ocklawaha timber men transporting cypress logs by raft to Palatka, late nineteenth century. Courtesy of State of Florida Photographic Archives, Tallahassee.

made entry by rail all but impossible. Logging crews lived on racially segregated houseboats on the river, and giant rafts of cypress logs, up to twenty-five feet wide and thirty-two feet long, became common sights along the Ocklawaha in the first two decades of the twentieth century. Once these improvised rafts reached the St. Johns, crews lashed up to six of them together to form a larger raft, and then towed them to the Palatka mill. Often workers even cut small canals through the swamps in the Ocklawaha bottomlands to gain greater access to large stands of swamp cypress. Such innovations paid off. By 1920, the Wilsons employed over three hundred timber men in their Ocklawaha operations, and the company produced over 400 million board feet of lumber. With its Palatka mill alone, the Wilson Cypress Lumber Company became the largest employer in that city, with close to six hundred mill workers on hand.

While steamboat tourism and cypress logging moved slowly and tentatively toward inclusion in a national market economy, a different pattern of economic development occurred on the western side of what would eventually become the Cross Florida Barge Canal. Unlike the Ocklawaha, the Withlacoochee experienced a much more rapid and profound transformation. With the accidental discovery of high-grade rock phosphate along the river in April 1889, this region underwent a tumultuous period of rampant exploitation and development, what the muckraking journal *Cosmopolitan* called "The Great Florida Phosphate Boom." Soon the community of Dunnellon, located on the Withlacoochee in western Marion County, became the center of America's largest phosphate mining operation, and the headquarters of the Dunnellon Phosphate Company, which eventually owned ninety thousand acres of land on which to conduct its operations. In 1890, the *New York Republic* marveled at the importance of the hard rock, a necessary component of agricultural fertilizers: "Within a year, Dunnellon had become famous on the two continents, owing to the discovery in her midst of what are probably the richest phosphate deposits in the world. This discovery . . . ranks with that of California's gold and Pennsylvania's oil."[31]

The discovery may have been accidental, but the extraction and marketing of Florida's phosphate exploded overnight as a result of three local entrepreneurs: Albertus Vogt, John Dunn, and John Inglis. Vogt was a thirty-nine-year-old transplanted Georgian struggling to make ends meet with a variety of business and real-estate ventures when his hired hand stumbled upon the rock while digging a well on his property near Dunnellon. Knowing immediately that he was on to something valuable, Vogt hurried to Ocala, the county seat and commercial center of the area. There he shared his find with John Dunn, an influential banker, lawyer, railroad organizer, and real-estate developer of the recently platted but as yet undeveloped town of Dunnellon. Dunn validated Vogt's hunch and, along with Vogt, immediately began purchasing property in the vicinity of the find. Having established a partnership called the Dunnellon Phosphate Company, Vogt quickly cashed in his shares for $191,000. Vogt would spend the rest of his life as the "Duke of Dunnellon," a quintessential Gilded Age robber baron who allegedly threw twenty-dollar gold

The great Dunnellon phosphate boom of the 1890s opened the Withlacoochee River valley to large mining operations. Courtesy of State of Florida Photographic Archives, Tallahassee.

pieces into Silver Springs just to see what they looked like at the bottom. Dunn, hardly content with his newfound wealth, expanded operations significantly in the next five years. By 1893, Dunnellon Phosphate, now under the directorship of local banker and developer John Inglis, had offices not only in Dunnellon and Ocala, but also in Boston and New York.

The phosphate boom erupted as Florida newspapers from Ocala to Jacksonville spread the word of the discovery. As a result, *Cosmopolitan* noted, "ten thousand feverishly eager prospectors overran the woods, and every man turned prospector for his own forty acres." All of these adventurers, remarked the *Florida Times-Union*, had "sand in their shoes, pebbles in their pockets, and a keen interest in prehistoric bones." As with most nineteenth-century enterprises, the industry quickly consolidated as small competitors and wildcatters overextended themselves and either went into bankruptcy or were bought out by Dunn's company. By the end of 1891, nineteen other phosphate companies, capitalized at over $21 million, mined the area. And through the first half of the 1890s, Dunnellon dominated the world's phosphate production. Reveling in the "glad song of the pick and shovel and the whir of the wheels of busy industry," the *Ocala Banner* hailed Dunnellon as a city "known all over the world. . . . It is the richest mining district on the globe."[32]

By 1893, Dunnellon had produced more than 216,000 tons of high-grade rock phosphate for foreign and domestic consumption. The rock would have been worthless, however, without a way to get it to market. Phosphate originally left the area by rail and crossed the state to Fernandina until John Inglis cleared the Lower Withlacoochee and developed Port Inglis at its mouth on the Gulf of Mexico. Even with the building of two large wharves at this seaport, however, railroads remained crucial to the transportation process. Shipping hard rock down the Withlacoochee from Dunnellon to the Gulf was prohibitively expensive. It was both cheaper and more efficient to move phosphate by rail to the town of Inglis, and then by barge down the river itself for seven miles to Port Inglis. As the major terminus for Dunnellon's leading industry, Port Inglis immediately became a thriving entrepôt for international trade. By 1905, the Dunnellon Phosphate Company shipped nearly 600,000 tons of the rock around the world. Ship manifests from a single month in 1905 vividly illustrate the global reach of the Dunnellon Phosphate Company: "the *Heronspool*, 4630 tons of phosphate to Stettin, Germany; the *Aislaby*, 3933 tons to Landskrona, Sweden; the *Kennett*, 2202 tons to Braila, Rumania; and the *Miramichi*, 3513 tons to Dunkirk, France." Thus, unlike the Ocklawaha, the Withlacoochee early on became a dynamic nexus of commerce and industry, fully engaged in not just a national but an international economy.[33]

By 1909, the phosphate industry had grown so large that, to meet the needs of mining operations in and around the town, the Florida Power Corporation built an earthen dam and hydropower plant on the river thirteen miles west of Dunnellon. The dam created an impoundment appropriately called the "Withlacoochee Backwaters," which would eventually be renamed the more prosaic "Lake Rousseau." But it also halted navigation along the river, as no lock was built to allow passage to Dunnellon. In 1916, however, the power company rectified the situation and began modifying the dam by building a 134-foot lock that would allow boats drawing four to five feet of water to travel from Dunnellon down to Port Inglis. That same year, the Marion County Board of Trade endorsed a proposal for the Army Corps of Engineers to dredge a ten-foot-

In the 1890s, Dunnellon mining interests used both barge and rail to ship phosphate all over the world. Courtesy of State of Florida Photographic Archives, Tallahassee.

deep channel in the river from Port Inglis to Dunnellon to enable the movement of phosphate, wholly by river, all the way to the Gulf. Boosters claimed the operation would provide "a lasting benefit to the whole country and one of the greatest factors in the development of Dunnellon and her interests." Prohibitive costs blocked the effort. The Corps did, however, recommend clearing water hyacinths and obstructions from the channel behind the dam all the way to Dunnellon. This could be accomplished at the modest cost of approximately one thousand dollars per year and, with the completion of the lock, would allow vessels drawing up to six feet of water to travel to Dunnellon. Though hardly a major river of commerce, the Withlacoochee therefore witnessed enough alterations that it would be difficult for anyone to claim it as a pristine wilderness waterway. From the standpoint of 1960s environmental activism, when one looked at a map of proposed canal construction, only one relatively untouched natural treasure remained in the path of ecological destruction—the Ocklawaha.[34]

As with most extractive industries, Dunnellon's phosphate boom proved short-lived. A combination of shoddy business practices, overextended capital, the depression of 1893, and the great citrus freezes of 1895 and 1896 left the industry in shambles. The discovery of lower grade, but significantly easier to extract, pebble phosphate in Florida's Peace River valley around Bartow, combined with the collapse of the world phosphate market as World War I broke out, completed Dunnellon's demise. By 1918, the city's mines produced only 18,000 tons of phosphate, compared with 2 million tons of the pebble variety mined further southward. With the end of the phosphate industry, Dunnellon, Port Inglis, and the Withlacoochee River corridor reverted to a sleepy somnolence, only broken by dreams of boomtimes once again with the coming of the ship canal in the 1930s, and the barge canal thirty years later.

As phosphate rapidly boomed and busted on the west coast, Hubbard Hart, in a manner similar to John Dunn and John Inglis, consolidated his enterprises on the Ocklawaha. In 1895, faced with falling revenues and rising competition from the newly established line of Captain Edward Lucas, Hart tried to continue dominating the tourist trade by merging operations with his rival. Two years earlier, Lucas had placed a new boat on the river. The *Metamora*, the largest and most luxurious boat then in service, forced Hart to overhaul and upgrade his own two vessels simply to compete. The new line, named the Ocklawaha Navigation Company, was in business less than a month when tragedy struck. In December 1895, Hart was killed in a trolley accident in Atlanta while attending a business meeting. According to the *Palatka Times-Herald*, "the death of Colonel Hart cast a gloom over our entire community, as no man who ever lived here was more universally esteemed and loved." After his death, the combined business continued its tourist trade up the Ocklawaha to Silver Springs, now reorganized under the management of Lucas and R. H. Thompson, Hart's brother-in-law. Other aspects of the business changed considerably, however. The establishment of rail lines throughout the Ocklawaha Valley made the steamboat less important as a carrier of freight and supplies. In addition, Hart's profitable orange groves were decimated by the "Big Freeze" of 1895. By the start of the 1897 tourist season, Thompson dissolved the partnership and reestablished the Hart Line as a competitor of Lucas. Revenues continued to decline for both companies, however, as fewer and fewer tourists took the trip. And those who did seemed more and more inclined to take it only one way, returning to Palatka from Silver Springs by railroad.[35]

The *Metamora*, pride of the Lucas Line, docked at Silver Springs, ca. 1900. Courtesy of Special Collections, Putnam County Library, Palatka.

In spite of, or maybe because of, the falling passenger numbers, Thompson's publicity agents continued to try to drum up business. The line's 1904 brochure extolled the virtues of "this wonderful, unique, fascinating, and romantic trip." It also included testimonials from northern tourists like Helen Pomeroy of Cooperstown, New York. Calling the journey "a weird and exquisitely beautiful excursion through the heart of Florida," Pomeroy praised both the scenery and the steamboat's wonderful crew and cuisine. For her, the danger of the voyage, with its tight turns and overhanging branches, remained an important part of its charms. "The steamer begins to wander here and there, first towards one bank and then the other," she wrote, "trying to find a place to get out of the maze of sudden turns, fallen trees, and sunken logs." Yet, she felt safe, as "the well-trained pilots are equal to their task, having the experience of many years to guide them." In March 1903, passengers on the *Metamora*, the pride of the Lucas Line, were not so lucky; it sank in the middle of the night while traveling on the Ocklawaha. The *Florida Times-*

Union, Jacksonville's morning paper, described how, "in all the years of travel on the famous Ocklawaha, this is the first serious accident." More significantly, however, the *Palatka News and Advertiser* recognized "the loss [as] almost a crushing blow to Captain Lucas." He never rebuilt the *Metamora*, and boats of his line never again made the passage to Silver Springs. Seeing an opportunity to regain the lion's share of the tourist trade, the Hart Line launched its last boat, the *Hiawatha*, a year after the *Metamora* disaster. Larger and more luxurious than the line's other remaining vessels, the *Okeehumkee* and the *Astatula*, the twin-decked *Hiawatha* carried up to eighty passengers on its trips on the river. Eighty-nine feet long, the boat displaced 129 tons and had twin smokestacks, as opposed to the single stack of Hart's previous vessels. The boats left Palatka three times a week, leaving the River City at 12:45 p.m. and arriving at the Springs around noon the following day. Departing the Springs at 2 p.m., the voyage ended back at Palatka early the next morning. This round trip cost twelve dollars, which included both sumptuous meals and a cramped stateroom berth. But even with the posh *Hiawatha* as its flagship vessel, the Hart Line struggled to remain solvent as Florida vacationers increasingly used Henry Flagler's Florida East Coast Railway to reach luxurious beach resorts along the Atlantic coastline.[36]

In early 1912, after traveling on the *Hiawatha* from Palatka to Silver Springs, the American musician Maud Powell wrote what might be seen as an epitaph for this period of steamboat tourism. Powell, considered by many as the leading American concert violinist of her time, was so moved by her Ocklawaha experience that she wrote a poem entitled "Up the Ocklawaha (An Impression)." She then had her friend and musical protégé, the young and talented female composer Marion Bauer, turn that work into a piece of music for violin and piano. Bauer completed the score by May, and it debuted in San Francisco in December with Powell playing the violin part. Powell seemed exceedingly pleased with the work, calling it "a tone picture, taking into consideration the story that called it into being." The piece remained a part of Powell's concert repertoire until her death in 1920. Powell's poem on which it was based reflected that sentimental vision of wilderness, much like that of Sidney Lanier, that Marjorie Carr would tap into fifty years later (though there is no evidence that Carr had either read the poem or heard the music) in her struggle to save the Ocklawaha. "Softly speeds the Hiawatha, search-

The *Hiawatha*, largest and most luxurious of the Hart Line steamboats, plied the Ocklawaha River until 1919. Courtesy of Special Collections, Putnam County Library, Palatka.

ing her way through the haunted swamp," Powell wrote. "Peace at last, Up the Ocklawaha." Amid the stillness, however, lay an undercurrent of gothic horror. For Powell, the trip had unveiled eerie, frightening feelings, as if the wilderness held barely repressed demons. She called the river and its environs "rank, dark, malarious, fearsome . . . a forest of doom," and concluded that "there is no solace in the mirrored depths of the Ocklawaha." Even the wondrous night fires, the very signature of the Hart Line, issued signs of dread when she wrote, "The pine-knot fires, in lurid relief, Double the curse in the ink-black waters."[37]

Little did Powell realize that, by decade's end, the steamboats that had provided a "Grand Illumination at Night," revealing "the Ocklawaha by torchlight," would be a thing of the past. The same year Powell had traveled on the river, the Silver Springs Company, which had purchased the springs themselves in 1909, launched its own boat, the *City of Ocala*. Much smaller than the Hart vessels and built with an internal combustion engine, it was designed for fast travel. With cheaper fares (round trips cost only ten dollars) and only one day of travel time, the boat cut even further into the Hart Line's declining passenger numbers. The company also raised landing fees at Silver Springs for its competitors, essentially prohibiting them from continuing their trade. In a effort to modernize, the Hart Line even abandoned its signature night fires and replaced them with electric searchlights to illuminate the river's environs. These efforts proved to no avail, however. On December 9, 1919, R. H. Thompson reported that the Hart Line would "postpone its winter time schedule, till further notice." The days of the *Hiawatha* and the *Okeehumkee* were over.[38]

Increasingly tied to market forces beyond its control, the Hart Line could simply no longer compete. Though the Silver Springs Company had essentially driven the Hart Line off the river, it quickly turned its back on the Ocklawaha itself. Formerly an integral part of the entire springs experience, the river now became an obstacle. Over time, more and more people arrived at the attraction by train and automobile instead of steamboat. Silver Springs, for fifty years an appendage of Palatka, was becoming part of Ocala, "only six miles by auto." In 1915, Ocalans made one last attempt to tie their city to both Silver Springs and the Ocklawaha when they put up fifty thousand dollars for a proposed canal connecting their city to the Silver River. The waterway would allow boats to make a passage from Ocala directly to Palatka, thus providing an alternative to "the stranglehold the Railroads have on us." The Corps of Engineers quickly shot down the plans, confirming the growing belief that river commerce was no longer viable. In December 1916, army engineer W. W. Fineren asserted that "the only freight boat of any consequence on the river during the present year (a 39-ton boat) has lately been sold under bankruptcy proceedings owing to the fact that the over-head charges were too much for the amount of freight that could be carried on the small boat." Two months later, an official Corps report concluded that "the cost of the proposed canal would be out of proportion to its usefulness and to its value to the country at large." With that dismal assessment, Ocalans quickly forgot the scheme, and therefore the Ocklawaha itself.[39]

With the transition from steamboat to rail and eventually the automobile, the idea of the Ocklawaha as a piece of wilderness to be admired as a thing in itself was lost, seemingly forever. Yet Florida's steamboat era seemed more about the packaging of nature than its preservation. By

1904, the Raymond and Whitcomb Company offered tours of Florida and the "romantic Ocklawaha River" to northern tourists. For three hundred dollars, one could take the train from New York, Boston, or Philadelphia to Palatka or St. Augustine, where a Hart steamer awaited to transport vacationers to Silver Springs. The company's promotional brochure promised that "if the day is warm and bright, alligators will be seen now and then," but bemoaned the fact that while, "fifteen years ago 2,000 or 3,000 might have been observed, fifty or seventy-five make now a very good record."[40]

Silver Springs and the Ocklawaha River now seemed sui generis, natural phenomena divorced from nature itself, as both the steamboat and logging trades in turn became separated from them. By 1920, loggers moved into virgin stands of cypress deeper in the valley, where river transport was no longer feasible. Now shipped by the Ocklawaha Valley Railroad, the logs were taken to the St. Johns, where they were then floated to Palatka. Having once been exploited as a resource and turned into a commodity, the Ocklawaha was now quite literally being relegated to a backwater, considered as irrelevant by both timber barons and tourist mavens. This very irrelevancy would enable the renewed interest necessary to turn the river into part of Florida's ditch of dreams. At the same time, the collapse of trade allowed the river to revert to a more primitive state, one that 1960s environmentalists would claim as an area untouched by human contact and thus worth preserving.

The rise and fall of the Ocklawaha's steamboat tourism and trade was hardly unique in late nineteenth-century Florida. In addition to Hart, a disparate collection of visionaries and rogues viewed the state as unclaimed territory and thus an opportunity to gain fabulous wealth and power. Their goal was simple—to develop the state while lining their pockets. While they lauded the laissez-faire principles of market capitalism, they depended on the financial support of both Washington and Tallahassee to fulfill their ambitions. Florida's Internal Improvement Fund, established in 1855, provided a mechanism for businessmen to develop much of the state's interior. Administered by a board of trustees composed of the governor and four other state officials, the fund controlled almost 60 percent of the state's lands—an area of more than 22 million acres—after the Civil War. Much of this land demanded draining and dredging to become productive, and the fund's trustees seemed more than eager to deed thousands of acres to businessmen like William Gleason and Hamilton Disston in exchange for vague promises of land improvement. While the state government offered the land at bargain prices, the federal government provided, through surveys and improvements made by the Army Corps of Engineers, the wherewithal to enhance its value. Once again, the federal government became an engine of economic change. And as the Corps began to ply its trade across the state after the Civil War, renewed talk of a cross-peninsula canal became widespread.

Southern politicians, whether Democrat or Republican, or even black or white, stood united in their desire for canal construction. Memories of the Civil War may have been relatively fresh, but the lure of money tended to soothe any remaining tensions. In 1872, Alabama's legislature, hardly known for its fondness for federal programs, issued a memorial to Congress requesting another "thorough survey be made of the Florida isthmus, with a view to ascertain the practicability of the ship-canal." The rationales tendered for building a waterway seemed familiar enough to be repetitious. The canal would prevent "a long, tedious, difficult, and dangerous navigation" around the peninsula and would advance "the interest of the whole country." Two years later, Florida congressional representative Josiah Walls, a black Republican from Gainesville, seconded Alabama's request in a speech on the House floor. Laying out the argument for canal construction in voluminous detail, Walls proclaimed the project a foregone conclusion. Nevertheless, he argued that this could not occur "without the aid of the general government." Still mindful of "the poverty, increased debt, and depreciated credit of the Southern States," Walls saw a ship canal as a necessary part of a Reconstruction program "to harmonize the seeming conflicted interests of the several sections of country." After cataloging the by-now conventional benefits of canal construction, Walls suggested the route run from the Withlacoochee across the state to Fernandina or St. Marys. Aware of the limitations of Jacksonville's port facilities—though strategically placed along the St. Johns River, its persistent problems with shifting sandbars blocked easy access to the Atlantic—he claimed the other two more northerly ports should be "the natural terminus of any canal through Florida." Desiring greater access to transport the fruits of Florida's rich soil, he also recommended connecting the canal to "an inland passage for steamboats . . . from the

St. Johns River to Key West." Though Walls never secured funding and, indeed, was voted out of office in 1876 as Florida returned to white rule under the leadership of Democratic Redeemers, his dream remained. Politicians and entrepreneurs who sneered at the mere thought of African American political participation embraced Walls's vision as part of their plan for a New South. As if to echo the sentiment, Charles Jones, Florida's first postwar Democratic senator, submitted a resolution to the Senate in December 1881 "relating to the construction of a ship canal across the peninsula of Florida."[41]

The steady demand for a canal from such advocates as Walls and Jones finally convinced federal officials to authorize further study of the issue in the late 1870s. In 1878, Congress passed the Rivers and Harbors Act, which, among other appropriations, provided money for yet another survey of Florida to determine both the efficacy and location of a cross-state canal. The result was not just a single survey, but an overwhelming series of them, conducted by several teams of army engineers under the command of General Quincy Gillmore, one of the Union commanders at the 1864 Battle of Olustee. Hardly neutral in determining the location of potential routes, these reports often reflected the desires of influential entrepreneurs like William Gleason. Suggesting that his political leverage, rather than geographical considerations, would ultimately determine the conclusions of the army's upcoming canal surveys, Gleason explained to an investor in 1880 that a surveyor assigned to the project was "much impressed with our proposed route" from the St. Johns to the Withlacoochee and desired to "make it a success." Gleason presumptuously overstated his influence with the Corps, however, as Gillmore's final report, issued in 1882, recommended a route far different from the one he favored. Gillmore laid out a detailed plan for a 122-mile canal between the St. Marys and the Suwannee rivers. The Corps contemplated a twenty-five-foot-deep, eighty-foot-wide canal, requiring fifteen locks, to wind its way through the Okefenokee Swamp and across the peninsula. Though he included detailed plans for building the canal, Gillmore clearly stated the project was neither commercially viable nor, with its $50 million price tag, fiscally responsible.[42]

Despite his reservations, however, Gillmore did see the canal "as part of a comprehensive scheme for improving and cheapening our water transportation from the heart of our grain and cotton growing regions to foreign ports." As would so often be the case, canal proponents will-fully ignored Gillmore's negative assessments and saw a plan that gave them ample room for hope. In the end, the Corps' voluminous surveys and reports resolved little. All that would come out of Gillmore's recommendations were meager appropriations for still more surveys. In many respects, the stalemate illustrated the circumscribed power of canal boosters, who had enough strength to keep their dream alive but unful-filled. Surveys established momentum for further surveys, but only for validating canal construction along particular routes favored by particular boosters. Though private investors would be quick to deny it, government assistance remained key to successful canal construction. Without federal help, both in terms of funding and the institutional support of the Army Corps of Engineers, even the most determined and enterprising plutocrats could make little headway in their quest for a waterway across the state. The Corps continued to issue negative assessments, but businessmen remained undaunted in their desire for a canal. For example, the publication of another Gillmore survey dashed Hamilton Disston's plans for a southerly route across the state. However, in 1880, Congress, following Disston's lead, authorized the already busy general to determine the possibility of a canal link through Lake Okeechobee. With the return of his engineers from the surrounding wetlands, Gillmore dismissed the idea as costly and impractical. Undeterred, Disston ignored Gillmore's judgment and pursued his elaborate scheme. In the end, he succeeded only in a modest project tying Lake Okeechobee to the Caloosahatchee River along Florida's lower Gulf coast. Frustrated with this and other financial failures, Disston eventually died mysteriously in April 1896 at the age of fifty-one, as much a victim of the dream of sudden Florida wealth as of his own overarching ambition.[43]

Much like Disston, Robert Gamble saw dollar signs in Gillmore's surveys and proposed the grandest vision to date—a 550-mile waterway from St. Marys to the Mississippi. More than simply a Tallahassee businessman, Gamble embodied Florida's transition from frontier territory to Gilded Age prosperity. Not only a cotton planter and banker in territorial Leon County, he served along the Withlacoochee in the Second Seminole War, developed and lost a thriving sugar plantation in Manatee County in the 1850s, and then moved back to Tallahassee after the Civil

War to become a major player in postwar politics and business. Presented to the Senate in 1894 by Florida senator Wilkinson Call, Gamble's ambitious plan best represented the almost century-long struggle for a cross-Florida connection. Gamble was no stranger to canal dreams. Fully forty years earlier, he had lobbied *DeBow's Review* to promote such a project. Now, as he neared the end of a long and productive life, he saw the canal as the culmination of his dreams. For Gamble, the $50 million cost of the project seemed a small price to pay for something that "would be of utmost value to the United States Government." As in the case of Disston's plan, government support was not forthcoming. Thus Gamble's vision remained unfulfilled at the time of his death, at the age of ninety-four, in 1909.[44]

The lack of government funding seriously impeded the plans of Disston and Gamble, but equally significant was the lack of a decent harbor on Florida's Atlantic coast. In particular, difficulties with the constantly shifting sandbars at the mouth of the St. Johns River prevented Jacksonville, located ten miles from the Atlantic shoreline, from capitalizing on its location to become a major seaport and canal terminus. The predicament led many, especially the Corps of Engineers, to conclude that the Atlantic outlet of the canal must be Fernandina, or even farther north in St. Marys, Georgia. To resolve the issue in their favor, Jacksonville's canal boosters continually pressed for federal assistance for clearing a permanent channel through the coastal sandbars. For nearly three decades, army engineers pursued a variety of possible solutions to tackle the problem. They dredged tons of sand and built stone jetties hundreds of feet into the ocean—all to no avail. In 1874, the peripatetic General Gillmore expressed frustration with the failure of these projects. "The natural and proper port for the shipment of all freight arriving in east Florida is Fernandina," he reported; "I doubt the wisdom of expending any more money upon the bar at the mouth of the St. Johns River."[45]

Supporters of Jacksonville's port refused to follow Gillmore's lead and abandon their quest to clear the bar and improve their city. In 1876, Dr. Abel Baldwin, an influential former state senator, railroad president, and Confederate surgeon, penned an article exhorting Floridians to embrace the project. "All citizens interested in the St. Johns and sections tributary to it, which embrace the entire Peninsula," he wrote, "have an interest in this improvement, . . . for it is the key to all improvements in the interior, . . .[and] the establishment of cheap and convenient water transportation throughout the Peninsula." Demanding further federal support, Baldwin and others convinced Congress to allocate funds for the Corps' plan to open the river for good. In 1881, in spite of his previous concerns, Gillmore designed two huge jetties that stretched nearly two miles beyond the shore to delineate the river channel. Built of northern granite and Florida limestone and anchored by "mattresses" of huge logs, the jetties took until 1895 to complete, at a cost of almost $2 million. Though they would demand constant reinforcement and improvement, the structures finally succeeded in providing Jacksonville with a safe and effective outlet to the sea. With the elimination of the sandbars at the river's mouth, engineers could now turn to dredging adequate shipping lanes for the ten miles from the city's center to the ocean. Constant dredging improved the depth of the channel from eighteen feet in 1895 to twenty-four feet in 1906, and, finally, to thirty feet in 1918. Now, Jacksonville could serve not only as a seaport but as the Atlantic terminus of a future cross-state ship canal.[46]

On the eve of the twentieth century, opportunities for construction suddenly seemed brighter, for the national political mood was changing to allow for a reconsideration of waterways and their relationship to economic growth. As canal boosters spent much of the nineteenth century fighting among each other for routes that suited their self-interests, their railroad rivals developed an overwhelming nationally integrated transportation system that became the primary engine of America's industrial revolution. Between 1865 and 1900, track mileage more than doubled, from 103,649 to 221,864 miles. Moreover, the industry's influence permeated everything: railroads were the nation's largest carrier of passengers and freight, they were the nation's largest employer, and they were the nation's largest consumer of coal and steel. Reflecting on the rapid rise of railroads by the turn of the century, *Popular Science* pointed to the power of monopoly and its tragic result for America's inland waterways. "Population and riches beyond the imagination of the nation's founders followed," it noted with astonishment, "and were bound by the iron bands until great commonwealths bridged the continent, until it were easier to think of millions than of thousands before, until one seventh of our

By 1900, the railroad had reached Silver Springs and provided economic competition for Hubbard Hart's steamboat operations. Courtesy of State of Florida Photographic Archives, Tallahassee.

swollen wealth came to be railroad property and its ownership a factor in law-making, until every-day ideas of domestic travel and transportation came to connote railways alone." As the industry expanded, moving from competition to consolidation, "river traffic was virtually dead, the water-fronts of every river town . . . controlled by railroad interests, and the once resplendent river vessels reduced to rattletraps."[47]

Politically, the Populist and Progressive movements reflected deep concern over the power and influence of railroads and sought solutions to curb their increasing monopolistic control. Years of discriminatory freight rates and hidden rebates had bred resentment among farmers and small businessmen alike. Moreover, critics complained that, though railroads clearly dominated American commerce, the industry lacked the capacity to meet the needs of a growing economy, especially in the Mississippi Valley southward to the Gulf of Mexico. Not content with governmental regulation and even ownership of railroads, some Populists sought an alternative to this relatively new form of transportation. Hence the demand for canals and waterways across the nation took on a greater sense of urgency and meaning. Canals would remove, according to the Marion County Board of Trade, "the stranglehold the railroads have on us and . . . the weighty burden of high cost of living [and] high rates of transportation." To break this control, the board recommended improving "God's great arteries of commerce [through] the ingenuity of a great Engineering force to harness, dam, canal, lock, or otherwise control, and make due man's bidding."[48]

Thus the turn of the century marked a propitious moment, and yet, as influential as canal boosters could be on the local and state levels, their voices were often muted and diffuse when compared to national railroad interests. To be heard, they had to join with hundreds of like-minded small-town boosters and promote a national agenda of waterways enhancement. In 1905, a group of Texas and Louisiana businessmen formed such an organization to advocate their cause. Originally known as the Interstate Inland Waterway League, and later renamed the Gulf Intracoastal Canal Association (GICA), it called for a national system of navigable waters from Chicago to New Orleans, from Brownsville to Boston. Crucial to this integrated system of more than eighteen thousand miles of navigable waters would be a canal linking the Atlantic and Gulf coasts—the very dream Floridians had held for nearly a century. Unwilling to compromise their vision, the Gulf Intracoastal Canal Association became a formidable interest group: "from camping on the doorstep of the nation's Capitol, to prodding sluggish county governments, encouraging the donation of necessary rights-of-way, and the rebuilding of bridges," the GICA concentrated all of its efforts to establish a Gulf Intracoastal Waterway. Through sheer diligence, the group tenaciously lobbied local and federal officials to consider a system of inland and intercoastal waterways that could supplement, or even rival, railroads, all of which magnified the importance of any cross-peninsula connection. As a prominent political scientist asserted in 1908, "the full value of this coastwise system will not be realized until a canal is constructed across Florida."[49]

As part of a network of artificial waterways that not only reached into the interior but hugged the Gulf and Atlantic shorelines, a Florida canal promised to become the lynchpin of a national transportation system. Florida would of course be an immediate beneficiary, but such a project would also enhance economic opportunities throughout the United States, and especially the Deep South. In 1903, the *Atlanta Constitution*, an ardent promoter of a New South based upon manufacturing and commerce rather than cotton production, called for a convention in Atlanta to discuss building the waterway. Though nothing came of the meeting, the paper continued its support for the project. Indeed, fifteen years later, it again proclaimed its position without, at least in the eyes of its editors, a hint of exaggeration: "Nothing which has been proposed since the civil war means more to the south than the St. Mary's–St. Mark's canal project, and the dozens of other industrial, agricultural, and mechanical developments which will follow close upon the heels of that project."[50]

Organized as interest groups, and with the persistent demands of newspaper editorialists to back them up, canal boosters eventually gained enough clout to get the ear of a sympathetic President Theodore Roosevelt, who established an Inland Waterways Commission in 1907. A vigorous supporter of waterways transportation, Roosevelt—a former assistant secretary of the navy—charged the commission to prepare an exhaustive report for the "full and orderly development and control of the river systems of the United States." Unlike the haphazard lobbying efforts of Gilded Age entrepreneurs, the commission promised to merge "local projects and the uses of the inland waters in a comprehensive plan designed for the benefit of the entire country." Recognizing "that no development of the railroads in the near future will suffice to keep transportation abreast of production," it proposed "the development of a complementary system of transportation by water" as a remedy to increased traffic. Navigable waterways offered a solution in which "the whole nation will share the good results." Issued in 1909, the commission's report vigorously reinforced demands for an integrated water system and laid the foundation for a protected waterway from Texas to Florida. Simultaneously, the United States embarked on a boldly ambitious project several hundred miles southward—the Panama Canal.[51]

With both its extraordinary technological achievements and potential to increase commerce through transoceanic connections, the Panama Canal symbolized a new era in transportation. Suddenly canals, long associated with antebellum internal improvements, seemed less backward. Indeed, canals once again held the promise of economic progress. Newspapers large and small explicitly made the connection between the

Panama and Florida canals with rhetoric that matched the enormity of both projects. In 1903, the *Atlanta Constitution* eagerly editorialized: "the south needs this canal. . . . In fact, the assurance that we are to have a Panama canal makes it practically essential that we should have the Florida ship canal." That same year, the *Ocala Banner*, always thinking locally, announced with relish that "the Panama Canal is in sight. It is an era of canals. Let the Silver Springs Canal follow."[52]

Boom or Boondoggle?

The turn of the twentieth century held enormous promise for canal construction. The backlash against railroads, the vigorous lobbying campaigns of waterways associations, the consistent backing of local newspapers, and the support of prominent Washington officials created an
air of hope and possibility. Everything seemed to be falling into place.
Then again, that conclusion could be drawn only within a relatively narrow constituency of canal boosters and pork-barrel politicians. Though
there was much talk of a canal, nothing—with the exception of survey
after survey—ever seemed to happen. Indeed, in 1908, the prominent
historian U. B. Phillips could only mention that "there were also plans for
canalling [sic] in Florida, some of them contemplating a route for ships
across the peninsula; but none of them were of enough importance to
demand a place in our narrative." Nevertheless, conditions for success
were already taking shape.[1]

As the Panama Canal captured the public's imagination, water transportation would assume greater significance, especially with the considerable influence of Florida senator Duncan Fletcher. As former mayor of
Jacksonville, Fletcher had an obvious stake in promoting the economic
interests of his hometown. Elected to the Senate in November 1908,
Fletcher wisely chose a position on the Commerce Committee, where he
could set policies and steer funding to waterways development. Moreover, that same month, he won another election: president of the newly
formed Mississippi to Atlantic Inland Waterway Association. Fletcher's
service as both U.S. senator and representative of one of the region's
leading lobbying organizations virtually ensured that water transportation, especially involving his home state, would be at the top of his legislative agenda. For Fletcher, a canal across Florida became a major focus
of his political career. In 1910, Fletcher rhapsodized about the need for
such a project. "Towards this work," he wrote, "our people are turning

today with the enthusiasm that the Crusador [*sic*] once turned his face towards the Holy City." And until his death in 1936, Fletcher provided much of that missionary zeal for not only a canal, but one with an Atlantic terminus located in Jacksonville.[2]

The Mississippi to Atlantic Inland Waterway Association was one of a number of transportation lobbying groups that developed in the years around the turn of the century. Designed specifically to advocate for "construction of a continuous inland waterway across the State of Florida," the organization held its first convention in Columbus, Georgia. Recognizing the importance of government assistance, the group "favor[ed] the execution of the work under federal appropriations." To secure this funding, members saw the importance of collective action. "If we do not hustle eternally, vigilantly, and energetically," they argued, "we of the Gulf Coast will be left out of that $500,000,000 waterway bill. The work must be done now. Who will do the work? Will you Mr. Business Man, Mr. Shipper and Mr. Consumer? You know you won't and that the result will only be accomplished through organized effort. You can help by joining the organization." Fletcher provided that help, not only by addressing the Columbus meeting on the importance of a ship canal, but also by donating one hundred dollars to the group, by far the largest single individual contribution. Fletcher's presence in the organization also helped solidify Jacksonville's position as the potential Atlantic port for any future project. To underscore the city's growing importance, association leaders designated Jacksonville as the host city for its 1909 convention.[3]

Duncan Fletcher represented a new generation of Florida politicians supporting a cross-state canal. Like Richard Call and Josiah Walls before him, Fletcher saw such a project as beneficial to both Florida and the nation. Fletcher, however, added a local dimension by emphasizing its importance to the city of Jacksonville. Fletcher even reached out to his political opponents in an effort to win support for the project. Napoleon Bonaparte Broward, the dashing state governor whose passion for development led to his ambitious plan to drain the Everglades, ran a tough but unsuccessful campaign against Fletcher in the 1908 Senate race. Having won the election, Fletcher appealed to his rival, a fellow Jacksonville resident and civic booster, to join the fight for the canal. In September 1909, Fletcher wrote to Broward, requesting he "attend the Convention of the Mississippi to Atlantic Inland Waterway Association, and make us a speech." The next year, the Jacksonville Board of Trade sent Broward, now a newly elected United States senator, to Providence, Rhode Island, as its representative to the annual meeting of the Atlantic Deeper Waterways Association. Established to promote "the construction of a protected interior waterway from Boston to Florida," the group endorsed the cross-Florida canal by its alliance with Fletcher's organization, "with which the Atlantic Deeper Waterways Association sustains the most cordial relations." The support of these two influential Jacksonville politicians as Florida senators would have significantly increased the chances for full federal funding of a canal project. Unfortunately, Broward died suddenly of gallstones in October 1910 at the age of fifty-three, before he even took his Senate seat. Duncan Fletcher would have to champion Jacksonville's canal role on his own.[4]

Fletcher could not have arrived in Washington at a better time to fulfill his vision. On March 3, 1909, Theodore Roosevelt signed the Rivers and Harbors Act, which provided funds for yet another survey of a cross-peninsula canal. This time, however, officials viewed the project as a fundamental part of a much broader network. In tandem with the report from Roosevelt's Inland Waterways Commission, the legislation reflected a grand vision that funneled nationwide commerce down the Mississippi to the Gulf of Mexico, to then branch both eastward and westward between Brownsville, Texas, and Florida. It also recommended the establishment of an east coast waterway linking New England and Florida. The only thing that impeded progress toward a national water transportation network was the lack of a link across Florida's peninsula—thus the need for cost estimates of a barge canal across the state. The bill commanded the Army Corps of Engineers to once again conduct canal surveys. With due diligence, engineers evaluated five potential paths from 1909 to 1911, and submitted their final report to Congress in late 1913.

The five routes under consideration stretched from St. Marys, Georgia, in the north to Lake Okeechobee in the south. Quickly dismissing the southerly route as "so far south that it would have few advantages," and the northern passage as both cost prohibitive and lacking a suitable water supply, the engineers turned their attention to the three central routes with Jacksonville as their terminus. Concluding that two of the routes had insufficient water supplies at their summit pools, the engineers recommended the path proposed forty years earlier by Hubbard Hart. Using the lakes of central Florida to connect the Withlacoochee to the Ocklawaha, this passage relied upon two established port facilities—Jacksonville on the Atlantic and Port Inglis on the Gulf. Far from natural harbors like Tampa and Fernandina, these ports had a history of development by public and private interests in the past quarter century. This was to be no small project, as the Corps envisioned a fifteen-lock canal costing $16.5 million. At first glance, this seemed like a positive step for men like Fletcher. But as with previous surveys, this one came with the usual mixed message. Though appearing to support the idea of a canal, the engineers made it clear that their surveying was done only to fulfill the congressional mandate and not necessarily to endorse its construction. They asserted that "a canal, such as is proposed, would have no great value as a through route between Gulf and Atlantic ports." To emphasize its point, the engineering board concluded unanimously that "the project is not one worthy of prosecution by the United States."[5]

Once again, the movement for a canal ran aground on the ambivalent language of another army survey. As America became embroiled in World War I, the pressing demand for efficient transportation revived the issue. In June 1918, Georgia governor Hugh Dorsey convened a national summit of businessmen and politicians at the small coastal town of St. Marys. Calling for "the immediate construction, as a war measure, of a barge canal between St. Mary's on the Atlantic and St. Mark's on the gulf," Dorsey saw the meeting as a crucial step to establish national support for the project. As the eastern seaboard witnessed shortages of coal and other commodities, Dorsey, with the support of the Georgia legislature, asserted the proposed canal would provide an efficient and inexpensive means of shipping coal from the bituminous fields of Alabama and Kentucky to cities and naval stations along the east coast.[6]

Though the St. Marys conference produced no immediate results, Dorsey's initiative once again energized canal supporters in Georgia and Florida. Following the Georgia governor's lead, the legislatures of both states established canal commissions that flooded Washington with memorials and resolutions favoring the project. Georgia, of course, had an obvious reason for pushing for the northerly route. Less understandable was the wording of the 1921 enabling legislation for the Florida Commis-

sion. The legislature specifically charged the commission with constructing a "sea-level canal, capable of carrying the largest vessels, commenc-
ing . . . up the St. Marys River . . . to St. George's Sound on the Gulf of
Mexico." By the mid-1920s, boosters in both states hailed the project
as "the All-American Canal." Reminding policymakers that the "Panama
Canal was agitated for three hundred years, with all sorts of adverse reports, and France expended millions of dollars in attempts to complete
it," the Florida State Canal Commission argued, "it is a success, today,
and has changed the shipping routes of the world."[7]

The growing support for Dorsey's vision meshed well with an increasing interest in the nation's waterways during the 1920s. Sensing potential support from far beyond their border, Floridians now placed their canal proposal within the context of a larger integrated whole; in the words
of one Corps observer, "a multipurpose inland waterways program that
would encompass flood control, navigation, water power, and irrigation."
This plan culminated in the passage of the 1925 and the 1927 Rivers and
Harbors Acts—two crucial pieces of legislation in the history of a cross-
Florida canal. The 1925 law authorized construction of a 100-foot-wide,
9-feet-deep barge channel from New Orleans to Galveston. Recognizing
the importance of a continuous Gulf waterway, the act also called for
more Corps surveys in anticipation of expanding the project to include
a protected waterway east from New Orleans through Mobile, Alabama,
to Florida. With the federal government now firmly behind this Gulf
waterway system, Florida canal backers seized an unprecedented opportunity.[8]

The 1927 Rivers and Harbors Act contained language much more specific to the Florida canal. At the behest of the Florida Canal Commission,
Florida congressman R. A. "Lex" Green of Starke inserted a provision
for yet another Army Corps of Engineers survey for a St. Marys canal.
Though similar to the dozens of surveys over the preceding one hundred
years, this one would actually provide the impetus for construction. In
1936, Henry H. Buckman, an indefatigable supporter of a route stem-
ming from Jacksonville, recognized this when he wrote that the canal
"may be said to have had in its beginning the River and Harbor [sic] Act
of 1927 . . . [which] serves as a convenient and appropriate dividing line
between the earlier and later history of the project." In 1929, the new
survey reinforced the conclusions reached in 1913 and declared that the

more southerly route, using the Ocklawaha and Withlacoochee rivers, "has advantages which are believed by the board to be paramount." Yet, nothing in the final survey report was conclusive. Buckman seemed ecstatic, however, as he reported that the survey's "engineering considerations . . . confined the location of the project to the peninsula of Florida and eliminated the possibility of a route through both Georgia and Florida." Capitalizing on this moment of good fortune, he and other proponents also pushed for a deeper and wider canal cut—one that would accommodate larger vessels. Unwilling to accept the limitations of a barge canal, Buckman argued that a waterway on the scale of the Panama would now be economically viable because "the development of ocean-going traffic through the Straits of Florida had reached a point where the savings to . . . ocean-borne commerce indicated the justification of the more costly ship canal."[9]

Others built on Buckman's enthusiasm. In 1928, Sinclair Chiles, a Pennsylvania civil engineer, published a detailed pamphlet, entitled *A Florida Cross-State Ship Canal*, extolling the virtues of a project from Jacksonville to the Withlacoochee. Punctuating his argument with voluminous detail, Chiles traced the various twists and turns of nearly a century of Corps surveys and called for a moment of clarity—only a ship canal could meet the transportation needs of a growing nation. Dedicating the pamphlet to Duncan Fletcher, Chiles even turned to romantic verse to chart what would become the route of the Atlantic Gulf Ship Canal of the 1930s. "From the Atlantic to the Gulf of Mexico, by ship canal we soon shall go, via the St. John's, Oklawaha and Withlacoochee, when they're united from sea to sea," he began. Continuing with what seemed to be a Florida geography lesson, Chiles cataloged the various locales that would benefit from canal construction. In addition to Palatka, Ocala, and Dunnellon, commerce would reach as far as Alachua County to the north, and Lake County to the south. Announcing that "prosperity'll be showered on every hand," he concluded that "as the prophet foretold," this land will "blossom as the rose, where rattlers and moccasins and 'gaters [sic] now repose." Unlike the works of Sidney Lanier and Maude Powell, Chiles's stilted stanzas were no backward-looking lament for a paradise lost. Instead, he issued a clarion call for progress, as he saw Florida becoming "not only America's playground, but also her greatest workshop."[10]

The decade came to a close without the completion of the surveys authorized by the 1927 Rivers and Harbors Act. New legislation in July 1930 rectified that problem, by demanding the most comprehensive survey yet. Left unresolved, however, was the contentious issue of whether the waterway should facilitate barges or transoceanic shipping. Proponents for Jacksonville now had a chance to influence the decision, as government policy relied heavily upon "local interests to furnish technical evidence that the waterway when and if approved will serve as a real artery of commerce." Emphasizing broader economic considerations over the more narrow debate regarding location, the legislation also demanded that the project "must have a demonstrable investment value based on concrete facts and figures."[11]

Unwilling to let matters take their own course, Jacksonville's city officials seized the initiative. On December 1, 1930, the city recruited the engineering firm of Hills and Youngberg to investigate the economic feasibility of a ship canal across the state. As a measure of their commitment, the city budgeted fifteen thousand dollars to fund the study, a serious sum in the midst of the Great Depression. The firm picked managing partner Gilbert Youngberg to lead the effort—a wise choice for he had just retired from a long career with the Army Corps of Engineers, spending his last tour of duty as the commander of the Jacksonville District. With the assistance of the Jacksonville Chamber of Commerce, Youngberg spent ten months plowing through reams of statistical data generated by the United States Shipping Board, the Department of Commerce, and the Board of Engineers for Rivers and Harbors. Submitted in October 1931 to both the city and the Corps of Engineers, the resulting report made the case for a 38-foot-deep, 200-foot-wide ship canal as the most cost-effective means of transporting goods across the peninsula. Youngberg admitted that the initial cost would be high, but the net savings of over $47 million by 1945 would be well worth the investment. Confident in the strength of his economic arguments, Youngberg then answered the perennial question of location by declaring that "the general topography of the region forbids the construction of a ship canal farther north than the Suwannee and St. Marys rivers." Essentially eliminating Georgia from serious consideration, Youngberg saw Jacksonville as the eastern terminus of any cross-Florida canal. Especially when comparing Jacksonville to Fernandina, "the advantage is with the former on every point: . . . rail connections, trade, banking and manufacturing, wharves, fuel supply plants, shipping firms, [and] depth of harbor." Aware that his obvious preference for both Jacksonville and a ship canal would draw charges of rank partisanship, Youngberg averred that his "report is in no sense to be construed as mere promotion propaganda, but as a statement of forthright facts and fair deductions." Fernandinans, of course, would beg to differ, but their protests were soon drowned out by a flood of articles and speeches by Youngberg and his associates.[12]

While compiling data for his report, Youngberg tirelessly promoted the need for a Jacksonville ship canal in both engineering journals and general public forums. Decrying the inaction following dozens of Corps surveys, Youngberg lamented, "why, in a hundred years, do we not know whether or not we should have a cross-state canal of sizable dimensions?" Countering rivals who questioned the legitimacy of his report with claims that there was "some sinister motive in Jacksonville's activities," he asserted, "what this city has done is a contribution to the welfare of the nation, and particularly to Florida as a whole." Few could match the retired engineer's zeal for promotion. At a meeting of the Life Insurance Underwriters' Association in January 1932, Youngberg used every trick in a motivational speaker's handbook to raise enthusiasm for his vision. Seizing upon the word "opportunity" as the key to the project's impact on the River City, Youngberg explained that the word meant "Optimism, Purpose, Perseverance, Organization, Resourcefulness, Tenacity, Unity, Nerve, Initiative, Timeliness, and YOU (or YES—let's go!)." Moving past the "glittering generalities" of what he called his "pep stuff," he went on to address the more concrete benefits the canal would afford Jacksonville. In addition to the significant construction activities associated with the project, Jacksonville would inevitably become a world leader in naval traffic, which would in turn greatly stimulate manufacturing throughout the area. Such promise would not reach fruition, however, "unless energy is displayed and efforts made to have the canal completed."[13]

When combined with the authority of his initial report, Youngberg's campaign for the canal solidified support for a route with Jacksonville as the Atlantic terminus. Still unanswered was the question concerning the western outlet of the canal. While Sinclair Chiles had envisioned a passage through the Withlacoochee, Youngberg's report suggested a more northerly route that entered the Gulf through the Suwannee River. Once again, the efforts of local boosters would resolve the issue. Chief among them was A. F. Knotts of Yankeetown, a small settlement along Florida's undeveloped west coast north of Tampa. In the early 1920s, Knotts moved to the Sunshine State from the industrial heartland of northwest Indiana. A major economic player in the steel town of Gary, Knotts was a land developer and local Republican politician, becoming mayor of the neighboring city of Hammond from 1902 to 1904. After visiting the Gulf Coast on a fishing trip in 1919, Knotts was so impressed by its primitive beauty that he migrated to the isolated area around Inglis, where he purchased about four thousand acres. Far from moving to Florida's undeveloped west coast to simply enjoy its natural beauty, he dreamed, like so many others, of quick riches in land speculation. He established the town of Knotts in 1924—a municipality so identified with northern transplants that a year later locals demanded that it become officially named "Yankeetown"—and went on to become the town's mayor and preeminent booster. Seeing a canal as the guarantor of Yankeetown's economic success, he became a tireless crusader for the cause, giving speeches, writing newspaper articles, and lobbying government officials—all in the cause of favoring his city as the western terminus of the project.

In 1928, representing Florida at the annual convention of the National Rivers and Harbors Congress, Knotts espoused a grand vision of Florida's future. "It has been the dream of the engineering department of the United States for many years to develop . . . a canal not only around Florida but one across Florida. We have in the interior of Florida, thousands of beautiful lakes which could easily be joined together by canals, thus making the entire State of Florida a Venice." A year later, he established the Florida Inland and Coastal Waterways Association, an interest group designed specifically to secure a route through his town. As the organization's president, he had, according to the *Tampa Times*, "the ability, the courage, the leisure, the resources, and the will to keep 'everlastingly at it.'" Hardly content with simply advocating for his hometown route, Knotts went on the offensive and blasted the Florida State Canal Commission for its unwavering commitment to the St. Marys–St. Marks canal. He especially targeted the commission consultant Harry Taylor, former head of the Corps of Engineers. Taylor, Knotts claimed, had the audacity to now lobby the government for a "route he once declared to be impractical." Such a blatant reversal in judgment could not stand. Stating that "it is time for the citizens of Florida to get busy and try and head off [the plans of] this ambitious and expensive Canal Commission," Knotts demanded that waterways lobbying groups like his have equal time to champion their cause. Undeterred by Knotts's assault, the commission quickly shot back that "any activity [by Knotts's organization] . . . that is unauthorized by the Florida State Canal Commission concerning said project is out of order, and should cease at once." In the end, however, the persistence of Knotts and Youngberg would eventually carry the day in establishing the location of the canal.[14]

As local boosters worked to route the canal through their hometowns, national supporters established a new organization to press their case in Washington. In March 1932, they incorporated the National Gulf-Atlantic Ship Canal Association at a meeting in New Orleans. Organized to "promote the construction by the United States Government of a ship canal across Florida," the group presciently chose Jacksonville as its headquarters and included representatives from Florida, Texas, Louisiana, Mississippi, and Alabama. Well-organized and flush with funds, this regional association quickly became a formidable interest group. Floridians held considerable influence within the organization as A. F. Knotts and longtime Jacksonville mayor John Alsop secured positions on its board of directors. Henry Buckman assumed the important post of secretary, from which he could release reams of information touting the benefits of the canal to Florida and the nation. Walter Coachman, chairman of both the Florida State Canal Commission and the influential Consolidated Naval Stores Company based in Jacksonville, served as treasurer. Though significant, the presence of Coachman, Buckman, Knotts, and Alsop on the board was not the crowning achievement of the Jacksonville and Yankeetown supporters. That honor would go to the installation of General Charles Summerall as the association's president. A native Floridian, Summerall graduated from West Point in 1892. After a long and distinguished career, he retired in 1930 as a four-star general and army chief-of-staff. Summerall's notoriety, connections, and dogged determination made him a perfect choice to lead the organization. With the former general running this highly visible organization, it was no surprise that the *Wall Street Journal* reported that "Jacksonville is the logical location for the . . . outlet of the Gulf-Atlantic ship canal—a national project of the first magnitude."[15]

Wasting little time, the Ship Canal Association sent Summerall to Washington in July 1932 to lobby on its behalf. Asking for an estimated $175 million in government funds, he played up the economic benefits of the project. American shipping would save $20 million, he asserted, "by avoiding the long trip around the Florida Capes." Though impressive, such arguments had been heard before and seemed hardly sufficient to build more support for the canal. Thus Summerall forwarded an additional incentive. As the nation remained mired in the third year of an unprecedented economic crisis, the former general suggested that the "construction of the canal would mean employment for hundreds of men." The ship canal, in short, offered a solution to the most pressing problem of the Great Depression.[16]

Summerall's timing could not have been better. By 1932, unemployment had become so severe that it forced President Herbert Hoover to establish the Reconstruction Finance Corporation (RFC). An independent government agency capitalized at over $2 billion, the RFC represented the largest direct government economic program established to that point. Designed to provide federal loans to banks and other large corporate enterprises, the RFC also assisted businesses engaged in large-scale public works projects. Considering Hoover had already proposed over $700 million in public-works programs, it seemed like the perfect moment to secure government funds for canal construction. Thrilled by the prospect of a receptive audience from a freshly minted federal agency, Henry Buckman spent the summer preparing a comprehensive assessment of potential canal costs, finally arriving at $160 million for a sea-level ship canal. On August 1, 1932, the Ship Canal Association submitted a preliminary application for that amount to the RFC.[17]

Federal legislation establishing the RFC stipulated that all funded projects had to be self-liquidating in nature; in other words, they had to pay for themselves. In the case of Summerall's request, that meant the canal would have to be a toll project. Though this would allow the government to recoup the $160 million loan within ten years, it also meant that any savings to shipping would be significantly reduced with the implementation of payments to use the waterway. With transportation savings as a fundamental rationale for canal construction, the requirement for a self-liquidating project made the RFC proposal problematic from the start. No other canal in the United States charged a fee. As a result, the promise of RFC financing spelled problems for canal advocates. The *Wall Street Journal* saw nothing but gloom by October when it reported that "the Reconstruction Finance Corp. [sic] is believed unlikely to look with favor on the request for a large loan to a private corporation for the construction of a ship canal across northern Florida." In spite of its bleak prospects, the Ship Canal Association soldiered on and called for a formal hearing on their plan before the Engineers' Advisory Board of the RFC. Held on December 19, 1932, the meeting marked the last chance to convince the government agency of the legitimacy of the project. The association placed all its hopes with Buckman and Youngberg's voluminous and sophisticated charts, tables, and graphs, emphasizing the canal's long-term economic benefits over the more immediate need to alleviate the nation's unemployment woes. The frigid Washington weather was nothing compared to the cold reception Summerall and his fellow canal proponents received. Though it did not officially cut the project, the RFC offered little assurance they would provide assistance. Luckily for the Ship Canal Association, political winds were shifting in the nation's capital. The November election of Democratic presidential candidate Franklin Roosevelt somewhat eased the bitterness of the RFC's rejection and provided the possibility of federal support from a different agency.[18]

As Summerall made little headway in Washington, Florida officially backed the Jacksonville ship canal route. In early 1933, the legislature abolished the State Canal Commission and established a new government agency—the Ship Canal Authority of the State of Florida—with the broad authority "to acquire, own, construct, operate and maintain a ship canal across the peninsula of the State of Florida . . . [and] to exact and collect tolls and to prescribe rules for the privilege of passing through or along said canal." This agency replaced the private Ship Canal Association as the local organization designed to secure federal funding for the canal. Its leadership, however, showed little difference from its predecessor. Charles Summerall comfortably moved over to chair the new organization's board of directors. Walter Coachman also served on the authority's board along with members from Ocala, Palatka, and Cedar Keys, verifying the state's support of the Jacksonville–St. Johns–Ocklawaha–Withlacoochee route. Henry Buckman became the authority's engineering counsel, and continued to issue information supporting canal construction. With a state agency in full support of a Jacksonville ship canal and a new Democratic administration in Washington encouraging massive work projects to combat the Great Depression, the stars suddenly seemed aligned for canal construction. The only issue remaining was the final Army Corps of Engineers report on the best route across the state.[19]

The Depression year of 1933 ended on a high note for Summerall, Buckman, Coachman, Youngberg, Knotts, and their fellow proponents of the Jacksonville-Yankeetown route. A special board of the Army Corps of Engineers issued their supposedly conclusive survey report on December 30. The board "considered 28 different routes across both Georgia and Florida" and concluded that "the most practical and economical route for a ship canal has been determined to be that following the St. Johns River, Florida, from the Atlantic Ocean to about Palatka, Florida, thence in a generally westerly direction to the Gulf of Mexico, near the mouth of the Withlacoochee River." In terse bureaucratic terms, the Corps simply designated this course as "Route 13-B." Proposing a lock canal so as to avoid "seriously disturb[ing] the natural ground-water table," the Corps recommended a waterway 250 feet wide and 35 feet deep, at an estimated cost of $223,400,000. If a barge canal was to be built instead of a ship canal, the board recommended following the same path. Of course, lost in the excitement of the final determination was the usual Corps' caveat—such a waterway "is not economically justifiable at this time, and [the Corps] recommends that the construction thereof be not undertaken by the United States." Undaunted, Summerall and his compatriots embraced the new route and ignored the by-now usual decision that a canal was not a viable option.[20]

The recommendation for Route 13-B dovetailed nicely with the arrival of a new administration in Washington. Franklin Roosevelt's ambitious New Deal centered on unprecedented federal expenditures for public-works projects to alleviate the effects of the Great Depression. Such concerted action fit well with the plans of the Ship Canal Association. The fact that FDR was also a Democrat similarly pleased southern politicians, who were happy to be working with one of their own for the first time in a generation. Within weeks of Roosevelt's inauguration, long-time Democratic senator Morris Sheppard of Texas wrote the president, "commend[ing] to your attention the proposed Gulf-Atlantic Ship Canal across Florida." As the chairman of the Senate's Military Affairs Com-

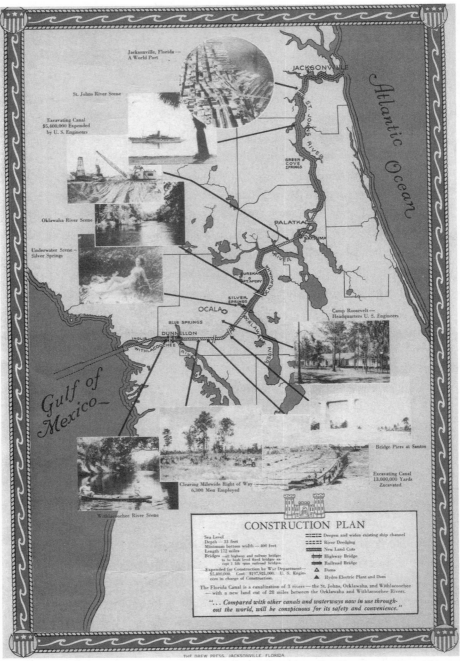

Army Corps of Engineers construction plan for the Florida Ship Canal, 1935. Courtesy of Department of College Archives and Special Collections/Olin Library, Rollins College, Winter Park.

mittee, Sheppard stressed that the project, among the usual laundry list of benefits, "afford[ed] an important element in the national defense." At the same time, Florida representative Lex Green, a member of the House Rivers and Harbors Committee, met the president to personally appeal for the project. Vowing "to leave no stone unturned in its behalf," Green seemed positive that the "actual beginning of this project may be realized in the near future" since Roosevelt expressed interest in the canal. Simultaneously, Florida governor David Sholtz provided an additional rationale, informing the president that "a very large amount of common labor can be used on this project." Noting that canal planning was "so far advanced that work can start promptly," Sholtz asserted that "its construction will necessitate large expenditures of private capital for auxiliary work, thereby causing additional employment of labor." Two months later, Florida's legislature sent a memorial directly to the White House. Echoing Sholtz's sentiments, it requested "the assistance and cooperation of every available federal agency in order to make possible, at an early date, commencement of construction work on a ship canal across the peninsula." Underscoring the shifting rationales for the project, legislators mentioned first, above all the fundamental economic and military considerations, the importance of giving "employment to a vast amount of human labor, thus greatly relieving the distress due to the unemployment crisis."[21]

The challenge of the Great Depression, rather than the merits of the project itself, became the major impetus for canal construction. Nearly one hundred years of surveys emphasizing the canal's economic, military, and transportation considerations had not been enough to ensure support for the project. With the country facing hard times, however, canal supporters saw an opportunity to convince the most skeptical of critics. Who could argue with a project that, in the midst of record unemployment, would put thousands of men to work? With a president in the White House willing to embrace expansive public-works projects, with a Democratic Congress willing to defer to executive authority in the midst of an economic crisis, and a public looking to Washington for bold initiatives, conditions seemed perfect to begin canal construction.

Bureaucratic impediments, however, delayed the project's start. Caught between the procedural technicalities of legislative funding and the transfer of projects from the authority of the RFC to the Public Works

Administration (PWA), the project's loan application, now sponsored by Florida's Ship Canal Authority, languished until an opportune moment could guarantee its success. Hopes for a canal diminished, however, as PWA engineers issued a report in October 1933 that was at odds with the Corps' cost estimates. To make matters worse, Secretary of the Interior Harold Ickes, who administered the purse strings of the PWA, had little enthusiasm for the canal. Frustrated by the lack of progress, nine senators from Gulf Coast states appealed directly to Roosevelt in March 1934. Pointing to the discrepancies between the engineers' reports, they called for the creation of a separate review board to resolve the fate of the Florida canal.

By early 1934, Roosevelt responded to their initiative with the establishment of a special review board consisting of five engineers—two appointed by the Corps, two appointed by the PWA, and a chairman representing the private sector—who tried to cut through the morass and reach a final, definitive conclusion. One of the army appointees was Major Brehan Somervell, who later managed the canal's construction. After visiting Florida to examine the route and determine the effect of a canal cut on the state's water supply, the board issued its appraisal directly to the president on June 28. The report arrived at a new set of conclusions concerning the entire conceptualization of the canal. Though in agreement with Route 13-B, the board took exception to earlier plans focusing almost entirely on the development of a lock canal. Instead, it called for a sea-level ship canal, which offered lower initial costs and easier construction. In addition to cheaper operating and maintenance expenses, a sea-level canal also offered greater ease and capacity for shipping. The only possible disadvantage would be the assumed negligible damage to local water wells along the right-of-way. Such a trade-off seemed worthwhile, however, when the price of the project came in at $142.7 million, considerably less than earlier estimates of close to $200 million.

Three months later, in a supplemental technical report, the special review board augmented its decision with voluminous tables and charts indicating the economic viability of the project. It also listed a series of "collateral and indirect benefits" that could "not be given a monetary value in a financial report." Besides the "reduction in hazard to shipping during the hurricane season," and the "value to shipping and possibly to the Army and Navy during a war," the board also affirmed the canal's importance in providing "relief in the present emergency." Canal boosters applauded the recommendation. The Army Corps of Engineers, now fully committed to the project, immediately designed a "tentative program for construction of the canal." On January 19, 1935, the chief of engineers informed Roosevelt that the Corps was "prepared to initiate the work with dispatch and to prosecute it with the greatest vigor." Both parties would only become frustrated, however, as Roosevelt delegated the final decision to Ickes. Ten days after the Corps articulated its eagerness to begin construction, Ickes validated his opposition to the canal by summarily rejecting the Ship Canal Authority's loan application.[22]

Having come so close to fulfilling their dream, canal advocates were unwilling to give up easily. They continually petitioned Roosevelt as they worked the halls of Congress, seeking legislation that might secure funding from other federal sources. In February, Congressman Lex Green tried to circumvent Ickes's decision by introducing a bill for a congressionally funded toll-free canal along Route 13-B. However, on April 8, 1935, Roosevelt signed the Emergency Relief Appropriation Act, which allocated federal funds to combat unemployment directly under executive authority, thus making Green's bill superfluous. Within the newly established Works Progress Administration (WPA), Roosevelt now had the wide latitude to grant money to the Florida Ship Canal Authority without congressional approval. Moreover, under the leadership of canal supporter Harry Hopkins, the WPA seemed much more willing to provide funding. By June 1935, everything seemed to be falling into place for the construction of a cross-peninsula canal.

For the most part, Roosevelt, like Hopkins, was receptive to the canal. The Atlantic Gulf Ship Canal was in keeping with the New Deal's effort to improve and modernize the nation's infrastructure. Yet, a primary motivation for the project was the pressing need for low-cost labor relief. As early as January 1935, the president suggested that he would allocate only as much money as could be spent in a year, with the condition that 50 percent of the funds must go to labor costs on a scale "somewhat below the local scale for common, semi-skilled and skilled labor but above [the] home relief scale." The WPA also had to employ people already on relief rolls. Canal boosters fully concurred with Roosevelt's conditions and continued to tout the project's potential to ease unemployment following the passage of the Emergency Relief Appropriation Act. Indeed,

Roosevelt's signature was barely dry before Florida's entire legislative delegation met with the president to seek his commitment to the project. Leaving the White House with the promise that the canal "will be one of the first things to be taken up by the works administration," they called a press conference and declared that "President Roosevelt was in full sympathy with the cross-state canal project connecting the Atlantic to the Gulf of Mexico." Amid the excitement, Duncan Fletcher's office, anticipating an announcement regarding the project's funding the following week, cautiously added that "the amount will depend largely on the number of men that can be employed—it may not be as large as the Senator wanted."[23]

As usual, Floridians would have to wait far more than a week for the government's decision. Once again, the proposal wended its way through yet another federal bureaucracy. On June 1, 1935, signs of success appeared as a WPA review board tentatively accepted a plan to allocate $25 million to cover construction costs for the first year of what it viewed as a three-year program with a total cost between $99 million and $119 million. That same day, Florida's legislature established the Florida Ship Canal Navigation District, a special taxing authority designed to provide the monies necessary to purchase lands for the canal's right-of-way and then convey them to the federal government, which in turn was responsible for the project. Though established as a separate bureaucratic entity, the Navigation District was in reality an extension of the Ship Canal Authority, with Walter Coachman serving on both bodies. In spite of the positive steps, a final decision, first from the Advisory Committee on Allotments, and then from director Harry Hopkins, dragged on into August. By that time, however, Roosevelt was publicly expressing doubts about the project. According to the *Gainesville Sun*, he concluded it would not be "right to proceed with construction of the canal without specific congressional authorization." Moreover, he balked at "the high cost—estimated at $146,000,000—mak[ing] the canal an unprofitable adventure under the emergency program."[24]

Such misgivings quickly alarmed the project's boosters, who saw their dream once again stalled. On August 21, Jacksonville mayor John Alsop sent a plaintive telegram to Duncan Fletcher complimenting him on his "eloquent and determined appeal to the president to make a reasonable allotment for the canal. . . . We beg of you Senator," he added, to "remind the President again that he promised you to make the first allotment this week." Alsop and his fellow supporters departed immediately for Washington to provide support for Fletcher's struggle to secure funding. For three days of closed-door meetings, Walter Coachman and Charles Summerall, with between twenty and forty proponents in tow, conferred with congressional officials and executive-agency representatives to determine how best to secure federal assistance. Supporters relied heavily on George Hills, the managing partner of Hills and Youngberg, the Jacksonville engineering firm that had produced the 1930 procanal survey. Hills was not only a competent engineer, but an ardent Roosevelt Democrat, labeled by *Time* as "the New Deal's dispenser of patronage in Florida." He pursued every possible political avenue to make the case for the canal. Taking a different tack than Alsop, Hills blasted Fletcher for his inability to persuade Roosevelt to fully support the project. In a letter to presidential secretary Marvin McIntyre, Hills declared that boosters "abused [Fletcher] as he had never been abused before. They told him that his failure to secure the Canal stamped him as a failure as Senator; that he had consistently supported the Administration and lacked back-bone in failing to demand and secure favorable action in the matter in return, and suggested to him that he should properly resign and make place for some one who could get results." He even went so far as to suggest that the unveiling of an honorary oil painting of Fletcher in the Jacksonville City Hall be canceled, since "if this project is not announced, . . . we feel it would be unwise to attempt to go through with that ceremony." Hill's screed was humiliating enough, but the senator was simultaneously barraged with telegrams from north Floridians demanding further action on his part. "Everyone is dismayed and cannot understand what has happened," asserted a typical message. Sumter Lowry, a prominent businessman and chairman of the state's Canal Committee, warned Fletcher that "as your friend and supporter, I strongly urge you not to let yourself be put in position where you will be greatly injured in Florida by the failure of the canal to be announced as promised."[25]

Shaking off the assaults, Fletcher recognized the depth of support for the project and redoubled his efforts to obtain funding. Moreover, Hills contemplated the dire political consequences to the Democratic Party if the canal was not constructed. He warned McIntyre that the "failure to announce a project in the substantial amount approaching $25,000,000 will seriously embarrass the position of the Party in this State, and bring about no end of punishment of the Administration's best friends here, including our Senior Senator." Piling on, he even brought up the Communist bugaboo, claiming they "are well organized in four of our principal cities. Each one of these groups is busily engaged in building up dissatisfaction with the Administration and they are not over-looking the failure of this project as a major argument in fanning the fire of discontent."[26]

In the end, Hills's persistent lobbying, combined with Senator Fletcher's renewed efforts, including a personal appeal to Roosevelt himself, proved successful. Both Hills and Fletcher stressed the economic benefits of canal construction, basing their argument on the need for work relief. The $25 million they requested from Roosevelt "would promptly absorb some 20,000 of our unemployed from North Florida and South Georgia. We need it badly [as] . . . we have had a number of disturbances by the unemployed through the past thirty days." On August 14, the president ordered Hopkins to prepare a report determining the number of people who could immediately be put to work on the project. At a press conference one week later, he remarked that "the government *probably* would go ahead and clear the right of way for the Florida Gulf-Atlantic Ship Canal" (emphasis added). The announcement revealed yet another moment of hesitancy on the president's part. Though impressed by the efforts of procanal forces, Roosevelt, always the consummate politician, was considering a more modest budget plan to cover only the initial phase of construction. By the end of August, Fletcher and Hills, buoyed by newspaper reports that the $25 million allotment was a done deal, still expressed optimism that FDR would provide that amount. However, on August 30, the president pulled some political chicanery that so often frustrated allies and foes alike as he finally freed up money for construction—but just $5 million instead of the full amount expected by supporters. Announced at a presidential news conference, the allocation was but a first step toward completion, and limited expenditures to clearing and excavation of the canal right-of-way as well as deepening the St. Johns, the Ocklawaha, and the Withlacoochee in anticipation of deep-draft ship traffic. With this simple bureaucratic decision, Roosevelt broke the logjam of political wrangling that had held up canal construction for decades. Summerall and Coachman, Buckman and Youngberg, Fletcher and Green—though disappointed by the meager initial allocation—found cause for celebration. Their dream for a bright Florida future, their belief in improving nature for man's benefit, was finally a reality.[27]

For Roosevelt, this decision was determined less by the grandiose vision of canal backers than the rough-and-tumble world of Washington. If the political winds blew in favor of the Ship Canal, as they did in August and September 1935, FDR would be an ardent supporter. Conversely, if the president detected even the slightest shift in those winds, he would withdraw his support without hesitation. FDR was particularly attuned to the political climate within the Sunshine State itself. Though Fletcher, Green, and Hills had done a good job of selling the canal to Roosevelt, other powerful forces within the state vigorously opposed construction. Railroad interests, a nascent conservation movement, representatives of the state's powerful and economically important growers, and even Miami and Tampa port officials concerned about the potential loss of revenue: all of these soon pressured the president to reconsider his position. In a June 1935 editorial, the *Tampa Morning Tribune* charged that "the whole state is awake to the menace." Even with the money allocated and with construction plans underway, the decision to complete the project still came down to politics. Thus canal boosters would have to keep up the pressure and convince Roosevelt he should stay the course and keep the funding spigot open.[28]

Of all things, a major natural disaster provided the politically astute president the chance to rally support for his decision. Not only political winds were blowing that fateful September. On September 2, an intense, tightly wound hurricane bore down on the Florida Keys. Dubbed the "Labor Day Hurricane of 1935" in a time before massive storms received such personalized monikers as "Andrew" and "Katrina," the hurricane was possibly the most powerful ever to strike the United States. Before smashing into Lower Matecumbe Key, the storm's nearly 200 mph winds forced the cruise ship *Dixie* to run aground on French Reef. Two days out of New Orleans and bound for New York, the flagship of the Morgan Line entered the treacherous waters of the Florida Straits as the hurricane approached the peninsula. For forty-eight hours, the passengers and crew endured battering winds, torrential rains, and surging seas before the Coast Guard rescued them and sent them to New York by rail. Though the *Dixie* suffered an estimated $500,000 in damages, no fatalities were recorded due to the professionalism of the crew and Coast Guard. The situation in the Keys was much different, as the storm destroyed a series of construction camps built to house workers building the new overseas highway to Key West. Predominantly populated by World War I veterans seeking work relief, the camps virtually disappeared in the face of the howling winds and deadly storm surge. The American Red Cross placed the death toll at 408, with many others unaccounted for in the wake of the tragedy.

As newspapers and magazines covered the story for weeks, much of the initial attention centered on the grounding of the *Dixie* rather than the tragic fate of the veterans. This played well into the hands of both boosters and the president. For Summerall, the *Dixie* incident provided an object lesson in why the canal should be built. "The canal," he remarked a few days later, "will enable shipping to avoid the hazards of the tornado [*sic*] area. This is forcibly illustrated at the moment of announcing the project by the wreck of the S.S. *Dixie* and the endangering, if not the loss, of lives incident to this calamity. The ship alone is probably worth more than the entire initial allotment to begin construction of the canal." Writing to Roosevelt a day after the "*Dixie* [was] wrecked on the Florida Keys, that maritime graveyard," Duncan Fletcher claimed the incident presented a "very forceful argument in favor of the Florida Ship Canal. The vessel was proceeding from New Orleans to New York. If the Florida Ship Canal had been available she would have escaped this disaster." Roosevelt concurred. Faced with the many media accounts of a ship being "bounced around like a log on a whirlpool," who could argue with the president's assertion that such a project "would forever make it unnecessary for sea goers to risk their lives in the circumnavigation of Florida's long, hurricane-blistered thumb." Indeed, the *Dixie*'s near-miss became inextricably linked with the canal's origins and lore for decades to come. Given all its play in the popular press, especially in a 1936 *The March of Time* newsreel, it is no surprise that much of the public gave the *Dixie*, rather than the prolonged lobbying efforts of men like Fletcher, Buckman, and Summerall, ultimate credit for clinching Roosevelt's support for the project.[29]

After a hundred years of countless surveys and bureaucratic foot-dragging, work began on a project of extraordinary scale. The 195-mile passageway would require the excavation of somewhere between 600 million

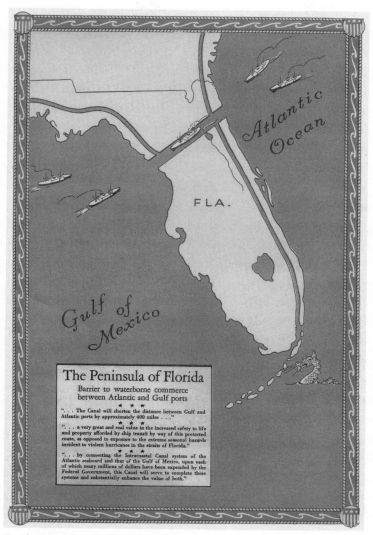

The Peninsula of Florida

Barrier to waterborne commerce
between Atlantic and Gulf ports

★ ★ ★

". . . The Canal will shorten the distance between Gulf and
Atlantic ports by approximately 400 miles . . ."

★ ★ ★

". . . a very great and real value in the increased safety to life
and property afforded by ship transit by way of this protected
route, as opposed to exposure to the extreme seasonal hazards
incident to violent hurricanes in the straits of Florida."

★ ★ ★

". . . by connecting the Intracoastal Canal system of the
Atlantic seaboard and that of the Gulf of Mexico, upon each
of which many millions of dollars have been expended by the
Federal Government, this Canal will serve to complete these
systems and substantially enhance the value of both."

For the Army Corps of Engineers, Florida represented an impediment to the nation's waterborne commerce that only a ship canal across the state could resolve. Courtesy of Department of College Archives and Special Collections/Olin Library, Rollins College, Winter Park.

and 900 million cubic yards of rock and dirt, significantly more than was removed to construct the mighty Panama Canal. Far from merely cutting a ninety-mile path directly through the Central Florida Ridge, the project also included significant alterations to the St. Johns, Ocklawaha, and Withlacoochee rivers. While initial designs recognized the need to preserve "the beauty of Silver Springs" as well the Ocklawaha and Withlacoochee rivers, project engineers quickly called for "much straighter cuts and the elimination of the sinuosities in the present channel" of the St. Johns River. Construction would similarly involve dredging a massive channel—500 feet wide at the shoreline and 1,000 feet wide at its mouth—nearly twenty miles into the Gulf of Mexico to make a navigable entrance for the cross-peninsula passage. Ancillary structures included four spillway dams and ten to twelve highway and railroad bridges with horizontal clearances of 300 to 500 feet and a minimum vertical clearance of 135 feet. The undetermined number of bridges underscored the flexibility of the Corps' planning, which even included the possibility of another feeder canal cut across the south Jacksonville peninsula.[30]

Those impressive plans for future canals were entirely dependent upon the possibility of increased funding. As it stood in September 1935, the Corps was restricted by the financial considerations of the relatively paltry $5 million and a demand to finish work within five years. With those considerations in mind, engineers followed the recommendations of the penultimate 1934 survey, redesigning the canal as a sea-level route rather than one with locks and dams. This meant simply cutting a 30-foot-deep, 250-foot-wide swath across Florida and its vulnerable freshwater aquifer. The project also entailed doubling the depths of more than 105 miles of existing waterways. Along the St. Johns, for example, the channel's bottom width would reach as far as 400 feet. The Herculean effort would be well worth the cost, however, as planners anticipated the canal accommodating 94 percent of the oceangoing commercial vessels that took the long passage around the Florida peninsula. With a transit time of roughly twenty-five hours, planners saw a new ship passing through the canal every hour of the day. Even in its narrowest sections, the canal's width would enable two cargo ships to pass by each other with relative ease. When compared to the carrying capacity of its predecessors, the proposed Ship Canal allowed for twice the traffic and nearly twice the tonnage as the Suez and Panama canals. This was to be no mere ditch,

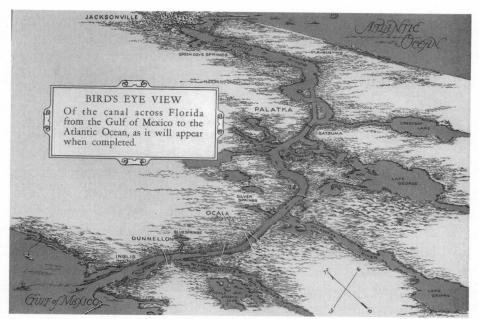

A bird's-eye view of the proposed Florida Ship Canal, 1935. Courtesy of Department of College Archives and Special Collections/Olin Library, Rollins College, Winter Park.

but the crowning achievement of an integrated American waterways system.

The Corps and canal boosters saw rivers not as distinct natural entities but as potential networks for a wide-ranging inland waterway system connecting the Mississippi Valley to the entire east coast. When completed, the Florida Ship Canal would be part of what the Corps consistently called "The Missing Link," the final connection between the Midwest and the Atlantic coastline. Fueled by the Corps' historic mission to facilitate internal improvements and helped by a federal government committed to public-works projects to relieve unemployment, such large-scale water projects as the Tennessee Valley Authority, the Grand Coulee and Hoover dams, and the Atlantic Gulf Ship Canal were particularly situated for success in the 1930s. Indeed, given those conditions, it is striking that the Atlantic Gulf Ship Canal was *not* completed, which underscores the importance of both local and national politics to the project's fate.[31]

Though boosters applauded the rapidity and decisiveness of Roosevelt's support, they would soon rue the relative lack of planning and forethought in making the project a sea-level venture. Cost-cutting measures may have guaranteed success among Washington's decision-makers, but in the long run such decisions led to the project's downfall. Ironically, a more modest lock-barge canal would have been more expensive in the short run but also would have been less intrusive to the Florida environment and less controversial to opponents both in Washington and within the state itself. For despite its proposed economic benefits, the audacious vision of the Ship Canal quickly galvanized a backlash opposed to the potential saltwater intrusion into the state's water supply, threatening both the life and livelihood of all Floridians.

While critics began thinking through the implications of Roosevelt's decision, the Army Corps of Engineers—even before the official groundbreaking—was hard at work on the project. Within a day after the *Dixie*'s grounding, the Corps established a separate Ocala District, organized specifically to manage canal construction. The newly appointed head of the project, newly promoted Lieutenant Colonel Brehon B. Somervell, proved to be an eager taskmaster. A competent, ambitious, and politically connected officer, Somervell arrived in Ocala the evening of September 6 and announced that he would immediately employ four shifts of workers to toil day and night. No mere military bureaucrat, he was already a rising star with significant experience dealing with water projects when Army Corps of Engineers chief Edward Markham appointed him to head canal construction. Somervell went on to a long and distinguished military and civilian career, heading the WPA in New York City from 1936 to 1940, playing a key role in the planning and construction of the Pentagon, leading the army's logistics branch in World War II, and participating in the Manhattan Project. Always an aggressive leader, Somervell seemed acutely aware that a fait accompli was perhaps the best argument to secure more governmental funding. As he began project operations, he boldly proclaimed, "we are going to push the canal right along as long as the money holds out. It's up to the other fellows to provide us with additional funds." Faced with the meager initial appropriation, he wasted little time in initiating construction. "We've got to clean up the relief rolls, that's the hurry," he announced. "The whole New Deal works program must be under way by November first and the Florida canal is one of the outstanding projects of the New Deal."[32]

In the shadow of an impending debate about both the efficacy of the canal and the lack of congressional authorization for funding, construction began at 6:00 a.m. on September 6, 1935, as thirty men began building a camp to house engineers and workers. Located about two miles south of Ocala on the Dixie Highway (present-day U.S. 441), the facility, officially named Camp Roosevelt, consisted of "quarters for officers and barracks for enlisted men and laborers, complete with canteens, mess-halls, and all the other appurtenances of an army post, including guard-house." Hailed as the future "show place of the state," the camp had already attracted more than five thousand visitors within the first ten days of construction. Initial plans called for an elaborate complex that included a school, hospital, baseball diamond, and other recreational facilities, as well as community gardens. Such amenities remained on the back burner, however, as the Corps scrambled to establish a much simpler base of operations. Within three weeks, the Corps put more than three thousand men to work in building the main camp as well as constructing portable bunkhouse sections for later use as satellite facilities established along the canal right-of-way. With hundreds of skilled carpenters and brick masons at their disposal, the engineers could guarantee results. "If an office building, a warehouse or anything else is needed," observed one reporter, "the location is staked off . . . and up it goes without a minute's delay. Work proceeds at night under flood lights and always under smooth army efficiency." By the middle of November, much of the initial infrastructure was completed, and the Corps turned its full attention to the primary task of digging the canal itself. Meanwhile, the state continued the process of acquiring land for the canal right-of-way. A special election was held in October that granted the Navigation District authorization to issue $1.5 million in canal bonds for land purchases. Showing significant support for the canal in the six counties it affected, the authorization passed by the overwhelming margin of 15,433 to 590. With this mandate in hand, the Ship Canal Authority had bought "either through option or condemnation, approximately 34 miles of right of way involving 21,173 acres" by the beginning of 1936.[33]

Few could contain their excitement amid the whirlwind pace of construction. Two weeks after the project began, a reporter announced that "more than 1500 workers on the Gulf-Atlantic ship canal buried the depression here today as clinking cash registers rang out a lively tune." As workers spent much of September building Camp Roosevelt,

Somervell began the process of clearing underbrush along the canal right-of-way by felling the first tree himself. Exuding confidence, Somervell more than lived up to his reputation as a no-nonsense administrator. When asked by a reporter if he was concerned that the government would fail to spend more than the $5 million already allotted, he replied, "If the War Department didn't believe it was a permanent thing, they could have sent a captain down here to handle the preliminaries." The reporter's follow-up vividly underscored the project's association with Roosevelt: "But suppose President Roosevelt is not re-elected. Would the canal have the same chance under somebody else?" Somervell answered laconically, "There will always be a president."[34]

Though construction began immediately after the Labor Day Hurricane, the official groundbreaking was held on September 19 at a location on the canal route nine miles south of Ocala. Touted as the greatest moment in not just Ocala's, but the state's, history, the ceremony was filled with pageantry and color. As if to underscore the importance of the event, Franklin Roosevelt himself, linked by telegraph from Hyde Park, set off fifty pounds of dynamite to inaugurate the project. With stores and schools in the region officially closed by noon, more than fifteen thousand enthusiastic supporters gathered to hear prominent Floridians extol the virtues of the project. Among them was the Ocala newspaper editor R. N. "Bert" Dosh, who saw the moment as the fulfillment of a dream in which "Ocala will be at the connecting crossway of the inland water courses of America." Dosh became to Ocala what A. F. Knotts was to Yankeetown—a tireless civic booster who saw the canal as the key to his city's economic improvement. He even went to Washington in August to participate in the last-minute lobbying efforts to secure funding.[35]

Though Dosh was a well-known local figure, his notoriety paled in comparison to the "bigwigs [that] packed Ocala for [this] historic occasion." State and national politicians came, as well as numerous business and civic leaders from Tampa, Jacksonville, and Orlando. Also on the dais that September day were the boosters and lobbyists who had long supported the canal. Their obligatory speeches highlighted the project's benefits to the state and the nation. Congressman Lex Green, recognizing the president's role in the construction, even suggested naming the canal in FDR's honor. Henry Buckman—who, more than anyone, appreciated the long struggle to get to this moment—announced that "the fight has only begun. . . . We should not relax our efforts until the ships are here at our doors." Seconding his sentiments, Sumter Lowry, a prominent Tampa businessman and the Canal Committee chairman, cautioned that "we have started but we must fight every foot of the way until ships are actually passing here."[36]

The honor of delivering the keynote address went to the elderly and ailing Senator Duncan Fletcher, credited by many Floridians with successfully securing the funds for the project. In his long and rambling speech, he boldly claimed the Florida Ship Canal would "make the Gulf of Mexico the Mediterranean of the western world. It will be an improvement for all the country. It will bring prosperity to Florida." The ceremony proved to be an auspicious start, yet those suspicious of omens had good reason to feel ill at ease. Unable to keep an eye on the clock, the long-winded Fletcher found his speech interrupted as Roosevelt triggered the dedicatory explosion at precisely 1:00 p.m. Just as the senator proclaimed "Ships from all nations will pass through this very site and they will avoid in the so-called graveyard of ships the fate of the *Dixie*," a deafening blast, "ignited by the President's touch, blew Florida sand and an under stratum of shell fifty feet into the air." The disruption immediately halted the ceremony as thousands began to scream and blow their car horns, rushing to the site of the new ten-foot crater. In spite of the blunder, boosters were confident that they were on their way to building "one of the wonders of the world."[37]

Duncan Fletcher's recognition of Roosevelt's role in canal authorization demanded a symbol of gratitude. In November, Fletcher sent the president "a token of appreciation." The gift was a silver tray on which was attached "a shell from the coast of Florida, encased in gold and containing a portion of the first earth excavated in the construction of the canal as a result of a blast set off by you." Engraved with the names of governors and senators, as well as members of the National Gulf-Atlantic Ship Canal Association, the tray praised FDR as the "builder of the Florida Canal." In terms reminiscent of Andrew Jackson, the tray's engraving represented the century of rationale for canal construction: "That our commerce be sped, our defense secured, and the unity of our country preserved." Fletcher assured fellow politicians that "the President was much pleased with our gift." With the ceremonies out of the way, it became paramount "to push the canal to completion as soon as possible."[38]

Along those lines, following the September groundbreaking, work began in earnest on both clearing the land and digging the canal itself. Among thousands of workers, crews of 80 to 120 men removed timber and underbrush by hand for eventual excavation. While project managers established portable camps from Palatka to Dunnellon, much of the work concentrated on the Central Florida Ridge, the nearly five thousand acres of land between the Ocklawaha and the Withlacoochee. Given the work-relief requirements of a WPA project, the Corps' excavation mixed modern technology and old-fashioned muscle: "Working alongside the modern, powerful excavating machines were men loading trucks with shovels and mule teams dragging old-fashioned scrapers. Huge tractor-scrapers, draglines, belt conveyers, tractor-hauled wagons, and trucks all played a major role in the excavation process, but always there were scores of men chopping and digging with shovels and trimming the slopes of the canal by hand." The use of relief workers came at a cost, however, as significant turnover resulted from "many of the relief laborers . . . lacking in physical stamina." Despite the preference for men over machines, the Corps made considerable progress by mid-1936, digging nearly ten miles of land across the Central Florida Ridge with no cuts into the underlying limestone.[39]

Canal construction, of course, brought a sudden burst of prosperity to Ocala as "money [was] easier to get and business generally [was] better." The project was so associated with the city that an *Orlando Sentinel* headline shouted, "Boom Awakens Sedate O-Can-Ala." Mayor B. C. Webb jokingly complained that the town was growing so fast that municipal officials now had to install large numbers of new traffic lights. Real estate changed hands so frequently that "telegraph messenger boys are being run to the point of exhaustion delivering messages." New restaurants, hotels, and theaters quickly opened as business increased between 25 and 50 percent. Native Ocalans recognized the economic importance of the project and conveniently looked the other way as bars and slot machines proliferated in their community. In one county meeting, ten applications for liquor licenses appeared on the agenda. Perhaps the biggest beneficiary of the boom was the proprietor of the Blue Bird barbe-

As a source of work relief, the Ship Canal counted on manual labor for excavation. Courtesy of Ocala/Marion County Chamber of Commerce, Ocala.

cue and filling station, located on the Dixie Highway near the entrance of Camp Roosevelt. Finding a "fortune dumped into his lap overnight," he had to employ "so many helpers they haven't room to turn around, and he [could] hardly get enough beer, sandwiches and soft drinks to supply the day and night crowds." His site was so lucrative, patronized daily by more than two thousand workers, that he was offered the then-exorbitant sum of twenty-five thousand dollars for a two-year lease of the business. Ocalans saw the Blue Bird's success as a portent for its own future prosperity. According to the *Tampa Tribune*, this fantastic growth "is the start of a boom that will make the city of 8098 population one of the greatest communities of the state." Camp Roosevelt, of course, provided the spark, and to some observers even posed a threat in overtaking the nearby city. The editors of the *St. Petersburg Times* ruefully noted that the work camp could maybe become a "permanent city, perhaps the future inland metropolis of Florida. All we old-timers, however, will hope that Ocala will always be at least its most important suburb."[40]

The canal had its biggest impact on the unemployed. Without a hint of exaggeration, the *Tampa Tribune* reported, "there is not a single unemployed person in Marion County today, and the jobless of neighboring counties as far south as Hillsborough are being rapidly put to work." Recruited from Florida's relief roles, more than six thousand men—a large number but far fewer than the twenty thousand the Corps had envisioned for completion of the project—had been put to work by mid-1936. In September 1935, a Pasco County judge wrote Congressman J. Hardin Peterson requesting assistance in "placing about six first class carpenters on the Canal job." These workers had been employed through the Federal Emergency Relief Administration (FERA) and were currently making only $4.50 per week. Peterson replied that all canal workers first had to be registered with the National Reemployment Service, but these carpenters could still end up on the canal as "100 carpenters were recently recruited from Tampa and sent to the canal project." African American workers also saw the canal as a chance to improve their condition. The president of Tampa's Urban League even requested a meeting with Brehon Somervell himself, "relative to jobs for Negroes at the Cross State Canal."[41]

By Depression standards, pay for canal work was good, with workers making thirty cents per hour. Laboring six days a week in six-hour shifts, the men cleared $10.80 weekly. With deductions for camp meals at fifty cents per day, workers brought home $7.80, enough to live on and spend freely in Ocala's burgeoning entertainment district. As Ocala boomed, officials of other Florida cities publicly denounced the ship canal for draining labor from their municipalities. On August 20, even before construction began, Tampa mayor Robert Chancey sent a blistering telegram to WPA head Harry Hopkins, complaining that "employables of Hillsborough County are to be assigned to Cross State Canal, taking them more than one hundred miles from their homes and depriving this community of its work man power in carrying on local work projects." Within Marion County itself, farmers and employers grumbled about hired labor, especially African Americans, being siphoned off by the project's lure of higher wages and shorter hours. One crate mill, for example, had to close operations because of the sudden labor shortage. Despite this and other problems, most residents of Marion County gladly accepted the canal, its workers, and the economic boost they provided.[42]

The single exception to all of this prosperity lay in the fate of Santos, a small, unincorporated, predominantly African American community located squarely in the projected path of the waterway. It would be naïve to presume that it was mere coincidence that this community would be the only one to disappear in the wake of the canal. Located seven miles south of Ocala, Santos emerged in the late nineteenth century as a free black settlement along the tracks of the Seaboard Coastline Railroad. Since that time, it remained a stark reminder of the strict color line of the Jim Crow South as well as a vivid example of the sense of community established by disenfranchised blacks. Cut off from the hustle and bustle of Ocala, it became relatively self-sufficient, with a local school and numerous black-owned businesses. As little more than a whistle-stop between larger destinations, Santos remained a small but vibrant community of hardworking and close-knit African Americans. Reflecting the tension between Saturday night and Sunday morning, the town simultaneously supported three juke joints and three churches that were at the center of black life. As with many black towns at that time, the community baseball diamond offered additional recreational opportunities as Santos became an integral stop on the local semiprofessional Negro League circuit. In addition to the usual agricultural activities, residents could work in a local rock-crushing mill and a factory that processed Spanish moss for stuffing pillows and mattresses. All of these activities came to an end with canal construction.

With the establishment of Route 13-B and the creation of the Ship Canal Authority in 1933, Santos quickly became little more than a memory. State officials under the auspices of that agency flooded the area, going house to house along the right-of-way, purchasing land for pennies on the dollar. Powerless African Americans, bombarded with rhetoric associating the project with FDR's New Deal, had little choice but to sell. In December 1935, for example, the state acquired the sanctuary and lands of the Calvary Baptist Church for a mere $750. The entire community was wiped out, with only a small Methodist church remaining. To add insult to injury, few of the town's displaced agricultural laborers found jobs associated with the canal. Even worse, many property owners, originally proud to participate in an endeavor that would ostensibly aid their country, became frustrated that they had lost their land and community for what would become an empty promise. Though no dramatic episode of racial hostility occurred in its destruction, the town disappeared just as completely as other traditional black communities like Ocoee and Rosewood.

As Santos disappeared, the rest of Marion County experienced rapid population growth. The Ocala area soon filled with more than nine thousand new residents, including "itinerant peddlers, preachers, medicine men, sooth-sayers, beggars, acrobats, and musicians," who crowded into "large and small side shows and tent meetings" in efforts to cash in on the project. Anticipating many "social problems caused by drifters and transients," Marion County Chamber of Commerce secretary Horace Smith asserted, "we are organized to cope with them when they arise." In spite of the carnival atmosphere of Ocala and Camp Roosevelt, few major disturbances occurred. Vagrancy became a considerable problem as transients, arriving with little or no money, put pressure on local relief rolls. Anticipating only 75 cases per month, the Salvation Army reported that it actually provided lodging for an average of 416 cases per month. Fighting, public drunkenness, and petty larceny became commonplace enough that the Marion County Sheriff's Office tripled its workload after canal work started.[43]

Local city and county law enforcement officials expanded forces, and the Army Corps of Engineers hired four officers, deputized by the county, to maintain order in the camp. In addition to guarding against illegal gambling, which proved difficult to prevent, camp patrols kept an eye out for confidence men on the prowl for easy marks among the workers. With so many laborers, prostitution became a perennial problem. "Questionable women" routinely drove to temporary camps looking for "prospects for their trade." African American prostitutes often lingered in nearby woods without fear of arrest "so long as they do not bring any liquor with them." While not legally sanctioned, prostitution was tacitly approved as community officials encouraged a local doctor to combat venereal disease at an established "disorderly house."[44]

While local officials and camp administrators overlooked minor legal transgressions, they could not ignore signs of what they considered a far greater source of disorder: union organization. Officials felt workers were well compensated and labor advocates were little more than troublesome intruders. Somervell especially had a well-earned reputation for union-busting and red-baiting; while director of the New York WPA, he crushed a major strike by construction workers and destroyed allegedly leftist murals destined for the walls of the administration building at Brooklyn's Floyd Bennett Airfield. On the canal, union organizers raised the specter of labor unrest that jeopardized timely completion of the project. In March 1936, George Timmerman, a thirty-year-old St. Augustine bricklayer, was found "nailed to a cross in a heavily wooded section near [Camp Roosevelt]. His lips were sewn shut and a heavy hunting coat was tied over his head to muffle his groans. . . . Officers said he had been engaged in labor difficulties on the cross-state canal." Instead of investigating the incident, local law-enforcement officials blamed Timmerman himself, claiming that he had staged the crucifixion to gain publicity for an ostensible sideshow career. Ocala Police Chief J. H. Spencer also tied Timmerman to left-wing causes, accusing him of "allowing himself to be nailed to the cross for communistic reasons." After taking him to the hospital for medical attention, officers forced Timmerman to immediately leave the area. He then disappeared from Ocala and the historical record. Workers were now warned: labor organization would not be tolerated along the canal.[45]

Labor unrest, though problematic, did not provide as significant an impediment to the canal's future as increasing statewide and national opposition. A loose coalition of railroad executives, citrus growers, central and south Florida shipping interests, and numerous municipalities raised a chorus of concern over the canal's long-term impact. At first glance, their efforts resembled a nascent environmental movement. While some of the anticanal forces, particularly the railroads, clearly pursued their self-interest, the opposition's objections to some degree presaged questions later raised with the construction of the Cross Florida Barge Canal in the 1960s. During the Depression, conflict over the canal was less a struggle of preserving Florida's environment than conserving a precious natural resource: freshwater. Without it, these opponents argued, the Sunshine State's preeminent industries—agriculture and tourism—would eventually come to ruin.

Criticism of the Ship Canal actually began long before its groundbreaking as a group of railway executives, in a February 1933 Army Corps of Engineers' hearing, leveled charges that a proposed canal would destroy Florida's aquifer. Canal excavation, they asserted, "may have a very decided effect on the underground flow in the Ocala limestone, and on the wells and water supply remote from the canal, and on the Silver Springs Run, as well as many of the streams that come to the surface" in central Florida. These criticisms, however, could easily be dismissed by canal sup-

The WPA also employed huge earthmoving equipment to dig the Ship Canal cut. Courtesy of Ocala/Marion County Chamber of Commerce, Ocala.

porters as the mere grumblings of discontented commercial rivals, embittered by government support for the canal. Four months later, however, Roosevelt received a personal letter from James Howe, his cousin residing in Daytona Beach, with no apparent ties to competing railroad interests. Calling for "a thorough investigation" of construction plans, Howe viewed the canal as a project that would bring "serious and irreparable injury" to the state. Laying out the framework of opposition that would hold for almost forty years, Howe listed his objections in a brief attached to his letter to the president. First, he reiterated the concerns of railroad men by proclaiming the canal's threat to Florida's water supply. The saltwater intrusion caused by the canal would make "living in the area a virtual impossibility." Calling the canal "a great ditch . . . being a very ugly feature in itself," he then attacked it for destroying the state's scenery. For Howe, this was not simply a matter of aesthetics, for "the preservation of natural beauty in Florida is doubly important because the State depends for its prosperity, to a large extent, upon its tourist business." He couched his final arguments in economic terms, calling the project "an unfair and impossible burden" and not even a beneficial work project since "the greater part of the construction would be performed by giant dredges."[46]

The assertions of railroad executives and Howe gained further credence with the release of a U.S. Geological Survey report in late August 1935, just as Roosevelt announced construction plans. The report concluded that the Ship Canal's deep cut would "inevitably drain enormous quantities of water from the limestone" and thus significantly lower the area's water table. More important, this sudden loss of freshwater would allow salt water to seep into the limestone, eventually contaminating the remaining deposits of freshwater, consequently corrupting underground water supplies across "a wide zone extending outward from the canal."[47]

By the end of the summer of 1935, as boosters from Jacksonville, Palatka, and Ocala seemed to secure Roosevelt's support for the canal, opposition from the central and southern parts of the state began an organized campaign to halt construction. Battle lines hardened as the issue pitted Floridians against each other. The Hillsborough Board of County Commissioners best summarized the opposition's argument: "Incited by selfish interests and from a purely mercenary motive, an effort is now being made, through the construction of a cross-state canal, to mar and at least in part to destroy" the region's "beauty, fertility, and health." Growers saw the project as a direct threat to their livelihood. The editor of the *Florida Grower* declared in June 1935, "in its pollution of our fresh waters, it would be a greater calamity than any freeze or hurricane which has come to this State." Indeed, the opposition portrayed the Ship Canal as evil incarnate. For if "Mephistopheles himself wanted to make Florida a part of the kingdom of the devil and to visit some cruel and lasting punishment upon its people," he would build a "big ditch" and poison the waters to leave rotting "oranges and carcasses on the parched sands of an empire once abundant in plant and animal life."[48]

The Army Corps of Engineers countered these critics with the appointment of a special board of geologists and engineers to further study the issue of Florida's water supply. In December 1935, Corps geologists issued a preliminary report arguing that the project's potential damage was negligible. Of the 195 miles of canal, only 27 miles of the cut—roughly 14 percent of the project—would have any "appreciable effect on the level of the ground-water table." While shallow wells had to be deepened along the right-of-way between Ocala and Dunnellon, the report claimed local agriculture and area vegetation would not be injured. With regard to the concerns of local officials in Tampa, St. Petersburg, Orlando, Sanford, Palm Beach, and Miami, the canal would have no impact on their water supplies whatsoever. Finally, while saltwater encroachment would take place at both ends of the peninsula, it would not pose a direct threat to the underground reservoir of the Floridan Aquifer. The assurances and authoritative tone of the report did little, however, to assuage growing concerns of canal critics.[49]

By late 1935, the opposition became so strident that many citizens increasingly feared that the Ship Canal's completion would reduce southern Florida "to the status of an island." In November, tempers were so hot that a committee met in Orlando to consider dividing "Florida into two states, one north and one south of the projected canal across [the] State." Taking issue with the Corps' report, one geologist complained to Harold Ickes that the federal government should not experiment with the state's water table. Drawing a comparison with another New Deal program, the Agricultural Adjustment Act, he remarked that "if killing pigs or plowing up every third row of cotton proves detrimental, the mistake can be corrected the next year." Damage caused by the Ship Canal, though, would be "irrevocable and there is no way in which atonement can be made." Another observer remarked how the tension between Floridians over the canal was "splitting the people of the state wide open." Likening the project to "pure dynamite from a dozen angles," it became "the hottest brick anyone ever picked up and if we don't have a civil war in Florida with secession of the Florida peninsula there'll be a trick in it."[50]

The largest outcry against the project came from agricultural interests concerned about the effects on Florida's valuable farms and groves. Farmers and growers, both individually and in their organizations, issued a flood of letters and telegrams to newspapers and politicians, decrying the canal's pernicious effects. Even before the canal's authorization, the Growers and Shippers League of Florida blasted the proposal. Attacking the canal on economic grounds, the league asserted the canal would provide cheap freight rates for competing citrus and thus "our grapefruit industry is doomed." It would also draw workers out of the fields and groves since the rumored twenty-six thousand men required to build the canal would be drawn from "pickers and other harvesting labor who would probably be attracted by long time permanent employment at government wages and hours." In February 1936, as construction continued, this opposition reached a head when the Central and South Florida Water Conservation Committee issued a thirty-eight-page brief against the canal. Located in Sanford, the center of Florida's valuable celery-growing area, the committee called itself a "Fruit Growers' and Truck Farmers' Protective Organization." Written by Frank Anderson, the brief echoed Howe's earlier private concerns, as it spelled out in voluminous detail the danger the canal posed to the state's underground water supply. It quoted a letter from state geologist Herman Gunter that concluded (all in capital letters—added by the growers so that no one could miss their point), "no one having the permanent, future welfare of the state at heart will want a cross-state canal started." It also attacked the Corps' science, accusing its December 1935 report minimizing the canal's effect on groundwater as "one more piece of propaganda on behalf of the canal, from a source which has shown itself to be anything but unprejudiced." Anderson also attacked the project on financial grounds, calling it "economically unsound" with "wildly exaggerated" estimates of shipping that would use the canal. Finally, the brief personally condemned Somervell for spending much too much time attempting to convince people of the canal's benefits. His "task of selling the canal to Floridians" led the Conservation Committee to telegram the Corps and request that Somervell "effect more engineering and less oratory and press agenting." All told, the brief galvanized canal opposition and provided a convenient place from which opponents could gather pertinent information to make their case against the project.[51]

Anderson and his fellow central Florida growers quickly found a friend in Washington in the person of J. Hardin "Pete" Peterson, a Democratic representative from Lakeland, in the heart of the state's citrus belt. In August 1935, the *Tampa Times* pleaded with the second-term congressman to take up the cause of stopping the canal. Complaining that "Senator Fletcher is spokesman for the Jacksonville gang that is ramming this thing through," the paper asked Peterson, "can't you demand that some guarantee be given to land owners and cities south of the Canal that their water supply will not be ruined?" Peterson was bombarded with letters and telegrams of support from agricultural interests, many organized around the Florida Water Conservation League and the Citrus Trade Produce Association. Peterson ardently took up their cause, becoming Florida's leading congressional opponent of the canal project.[52]

As growers jumped on the canal's perceived dangers to their water supply, other groups attacked the project for its destruction of Florida itself. Conservation organizations, often composed of upper-class women interested in protecting birds and flowers, sent newspapers and legislators dozens of letters and resolutions attacking the canal. The Winter Park Garden Club, representing the elite of that wealthy Orlando suburb, even established an anticanal committee to organize opposition. In a resolution widely distributed to Florida legislators, congressional representatives, and even FDR himself, they decried the "vandalism that would destroy two of the loveliest rivers in this great country," and "deplore[d] the blindness of those who would convert these quiet shores into huge ugly sand-banks." Making arguments that would presage issues raised by environmental activists of the 1960s, they considered "the temporary relief of a few thousand workers . . . is [an] insufficient reason for destroying the beauty, fertility, and habitability of the most beautiful section of Florida." Upon receipt of the resolution, Representative Peterson replied to the committee chairwoman that the canal would do "untold damage to the water supply of Florida." Concluding "it is my duty to oppose it," he asked the clubwomen to "contact other garden clubs and let them know of the possibility of damage to the water supply."[53]

Sensing growing resistance from across the state, as well as in the halls of Congress, Roosevelt cautiously backed away from the project by the end of 1935. As initial funding rapidly dwindled within months of the groundbreaking, Duncan Fletcher pressed the president for an additional outlay of $20 million to expedite construction. While Roosevelt had assured Fletcher that more funding would be available, by mid-December he stipulated that further support for the project—unlike the original grant that came directly from the executive branch—would have to come from "some kind of Congressional sanction." According to Ickes, who staunchly opposed the canal, Roosevelt's decision was less a matter of deference to the legislature than the expression of serious doubts "about the practicability of the canal." Unwilling to waste political capital on an increasingly controversial issue, the president withdrew his leadership on the project and opted to "let Congress handle the whole thing."

The administration did request more funding for the next fiscal year in the War Department's appropriations for rivers and harbors projects. However, canal boosters had to secure future support from an increasingly skeptical Congress.[54]

As Floridians remained profoundly divided over the supposed threat to the Sunshine State's water supply, opposition on the national level centered on the canal as a stunning example of pork-barrel politics. Canal critics viewed the project as "utterly without economic justification" and, perhaps more irritating, "built solely by Executive Decree." The latter point was hardly rhetorical, for opponents saw Roosevelt's funding provisions without congressional authorization as a constitutional issue concerning the "very process of orderly government." Leading the charge against the canal was Republican senator Arthur Vandenberg of Michigan. A hardworking, humorless, former small-town newspaper editor, Vandenberg had risen to a prominent place in the Republican Party by doggedly attacking the New Deal. Characterized by *Time* as "pompous, dull, a stuffed-shirt," Vandenberg was at his best on the Senate floor, where he "kept the Republican Party alive . . . from 1934 to 1939." There, his long-winded speeches found a resonance beyond anything he could express in a more public forum. To him, Roosevelt's initial support under the Emergency Relief Appropriation Act was a dangerous precedent, not only bypassing congressional authority, but in so doing committing the "treasury to vast long-time public works" that would transfer "the control of the purse from the Capitol to the White House." Moreover, what began as a $5 million appropriation was only the first installment of what Vandenberg saw as a massive drain on the federal coffers. While the canal's estimated cost was roughly $146 million, he claimed it could increase to well over $200 million before completion. Vandenberg hammered home this message as he tried to convince legislators of the wastefulness of the project that he tied directly to FDR and his New Deal.[55]

The canal debate shifted toward Congress in early January 1936. Writing to members of Florida's legislative delegation, Vandenberg expressed his doubts about the canal. In a letter to fellow senator Duncan Fletcher, he concluded that "a water-project of this magnitude should be passed upon by direct action of Congress." Two days later, writing to Representative Peterson, an ally in the anticanal struggle, he laid out his concerns about the canal itself, rather than its method of financing. "I [want] . . . to investigate the so-called Trans-Florida canal," he wrote, "with a view to determining whether it is feasible and advisable for the Federal Government to persist with this project." Using the by-now usual arguments— damage to Florida's groundwater, limited economic benefits, and the lack of support for the project among south Floridians—Vandenberg announced he was proposing a resolution calling for a full investigation of the project. The result, introduced on the Senate floor on January 6, was more than a mere partisan attack on another example of government waste and New Deal profligacy. Through a series of subcommittee hearings, Vandenberg raised doubts about the canal, questioning the legitimacy of the project's authorization as well as the safety of the state's water supply. Moreover, he asserted the savings in time and travel costs were marginal at best, providing letters from leading shippers who claimed they would not even use the waterway because of the "risk of collision and grounding that would be taken in navigating the canal."[56]

Vandenberg cleverly tied the ship canal to other examples of what he viewed as pork-barrel politics at its worst. Chief among these "White House pets" was the Passamaquoddy Bay Tidal Power Project, slated for construction in eastern Maine. Designed to harness the dramatic tidal surges of the Bay of Fundy, the project, informally known as the "Quoddy," anticipated expenditures of $37 million. Vandenberg charged the sum was a mere starting point, with a final figure approaching the $100 million mark. Associating the two programs with political horse-trading rather than economic necessity, the senator made the case that wasteful government spending was simply a manifestation of Democratic politics. The Republican-leaning *Boston Herald* best summarized Vandenberg's position: "the general suspicion is that the President wished to sweeten up the voters of Maine so that they would be in a Democratic mood at the September election." Vandenberg's campaign against excessive government spending ensured that the fate of the two projects remained intertwined throughout the congressional session. Following the senator's lead, national newspapers and magazines rarely failed to mention either project without the other. Vandenberg shaped the debate so successfully that on the other side of the Capitol, a subcommittee of the House Committee on Appropriations considered the canal and Quoddy as a single issue. While not abandoning them entirely,

House officials suspended appropriations for both until Congress, rather than the executive branch, authorized funding.[57]

For months, the fate of the canal was buffeted about as both houses of Congress debated a series of appropriations bills in the spring and early summer of 1936. Canal boosters placed their faith behind Duncan Fletcher in the legislative showdown over a $29 million appropriations bill, $12 million of which would go to the Ship Canal. Anticanal forces saw Vandenberg as their savior and contacted him with supportive telegrams and letters as if he were their own senator. The Seminole County Chamber of Commerce even invoked "the blessings of God upon Michigan's Senator Vandenberg." On the floor of the Senate, Vandenberg warned that the issue was "not just a little innocent amendment involving $20,000,000 . . . that is just the admission fee" for what would eventually cost taxpayers as much as $200 million. Pleading for support, Fletcher countered Vandenberg's charges with oratory, asking if the Senate would dare oppose "a mighty stride of progress, the greatest undertaking in this generation on the part of this Government. Is it possible that Senators will block the way of the greatest accomplishment achieved by the Government in this century?"[58]

In March 1936, the answer was clearly "yes" as the Senate struck down Fletcher's amendment by a single vote. The issue did not die, however, as Fletcher and other canal supporters relied upon a variety of parliamentary procedures to attach additional funding to a series of other legislative measures. Vandenberg fought back, working closely with Florida's anticanal forces that gathered petitions and resolutions against the project. The senator suggested that telegrams and letters from "every Chamber of Commerce and every Luncheon Club and every available political organization and every Woman's Club" would provide support for the struggle. Frank Anderson, author of the influential February 1936 brief against the canal, even threatened to organize a demonstration of "approximately 60,000 men, women, and children" at the canal's construction site within forty-eight hours of "due notice from metropolitan newspapers, news agencies, and newsreels. There would be no arms and no violence, only friendliness and jocularity; but digging operations can be forced into suspension until troops are called out."[59]

In the light of the Senate's vote, canal boosters attacked both Vandenberg and those who backed him in the fight. On April 5, Henry Buckman wrote a letter to the *Washington Post* calling the paper's support for Vandenberg "not completely informed," and proceeded to list reasons why the canal should be completed. Responding in kind, Vandenberg penned a rejoinder, published in the *Post* three days later. Calling Buckman's letter "an effort . . . to resurrect this thoroughly discredited enterprise," Vandenberg concluded that "the Florida Canal, in spite of all that may be conjured by its aggressive apologists, remains a shining example of the wrong way to spend the public money." Not to be outdone, Buckman wrote back to Vandenberg privately in scathing terms. Demanding an opportunity to "publicly debate this matter," Buckman accused Vandenberg of attempting to use the canal issue to gain political advantage. "You are deliberately attempting to destroy this really splendid national enterprise, regardless of the facts," he claimed. "Because the immediate responsibility for its construction lies in the lap of the President, you propose to sacrifice this great project, involving the well-being of more than two-thirds of the people of this country." For Vandenberg, since the vote had already been taken, the subject was closed. "I have already 'debated' it to a conclusion on the Floor of the Senate," he asserted. The final salvo in the war of words came not from either participant but from the Central and South Florida Water Conservation Committee. Vandenberg had kept his Florida allies informed of the battle and they sent a telegram to Buckman on April 11. "Your challenge to Senator Vandenberg to debate [the] Florida Canal [is] both inappropriate and in bad taste," they wrote. "If you desire to air this subject fully the people of Florida should hear you therefore we invite you to debate the subject with one of our engineers or water experts next week at any point in Florida you may select where radio transmission facilities are adequate." Though Buckman seemed eager to challenge Vandenberg, there is no record of his response to the committee.[60]

As the debate dragged on, Roosevelt cautiously sat on the sidelines. Indeed, in March the president had visited the Sunshine State for some fishing and had scrupulously avoided contact with both pro- and anticanal forces. That same month, Florida governor David Sholtz wrote Roosevelt to warn him of the political dangers in continuing the project. Sholtz particularly addressed the issue of the bridge at Santos, designed to carry the Dixie Highway well above the canal. In no uncertain terms,

Sholtz demanded that "under some pretext or other, work be immediately stopped on at least the construction of this particular bridge." Otherwise, if the project failed, pictures of a bridge over nothing would be used by Republicans "in the November Campaign under the title of 'Roosevelt's Master Folly.'" In response, FDR's personal secretary called the Corps to request a temporary halt in construction on the bridge piers. When the army asserted that contracts had already been let and that any impediment would be costly, Harry Hopkins intervened and demanded the work must cease immediately. Somervell ignored both Sholtz and Hopkins and continued construction. Maintaining "the unsightly effect of this construction, if work on the canal should be discontinued, has been exaggerated. . . . They could be removed at no great cost should a final decision be made to permanently discontinue the construction of the canal." Thus the remnants of the Santos bridge piers and abutments still remain today in the median of Highway 441; although gradually being taken over by second-growth forest, they stand as silent reminders of the power of the Corps of Engineers.[61]

Facing an upcoming election, Roosevelt sought to avoid alienating Florida voters by letting the controversy run its course. At a news conference on April 15, he announced that he would not forward any relief funds until Congress resolved the issue. At the same time, Roosevelt offered canal supporters a thin reed of hope by vaguely suggesting that he would consider modified plans to further finance construction. Though national papers characterized the situation as "very dark," noting that "a strong dash of magic would be necessary . . . for another substantial appropriation," Senator Fletcher remained ready for another fight, hoping he could provide the potion necessary for success. In poor health, the seventy-seven-year-old senator marshaled all his energies for another round of congressional wrangling.[62]

In late May, showing political skills that more than matched those of his adversary from Michigan, Fletcher succeeded in not only putting the project up for yet another Senate vote, but also in separating it from the Quoddy. With this maneuver, Fletcher forced a vote on providing $10 million in funding for the canal itself rather than a referendum on the perceived excesses of the New Deal. Vandenberg was prepared for a long struggle, threatening at least sixteen amendments to Fletcher's bill, part of a much larger work-relief appropriation measure. Taking to the Senate

After the 1936 congressional defeats, the abandoned cuts southwest of Ocala were all that remained of the Florida Ship Canal. Courtesy of Ocala/Marion County Chamber of Commerce, Ocala.

floor for over two hours to oppose both projects, Vandenberg asserted that "it's perfectly absurd to authorize another year's work unless we are prepared to carry them through. No amount of camouflage or weasel-worded argument can make it appear otherwise." Concluding that "we are back again boondoggling on Quoddy Bay and pipe-dreaming on the phantom Florida canal," the verbose senator from Michigan demanded that the Senate vote to kill both projects again. The vote occurred amidst the contentiousness of two days of arcane parliamentary procedure and the demand for a decision, as senators anxiously looked forward to adjournment. When the smoke cleared, Fletcher had his funding as the Senate voted 35 to 30 in favor of the canal. Showing the wisdom of his strategy, a second vote rejected the Quoddy project by a margin of 39 to 28. The Jacksonville Chamber of Commerce waxed poetic in its excitement: "If not now, sooner or later, Congress is bound to finish the job. The King of Spain's galleons will be safe for all time from Frankie Drake and his buccaneers in the Florida Straits. Old Hickory's dream is coming true."[63]

Fletcher may have won his battle in the Senate, but another struggle had to take place in the House of Representatives. In spite of Lex Green's admonition on the chamber floor that "the canal will be finished," the House dealt a crushing blow to Fletcher's proposal on June 17, after only forty-five minutes of debate. With a convincing vote of 108 to 62, the House dismissed further funding for the Gulf Atlantic Ship Canal. Moreover, House members made it clear their position was final—there would be no consultation or reconciliation with the Senate's position. For all intents and purposes, the ship canal was dead. That same morning, in what the *New York Times* called a "bold stroke of irony," Duncan Fletcher suffered a fatal heart attack at his Washington home. As the *Florida Times-Union* reported, "Florida's 'Grand Old Man' died here today as he hoped to die—'in harness.'" Colleagues were quick to make the connection between the arduous struggle for the canal and Fletcher's demise. Citizens from the senator's hometown of Jacksonville, mindful of Fletcher's longtime boosterism, expressed deep condolences for both him and the fate of the canal. The city commission ordered his official city hall portrait—the same one whose removal George Hills had demanded less than a year earlier—be draped in black for a month. The same act of mourning might have been appropriate for the Atlantic Gulf Ship Canal.[64]

The Dream Deferred

Canal work stopped with the June 17, 1936, House vote to deny further federal funding. As workers went home, Ocala's boom quickly came to an end. Only 3 percent of the project was complete, with approximately one-third of the land clearing finished. For all the money and time spent, the only visible reminders of all the year's efforts were four thousand acres of land cleared along the right-of-way, almost 13 million cubic yards of excavated soil, and the four concrete stanchions in Santos marking an incomplete highway bridge over a phantom waterway. In November, the *Christian Science Monitor* asked the question on everyone's mind, "Just what use will be made of the bridge foundations or the miles of canal excavated to bed-rock?" As for the ninety-seven buildings on the 215 acres of Camp Roosevelt, "until a few weeks ago, a thriving symbol of Rooseveltian philosophy," they now became "but a ghost of a community." But the New Deal's mark on north central Florida would not be erased. Within a year, the work camp became the staging ground for other projects, first as an adult extension service of the University of Florida, then as an aviation school through the Florida WPA, and finally as a National Youth Administration camp. At the conclusion of World War II, seventy individual houses and two dormitories at the camp built to house canal workers were put to use as housing for veterans attending the University of Florida in Gainesville.[1]

Still, project supporters refused to give up. In July, *Southeastern Waterways*, a Jacksonville trade paper, published a special "Canal Edition." Admonishing readers that "once more the fight is on," the editors encouraged them to "become a booster." With articles by the usual advocates—Youngberg, Coachman, Lowry—the special issue again summarized the considerations in favor of canal construction. But a new, and potentially more persuasive, argument started to surface. With war clouds looming on the European horizon, canal supporters made the case that the project was a matter of national defense. Sumter Lowry summed up this theme when he wrote that the canal "could easily be the difference between defeat and victory in the event of war." In November 1936, this argument achieved resonance with the Army Corps of Engineers as it initiated yet another reevaluation of the project. Much like the reports of the nineteenth and early twentieth centuries, this review of economic projections, construction costs, and the contentious groundwater issue went through several drafts. While the costs of the canal seemed to shift with each preliminary report, engineers concluded that "construction of a sea-level canal . . . at an estimated cost of $157,585,000, exclusive of lands, in addition to $5,400,000 already expended, is justified in the public interest." The Corps also announced that public hearings on its recommendation were to be held in Washington on December 16. As the *New York Times* put it, "the temporary armistice between Northern and Southern Florida has been terminated, and the battle of words is being resumed."[2]

The hearings centered around two major controversies: the effect of the canal on Florida's groundwater and opposition by the nation's railroads. Army engineers had avowed the canal would not endanger the state's water supply, but opponents were not willing to take this position at face value. "It is possible . . . that Army engineers are correct in their conclusion that the fresh water supply of this section of the State will not be endangered," editorialized the *DeLand Sun News* on December 9. "[B]ut it is likewise possible that they are incorrect . . . It doesn't take any great reasoning to figure that DeLand would suffer irreparable damage from construction of the big ditch." At the hearings, the newly formed Florida Water Conservation League presented a brief that built on the concerns of the DeLand paper. Citing the 1935 U.S. Geological Survey report, the league saw the project as a disaster for Florida farmers. Rebutting those assertions, Charles Summerall testified that every study and investigation (conveniently forgetting the one used by his opponents) concluded the canal would not imperil Florida's water. Corps chief General Edwin Markham, the presiding officer at the hearing, listened politely to both sides but ultimately stated that "there will be no need to worry on the part of Floridians concerning their water supply." Railroad interests focused on economic concerns, maintaining that the army estimate of $162 million for construction was artificially low. J. C. Willoughby, representing the Atlantic Coast Line Railroad, estimated that a more accurate figure would be $366 million, with approximately $4 million annually in maintenance costs. Proponents scoffed at these figures, concluding that railroad objections had little to do with costs, but instead were made "on purely selfish grounds, pointing out that such a waterway would take business from them." This argument would continue for decades. Canal supporters accused opponents, even those whose rationale rested on the canal's destruction of Florida's water supply, as at best manipulated by railroad interests, and at worst under their control. John Small of the Atlantic Deeper Waterways Association testified that "no intelligent person would contend that the completion of this interior inland system of transportation would not affect existing instrumentalities of transportation. Of course it will. But if that objection had prevailed during the last 300 or 400 years . . ., we would today be as static as China." In the end, the hearings resolved little, as both proponents and opponents returned home for the holidays more determined than ever to make their case. As the new year approached, a *New York Times* headline best summarized the mood: "Renew[ed] Warfare on Florida Canal."[3]

Though the November preliminary Corps report unabashedly supported canal construction, another Corps engineering survey submitted in February 1937 came to a more tentative conclusion. While it still defended the venture, the new report, this one under the auspices of Major General George Pillsbury, addressed the concerns of many Floridians about the project's impact on their water supply. "In view of the undeterminable possibility that the excavation of a sea-level canal might [cause] . . . extensive damage to ground-water supplies," the report recommended that contingency plans be developed "for the construction of a lock canal . . . in the interest of protecting the ground-water supply." Pillsbury also asserted that the supposed benefits of the Ship Canal would eventually decrease as the size and speed of ships continued to expand. Therefore, he recommended that if the canal be built without locks, it had to be 35 feet deep and 400 feet wide to accommodate these newer ships. This was 2 feet deeper and 150 feet wider than the canal that had been started in 1935. He estimated that this new canal would cost $263 million. Echoing Arthur Vandenberg, however, he conceded that the "benefits from a canal across Florida do not establish the economic justification for the large expenditures necessary for its construction."[4]

Such hedging did not deter procanal forces at either the state or national levels. In April, Chief of Engineers Major General Edward Markham rejected the new board's February report and issued his own analysis. The general came out squarely for further construction in the name of both work relief and navigation improvement. He recommended building the canal with a compromise depth of 33 feet, which he estimated would cost close to $200 million to complete. He also attempted to put to rest concerns about the negative effects of the project on Florida's water supply. "I do not share the apprehension expressed in the report of the Board of Engineers [Pillsbury's February findings], as to the possible adverse effect on ground water supplies," he wrote. "A sea-level canal," he concluded. "will not . . . result in serious intrusion of salt water." This back-and-forth wrangling among Corps officials led *Business Week* to make the literary allusion that "Banquo's ghost had nothing on the Florida ship canal, which is back in the news again after being killed off by heavy oratory during the last session of Congress." Markham's analysis kept the idea of the canal alive. By reopening the question, the Corps once again gave hope to boosters. This in turn stiffened the determination of canal opponents, especially Arthur Vandenberg. His midwestern mouthpiece, the *Chicago Tribune*, laid down the gauntlet in early 1937. "We do not doubt that construction engineers can build the canal—and want very badly to build it," the paper editorialized. "But to rely on their opinion of its commercial value is too much like asking an automobile dealer whether we need a new car."[5]

Markham's proposal reenergized canal supporters, but without congressional authorization for further funding, it had little impact. Charles Summerall recognized the need to continue to lobby Congress. In June 1937, he wrote to the Ship Canal Authority, noting that "the hope of early resumption of construction work on the Florida Ship Canal depends on securing the passage through Congress of a bill approving the project." He asked the agency to be "prepared to wage an intensive educational campaign . . . essential in fighting for congressional recognition and approval of this meritorious project." The death of Duncan Fletcher seemed like a serious impediment to securing such congressional approval, for few could rival the tireless advocacy of Florida's senior senator. And yet, by 1938, Fletcher's successor, Claude Pepper, had become as associated with the project as Fletcher himself.[6]

Born in 1900 amidst the grinding poverty of southern sharecroppers, Pepper escaped his Alabama upbringing to attend Harvard Law School. Full of political ambition, he migrated to Taylor County, Florida, a sparsely populated area within the Big Bend region that closely resembled his rural roots. By 1934, he had made enough of a name for himself that he challenged incumbent senator Park Trammell in the Democratic primary. Though unsuccessful, Pepper's strong showing made such an impression with Florida's political establishment that they encouraged him to run in the November 1936 special election to fill Fletcher's seat. Arriving in Washington after winning the contest, the jug-eared, pock-marked young politician quickly became one of the most ardent supporters of Franklin D. Roosevelt and the New Deal. In spite of his strong partisan affiliation, Pepper faced a hotly contested three-way race for the Democratic nomination for reelection in 1938. Opposed by sitting governor David Sholtz and Representative Mark Wilcox of West Palm Beach, Pepper criss-crossed "Florida's sticky villages and sun-blistered swamp towns, its resort cities and its inland flatwoods, to an accompaniment of loud speakers, flood lights, bad cigars, and baby-kissing." Pepper's initial support for the canal seemed more implicit than explicit, as he could not afford to alienate those in south Florida still vigorously opposed to construction. Verifying Pepper's need for caution, the *Miami Herald* reported that a Pepper victory in the primary would be a "sad day for Florida if New Dealism should go down as the political movement that ruined the peninsula." But as the campaign drew to a close, Pepper began to distinguish himself from the field by speaking openly about his support for the canal, especially as his two opponents scrupulously avoided the issue. According to a *Time* magazine article that featured Pepper in full campaign regalia on its cover, the senator, in "deciding most of his votes will come from north Florida anyway, told citizens of that section he was strong for the canal" and accused his opponents of "pussyfooting" on the issue. In May, Pepper won a convincing primary victory over his two well-known opponents without even the need for a runoff. In a much less contested November general election, he "did not camouflage, sidestep, nor attempt to carry water on both shoulders. He was an out-and-out exponent of the cross-state canal, and the people of this state, from the high gentry of the press to the lowly ditch digger, overwhelmingly endorsed this platform." Pepper handily defeated his Republican opponent, and retained his Senate seat in what national observers saw as a referendum on the New Deal. To many Floridians, that meant ardent support for the canal.[7]

Unlike Duncan Fletcher, whose commitment to the project stemmed from his Jacksonville roots, Pepper's backing seemed less personal and more the result of political expediency. A brief notation in his private diary from early 1937 revealed that "this canal is a bad decision but I have committed myself to it and must go ahead." Full of confidence, he added, "I believe I can put it over." His impressive senatorial campaigns of 1938 allowed him to overcome this ambivalence and become as true a believer as Fletcher. From then on, he assumed the role of the primary senatorial proponent of the project. As one constituent explained in a letter to the *Miami Herald*, Pepper was "100 percent for carrying out the policies of President Roosevelt . . . and he was 100 percent for completing the cross-state canal." Underscoring the sense that the election of 1938 was a turning point, he added "the people of this state, by an overwhelming majority, gave Senator Pepper their mandate to push through this project."[8]

Pepper faced a daunting task with each congressional session, as considerable opposition from both Republicans and conservative Democrats stymied the passage of bills to provide construction money. As early as the spring of 1937, the ardently procanal congressman Lex Green of Starke introduced a canal bill that produced weeks of hearings before the House Committee on Rivers and Harbors. Though the committee favorably reported the bill to the House floor by a vote of 12 to 5, Chairman Joseph Mansfield of Texas stopped action on the measure on June 16. With this defeat, the issue lay rather dormant for the next year and a half, as Congress refused to take up the measure again. In spite of this inaction, boosters remained curiously optimistic, so much so that Henry Buckman could confidently state in December that the project "has enough pledged votes in Congress to guarantee its authorization." With Pepper's victory in the May 1938 Democratic primary, which virtually assured his reelection to a full Senate term in November, Buckman's appraisal seemed more realistic.[9]

Pepper's willingness to stake his reputation and career on the canal impressed true believers like Buckman, who was already lobbying the senatorial novice. Always attuned to the ins and outs of Washington politics, Buckman saw a particularly opportune moment in the spring and summer of 1938 to fight for the canal. In late May, Pepper pushed the distribution of a movie produced by the State of Florida extolling the importance of the canal. Pepper wrote Roosevelt's secretary about the film, imploring him to get the president to "see the picture so he can see the way in which we are trying to carry out his idea of selling the national importance of the canal to the country." Boldly entitled *Straits of Destiny*, the fifty-minute feature linked the Florida project to the earlier Suez and Panama canals and stressed the value of the project to national defense, an increasingly important rationale during a time when the world appeared headed for war. A month later, the death of New York senator Royal Copeland, "a bitter, determined and effective" canal opponent, offered Pepper the chance to realign votes on the influential Senate Commerce Committee. For years, Copeland, as committee chairman, impeded procanal legislation in conjunction with fellow committee member Arthur Vandenberg. Buckman advised Pepper, as a member of the committee, to support the appointment of Senator Josiah Bailey of North Carolina as Copeland's replacement. While Buckman said that Bailey "always voted for the canal," he cautioned, "it is possible that he did so more out of his friendship for Senator Fletcher than any other reason." Furthermore, as one of the leaders of the anti–New Deal conservative coalition, "his very bitter opposition to the Administration may cause him to be confused as to the real merits of the project." If there was to be any promise of legislative success, Pepper would have to "begin at once the cultivation of Senator Bailey's friendship for the canal."[10]

Pepper spent the remaining months of 1938 "cultivating" not only Josiah Bailey but anyone else in Washington who might provide support. He forwarded Roosevelt voluminous statistical data showing ostensible benefits to commerce and national defense. The president passed the information on to the National Resources Committee, a federal planning agency established in 1935. The conclusions of its chairman, Frederic Delano, reflected the ambivalent stance of most objective observers. While conceding that "shortening the sea route . . . in time of war might be desirable," Delano concluded that "we can hardly conceive of any qualified man endorsing the project without many reservations. We should say that the obvious arguments against the canal are so strong that the burden of proof rests on those who are advocating it." Pepper, along with

Florida congressmen Lex Green and Joe Hendricks, continued to fight in the halls of Congress to provide that proof.[11]

While Pepper, Green, and Hendricks worked in Washington, local boosters pursued the issue back home. Led by Florida's Ship Canal Authority, these advocates tirelessly sought to keep the canal issue on the front pages of newspapers and thus on the minds of average citizens throughout the state. Since construction had halted in 1936, the state agency had no real mission—it legally could not acquire land for a canal that was not authorized to be built. Rather than shut down, the authority took on a new role—public cheerleader for a quick renewal of canal building. That new task took significant amounts of money—cash the State of Florida would not provide. The authority instead turned to other sources to fund its lobbying and public-relations efforts. From 1938 to 1943, commissioners of Duval County, which included the city of Jacksonville, annually allocated the large sum of ten thousand dollars to be used by the authority "to expedite and advance the early resumption of construction work on the Canal." The authority then entered into a contract with the National Gulf-Atlantic Ship Canal Association to provide "the services of skilled engineers and economists . . . in order to furnish Congress with technical information and in order to overcome the attacks to be expected from opponents of the project." The private advocacy organization would also use the funds to "inform the country generally of the great merits of the project."[12]

What supporters and opponents thought was the final canal battle dawned with the year 1939. Pepper and Green marshaled their forces and organized two official congressional inspection tours of the proposed canal. From January 18 to 23, members of the House Committee on Rivers and Harbors were the guests of Henry Buckman, Walter Coachman, and the Florida Ship Canal Authority. Designed to provide congressmen with both "the maximum of pleasure and recreation" while simultaneously furnishing "a comprehensive 'on the ground' inspection of the Canal," the junket gave officials an opportunity to visit Jacksonville, Palatka, Silver Springs, Camp Roosevelt, the Santos bridge piers, and the Gulf canal entrance at Yankeetown. Five weeks later, the authority hosted a similar gathering of senators. Led by Pepper, twelve senators met with Florida governor Fred Cone and "drove to parts of [the] canal where digging occurred in 1935." While congressmen and senators enjoyed the Florida sunshine, Franklin Roosevelt shored up support for the canal from the White House. On January 16, Roosevelt wrote a letter to both Josiah Bailey, the new Senate Commerce Committee chairman, and Joseph Mansfield, chairman of the House Committee on Rivers and Harbors. In it, he asserted that "it has long been my belief that a Florida Ship Canal will be built one of these days and that the building of it is justified today by commercial and military needs. As you know, it has been my thought that the government should continue its construction." Once again tying the project to unemployment relief, he added, "I would not personally object to a construction period lasting as long as ten or fifteen years."[13]

Roosevelt's renewed interest certainly delighted canal advocates. Representing the Florida-Ship-Canal Club (Chapter No. 1) of Belleview, Florida, Mrs. W. T. Walters wrote the president to commend him for his position, adding "we trust that it may never have to be used for war purposes, but will be a monument of defense." With Europe on the threshold of war in early 1939, national security considerations were now becoming a major part of the procanal argument. Echoing Walters's concerns, Governor Cone went so far as to suggest that the canal's access to interior Florida "would make an ideal base for building and repairing submarines and destroyers which would possibly be needed in time of war."[14]

South Florida, however, remained skeptical. In January 1939, the Miami Beach Chamber of Commerce declared in a letter to the president that it was "still against the construction of the Florida Cross-State Canal." The organization gently suggested that FDR "carefully study the other side of the story and learn that the issue is a serious one to us." The attached editorials from two Miami dailies struck a much different tone. Denouncing the president's actions as a betrayal, editorial writers blasted the canal as "unwanted, unneeded, undesirable, and dangerous." The *Miami Herald* asserted that the commercial rationale for the canal was "slightly goofy, . . . [and] the idea that the canal would have any military value is even goofier." The *Miami Daily News* seconded its rival's conclusions by declaring that "the military question is not a significant factor." After blasting FDR's backing of the canal, the editorial writers expressed regret that "our Florida Senators cannot realize that all of Florida is by no means united in this project," and felt the need to look far beyond the state line for allies. As the *Herald* put it: "Florida is delighted to know that it does have at least one representative in the Senate. That representative is Vandenberg and not Pepper."[15]

Letters, petitions, and telegrams filled the mailboxes of Claude Pepper and Charles Andrews, Florida's other senator, demanding that they withdraw their support for the canal. In writing to Pepper, for example, a Miami Beach constituent excoriated the senator for his "betrayal of southern Florida and the growers of citrus fruit." In spite of the spate of mail, Pepper and Andrews remained steadfast. Anticanal forces thus turned more and more to Arthur Vandenberg to represent their interests. They commended the Michigan senator for his genuine concern that "the canal will seriously endanger our water supply." Indeed, south Floridians assured Vandenberg that he would "receive a royal welcome to this greater Miami area" for "his active interest in opposing the said canal." On January 19, Vandenberg took to the Senate floor to ask, once again, for another study of the potential adverse effects of the canal on Florida's water supply. Pepper responded by caustically reminding the Michigan senator that Florida was already "represented by two Senators and was doing fairly well without help" from Vandenberg. Continuing in that same vein, Pepper declared that he was "somewhat chagrined" that Vandenberg should "take upon his shoulders the representation of the people of Florida about their local water supply." Vandenberg did not respond to Pepper's charges, and instead reverted to his usual concerns about the canal and economic profligacy. "Inasmuch as the State of Florida is asking for upwards of $200,000,000," he intoned, "of which my State of Michigan has to pay . . . at least . . . $10,000,000, I shall continue to exercise at least $10,000,000 worth of interest in what Florida is undertaking to do."[16]

By February 1, 1939, another epic battle for canal funding began to take shape. Two Texas legislators, Senator Morris Sheppard and Congressman Joseph Mansfield, introduced bills in both houses authorizing the "construction of the Atlantic-Gulf Ship Canal across Florida." Vague in their formulations but similar in their language, the bills raised the long-dead issue of "the advisability of levying tolls on vessels . . . transiting the canal . . . providing for the maintenance and operation of the canal and for the amortization of the cost of its construction." Both bills were referred to the appropriate congressional committees, where they would undergo weeks of hearings. Much of the debate was tediously repetitive, with both sides marshaling the same arguments they had made four years earlier. Proponents continued to tout the commercial benefits of the canal with increased attention to the project's importance to national defense. Opponents railed against government waste and the potential damages to Florida's vulnerable water supply. There was more than a little self-interest in their positions. Commercial maritime organizations like the National Rivers and Harbor Congress and the Atlantic Deeper Waterways Association touted "the benefits to shipping which will result from the opening of the canal indicating an increase . . . of over 50 percent in the existing ocean-born tonnage." Representatives of the Association of American Railroads fired back with a blatant complaint that "the railroads, particularly the Florida railroads, are not in a position to withstand the loss of revenue on traffic which [the proposed canal] might divert from them."[17]

As the hearings dragged on, congressmen found themselves plowing the same ground they had for decades. As the House Committee on Rivers and Harbors noted in April 1939, "few, if any, river and harbor projects had ever been accorded more exhaustive study and discussion. There is available unusually complete data and information for judging all phases of the enterprise." By the end of the month, the committee submitted a report to the floor with solid support for authorization and construction. The Senate Committee, however, remained divided. Even with Pepper's influence as a committee member, the body simply forwarded the bill to the full Senate without comment. Though the ensuing debate began with the particulars of the canal controversy, it increasingly moved toward bigger issues. Arguments over saltwater intrusion and the destruction of Florida's citrus industry were replaced by larger questions of national defense, government spending, and the very nature of the New Deal. The specter of war added new urgency to proponents' arguments. The House Committee on Rivers and Harbors succinctly summarized them: "[B]y shortening the distance . . . for sea-going vessels between the Atlantic Ocean and the Gulf of Mexico, the Atlantic-Gulf ship canal would be of definite value during war in the shipment of necessary material for both military and commercial use. . . . [Thus] the canal is . . . essential to the adequate defense of the United States."[18]

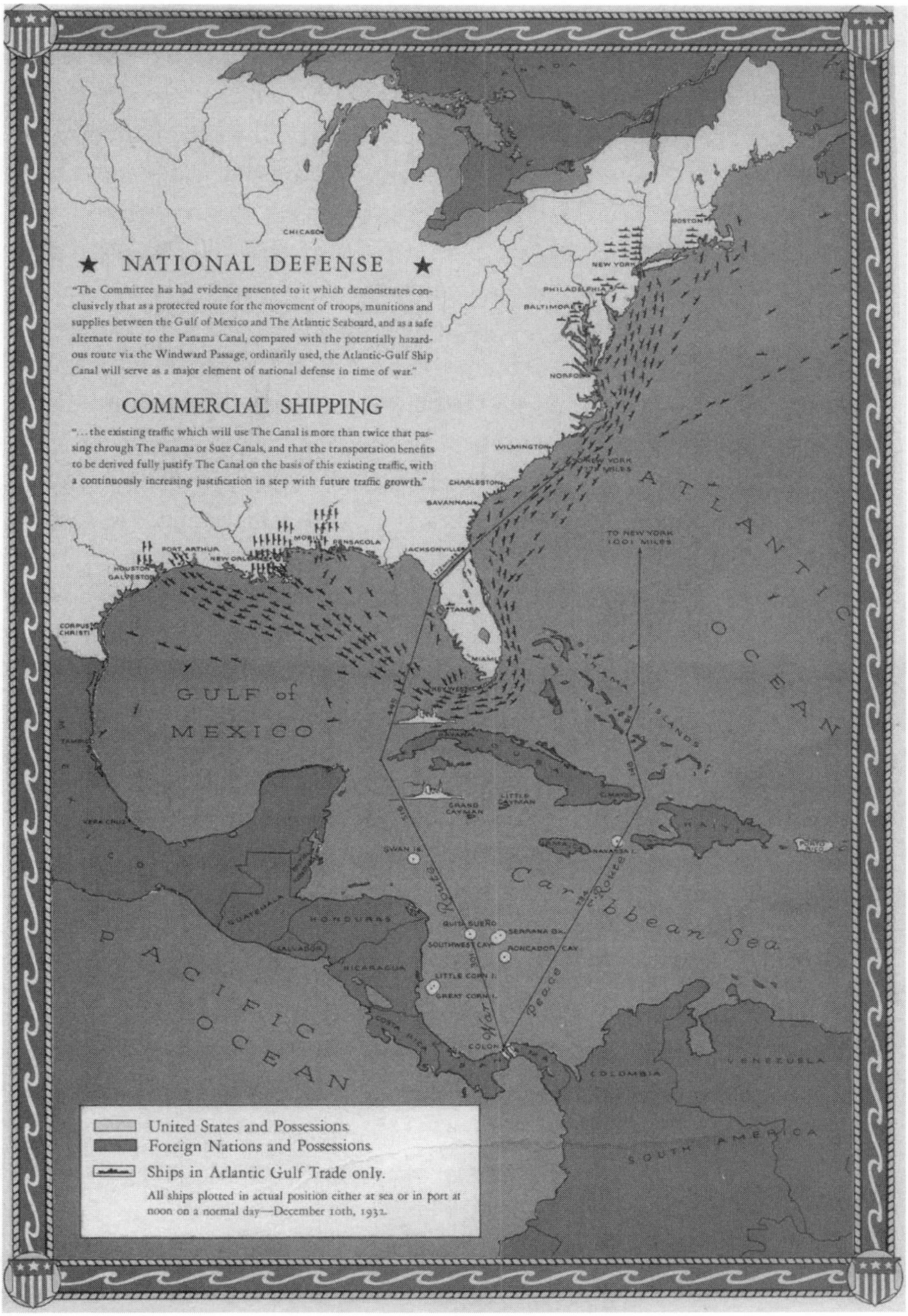
★ NATIONAL DEFENSE ★
"The Committee has had evidence presented to it which demonstrates conclusively that as a protected route for the movement of troops, munitions and supplies between the Gulf of Mexico and The Atlantic Seaboard, and as a safe alternate route to the Panama Canal, compared with the potentially hazardous route via the Windward Passage, ordinarily used, the Atlantic-Gulf Ship Canal will serve as a major element of national defense in time of war."
COMMERCIAL SHIPPING
"... the existing traffic which will use The Canal is more than twice that passing through The Panama or Suez Canals, and that the transportation benefits to be derived fully justify The Canal on the basis of this existing traffic, with a continuously increasing justification in step with future traffic growth."
CHICAGO
BOSTON
NEW YORK
PHILADELPHIA
BALTIMORE
NORFOLK
WILMINGTON
CHARLESTON
SAVANNAH
JACKSONVILLE
PENSACOLA
MOBILE
PORT ARTHUR
HOUSTON
GALVESTON
CORPUS CHRISTI
TAMPA
MIAMI
TO NEW YORK 1001 MILES
ATLANTIC OCEAN
GULF of MEXICO
GRAND CAYMAN
LITTLE CAYMAN
SWAN IS.
QUITA SUEÑO
SERRANA BK.
SOUTHWEST CAY
RONCADOR CAY
LITTLE CORN I.
GREAT CORN I.
Caribbean Sea
War Route
Peace Route
HONDURAS
NICARAGUA
GUATEMALA
COLON
COLOMBIA
VENEZUELA
SOUTH AMERICA
PACIFIC OCEAN
United States and Possessions
Foreign Nations and Possessions
Ships in Atlantic Gulf Trade only.
All ships plotted in actual position either at sea or in port at noon on a normal day—December 10th, 1932.

By 1940, canal supporters relied on national defense as a major rationale for construction. Courtesy of Department of College Archives and Special Collections/Olin Library, Rollins College, Winter Park.

Arthur Vandenberg countered that contention with an earlier report from the secretary of the navy that claimed the canal would add little to national defense. The Michigan senator suggested the argument was simply a cover for profligate New Deal spending. "The national defense is used as an excuse or a justification for anything that anybody wants to do," he declared from the floor. "Patriotism goes on parade in the direction of the Treasury with a flag in each hand." In this assertion, Vandenberg echoed the claims of the *Baltimore Sun* columnist Frank Kent, a longtime Roosevelt nemesis. "The project hitchhikes on the war scare," Kent wrote in May 1939, "and there is a great deal of ballyhoo about how necessary the canal is for national defense purposes. The war and navy departments now say they want the canal . . . but the significant thing is that two years ago when there was no war scare, [they] gave the country a very emphatic opinion against the canal." Roosevelt's enemies were not the only skeptics. On March 12, Harold Ickes wrote in his diary that, "The President seems to justify it to himself now as an instrument of national defense, but I suspect that there is some rationalizing in that."[19]

For Vandenberg, stopping the canal was part of his concerted conservative strategy to curtail big-money projects associated with the patrician in the White House: "At a moment when the country is deeply anxious about the solvency of the federal government, the passage of this bill would cruelly disillusion even those brave souls who still think that the congress intends sooner or later to protect the public credit. If this thing can be permitted under the circumstances, then the country is indeed helpless." Even some Democrats were willing to cross party lines to take a stand against excessive government spending. Senator Millard Tydings of Maryland expressed frustration with the canal's exorbitant price tag: "The good that would come to shipping from the construction of the canal would not be a drop in the bucket to the harm that is being done the entire country by constantly appropriating money which we have not got [*sic*]." Fellow Democrat Alva Adams of Colorado hewed to this same line, complaining that "if Congress does not restrict its spending, the disaster to the country is inevitable; and of all the expenditures we are asked to make, the Florida ship canal is the one above all others lacking emergency character."[20]

As early as May 11, Pepper recognized that he may not have had the votes to win, and he called on Roosevelt to work his legendary powers of persuasion on vacillating Democrats. When the vote was called on May 17, Vandenberg's argument held the day by a tally of 45 to 36. Twenty-three Democrats, in addition to the Independent Robert LaFollette of Wisconsin, aligned with Republicans to kill the project. Florida opponents seemed ecstatic. A Florida grower wrote appreciatively to anticanal congressman "Pete" Peterson of Lakeland that "the vote in the Senate on Wednesday, on the Florida Ship Canal, was the best news we have received in a long time." With the Senate victory, canal opponents pushed to kill the measure in the House as well. On May 19, the Chase Packing Company of Sanford, a major citrus distributor, sent a memo to "all our good friends" reminding them that "the results in the Senate yesterday show that your efforts did good," and urging them to lobby their "Representatives to vote against it." Faced with the Senate's negative vote and an intense lobbying effort, congressional supporters refused to move on Mansfield's version of the bill. The House's inaction showed the waning support for the canal. Representative Peterson crowed in a letter to one Bradenton supporter that "I am pretty well convinced I could have beaten it in the House by around a hundred votes."[21]

The canal was seemingly defeated again. Conservative opponents reveled in their victory over an element of Roosevelt's New Deal at a time when such wins were hard to achieve. Much of the credit went to Arthur Vandenberg, whose star was clearly rising. From the Miami Beach Chamber of Commerce, which issued the senator an honorary membership, to the Chicago Tribune, which lauded him for his "single-handed opposition" that "kill[ed] the New Deal Florida Canal," Vandenberg stood as the conservative hero of the moment. By October, his rotund visage even graced the cover of Time as pundits began to see him as a viable candidate for the presidency in 1940. Ironically, the last sitting U.S. senator to appear on the magazine's cover was none other than Florida's fervently procanal Claude Pepper.[22]

The defeat meant little to canal hard-liners. As early as June, Henry Buckman advised Pepper not to give up and "take this one step at a time." They considered a number of alternatives, including provisions for self-liquidating tolls and even a redesign as a shallow-draft barge canal. These potential changes would undercut the major concerns raised by opponents like Peterson and Vandenberg. No longer could these critics claim that the canal would threaten the Floridan Aquifer, as the barge canal would be designed with a depth of only twelve feet. The issue of profligate government spending would similarly be mitigated by user fees that would eventually ameliorate construction costs. To build popular consensus for these new proposals, Claude Pepper arranged for the Army Corps of Engineers to hold a public hearing on the efficacy of a barge canal, with "all interested parties given an opportunity to express their views on the . . . public benefits of such a connecting channel." Held in April 1940 at the Dixie Theater in downtown Ocala, the three-and-half-hour meeting featured proponents hailing construction as an economic benefit to the region. In the end, the meeting accomplished little other than showing that north central Floridians still felt positive about the project. But it would also be the last time citizens could comment on the project in a public forum for more than a quarter century.[23]

Proposals for a redesigned canal may have altered the shape of the debate, but at bottom the issue remained a political one. No one knew that more than proponents. In early August 1939, Charles Summerall addressed a joint luncheon meeting of the Jacksonville Rotary Club and Chamber of Commerce in the downtown Mayflower Hotel. Giving a speech that centered on the present status of the canal, Summerall quoted selectively from government reports to bolster his case that "the canal is necessary and economically justified." Charging that the canal was defeated by false statements and propaganda, he asserted that "chances for passage were much more favorable than before." "Eight years ago the Florida Ship canal was scarcely an idea," he noted. "An idea once born never dies, and the better the idea the more opposition it will have." Undaunted by failure, Summerall asserted that canal supporters will work on "converting defeats into victories." For example, he turned Vandenberg's argument about government waste on its head, announcing that "to quit now would be to lose all of our investment of $300,000 in the right-of-way." He concluded by brashly announcing that "the canal is going to be built, and it will be built by fighting men." But in tacit recognition of the tenacity of his opponents, Summerall admitted that "when the complete victory will be, no one knows."[24]

World War II offered the perfect occasion to fulfill Summerall's prediction. As war spread across Europe, canal proponents increasingly tied the fate of the canal to America's national defense. Solving the issue of saltwater intrusion by now advocating a shallower-draft barge canal with locks, these advocates saw the war as an opportunity to press their case. In April 1940, the Senate printed a report by Henry Buckman entitled *Defense Coordination of the Panama and Florida Canals—A Preliminary Study*. In this detailed forty-nine-page report, replete with charts, maps, and reams of statistical data, the influential Jacksonville lobbyist made his case for canal construction as an important component of national defense. While Buckman maintained that "there are other defense values in the canal which are not directly translatable into money savings," he overlooked these and focused directly on the numbers. He concluded that the canal would reduce the number of tankers sunk by enemy ships and save millions in both tanker-construction costs and transportation. In the event of a two-year war, these total savings would add up to a minimum of $718,995,000. To Buckman, this figure represented a major bargain, as he calculated the cost of building the canal at $160 million.[25]

That same month, Pepper took to the Senate floor to second Buckman's ideas: "We could not be the victor in any major struggle forced upon us in which we could not adequately defend our merchant shipping." Recognizing that fact, Pepper asserted that delay in building the canal "would be a grave and most imprudent omission in our plans for national defense." In the first six months of the war, the allies and neutrals have "suffered the loss of 401 merchant ships and cargoes with a total value of $900 million." Without the canal, "we could not expect to escape the hazard of merchant ship losses. . . . To ignore the necessity . . . is to court disaster." Though Pepper's words received positive play in the Jacksonville and Ocala newspapers, the *Orlando Sentinel* charged that Pepper was overplaying his hand. In an April 11 editorial entitled "Why Dig This Ditch?" the paper argued, "we don't believe the ship canal is needed as a defense measure." Seeing this new rationale as simply reheated rhetoric for canal construction, the *Sentinel* declared, "Florida needs broad roads more than she needs a canal, and would do well to take a stand for the things she should have that would advance her further towards her destiny." This skepticism continued to bedevil proponents. If the state was not unanimously supportive, even with the threat of war, how could Claude Pepper and Lex Green convince their congressional colleagues to adopt a measure for construction?[26]

Correspondence from Floridians reflected this continued lack of consensus. Throughout 1939 and 1940, Pepper received dozens of letters decrying the lack of progress on the canal. A Jacksonville resident noted, "Almost every time I go down the street, somebody stops me and wants to know why, with national defense uppermost in the minds of the nation, we don't make some headway in connection with the canal." For every one of these procanal letters, however, Pepper received another attacking his stance. In September 1940, a Miami attorney railed against both Pepper and the Jacksonville Chamber of Commerce for having the "affrontery to publicly proclaim to the intelligent people of Jacksonville, and other parts of Florida, that 'the Florida canal is the very keystone of our national defense.'" "This keystone claim," the writer continued, "is therefore the ACME of absurdity and insincerity."[27]

In his national newspaper column, Frank Kent saw the argument for national security as an example of "whipped up patriotism to cloak a political grab." Observing that with "the newspapers full of war headlines and public attention diverted by national defense, neutrality act debates and other matters, this thing has been quietly slipping along." Kent derisively concluded that "the effort to jam this scheme through under the guise of a national defense measure is inexcusable."[28]

The *Chicago Tribune* explicitly referred to the project as the "New Deal Florida Canal" and Roosevelt's "Pet Scheme." In the election campaign of 1940, the Republican Party commissioned a film entitled *The Truth about Taxes* to support the presidential campaign of Wendell Wilkie. The ten-minute screed vigorously attacked Roosevelt's profligate spending while letting the nation's defenses slip. It particularly singled out the Florida Canal as "another wasteful boondoggling failure into which the New Deal has squandered more millions instead of spending money for necessary additions to our national defense." Another virulent critique of the project came from a 1941 letter to the editor of the *Wall Street Journal*. Explicitly tying the canal to other New Deal policies, the writer asserted that "we have in power those who espoused such things as devaluing the dollar, plowing under crops, killing pigs, burying gold, buying silver at outlandish prices, condoning sit-down strikes, baiting employers, advocating construction of the St. Lawrence Waterway, the Passamaquoddy fluke and the Florida Ship Canal and many other equally ridiculous projects." Even in early 1941, such arguments made headway as Americans remained profoundly divided over the efficacy of entering the war. Arguments about national defense could cut both ways.[29]

Florida's canal boosters shied away from the broader issues raised by the New Deal and focused instead on the importance of Washington's political battles for the ultimate disposition of the project. These men often saw members of their congressional delegation, even the sainted Claude Pepper, as less than stalwart defenders of the project. In a March 14, 1941, letter to Bert Dosh marked "personal and confidential," Henry Buckman complained that "our senators are simply not putting their shoulders to the wheel in this matter. They must be caused to alter their attitude at all costs. . . . Neither one of our senators have done a lick of work since you left here. . . . [and] not a word in defense of our project has been offered by Senator Andrews or Senator Pepper."[30]

While supporters bemoaned the weakness of their legislators in pushing the project, opponents feared the arguments over national defense put forth by Pepper would overcome local issues, particularly that of destroying south Florida's water supply. In February 1941, a prominent Sanford grower wrote a letter to both Andrews and Pepper that was published as a letter to the editor of the *Orlando Sentinel*. The canal would "would perform a surgical operation on Florida," he ranted, "and mutilate and destroy its most important asset, its underground water supply." Assailing the ostensible military justification for the project, he concluded that "the defense feature being emphasized is only an excuse to take advantage of emotionalism and patriotism to put across something that is not needed." The arguments of both Buckman and the growers reflected an atmosphere where each side was determined to gain maximum political advantage from a deteriorating international situation beyond their control.[31]

The events of December 7, 1941, and America's entry into a two-ocean war put the canal controversy in a new and infinitely more urgent idiom. Canal opponents no longer assailed the project as an example of the New Deal's leftist tendencies. Their opposition would increasingly be tied to the issue of national defense. Whichever side could manage to "prove" its case about the canal's contribution to the national war effort would carry the day. Newspaper headlines throughout the state echoed this contentious debate. On December 31, a *Miami Herald* headline blared, "Florida Canal Unnecessary, Survey Shows." The article reported the results of a survey that concluded the canal was "not justified as a national defense project and should be abandoned." Five weeks later, a *Tampa Tribune* editorial proclaimed "Canal Out for Duration," as it announced that the House Rivers and Harbors Committee voted to not spend any funds on projects not tied directly to the war effort. Since the canal could "in no sense be tied to the war effort," the paper asserted that "surely this will eliminate the Florida Ship Canal." Yet only five days later, a headline in the same newspaper boldly announced that "Florida Barge Canal Urged to Offset U-Boats." The article quoted Joseph Mansfield, the chairman of that same House Rivers and Harbors Committee, as saying that the canal would protect American shipping "from the gulf coast to the east" and would prevent the "danger of a rise in the cost of gasoline in the east."[32]

Submarine warfare provided a rare opportunity and enormous leverage for canal boosters. Throughout the winter and spring of 1942, German U-boats initiated Operation Drumbeat, a mission designed to cripple the American merchant fleet in its home waters. In the campaign, the submarines destroyed nearly four hundred merchant ships off the Atlantic and Gulf coasts of the United States. Particularly striking for Floridians was the April 10 torpedoing of the SS *Gulfamerica*, a tanker carrying ninety thousand barrels of oil on its maiden voyage from Port Arthur, Texas, to New York. Sunk four miles off the sands of Jacksonville Beach by the torpedoes and deck gun of *U-123*, the incident took place within full view of horrified spectators on the shore. Walter Coachman, executive director of the state's Canal Authority and one of the most stalwart of canal advocates, quickly recognized the importance of this U-boat offensive to the campaign for the canal. As early as January 24, 1942, Coachman issued a statement to the members of the Executive Committee of Canal Counties: "The war has . . . given us an opportunity to render a real national defense service to the nation." Nine days later, he elaborated on this point in a confidential letter to the same group. "The submarines and Adolph are aiding us," he boldly announced, adding cautiously, "but the opposition is also intensely active." Later in February, writing to M. C. "Doc" Izlar of Ocala, he commended Congressman Mansfield for his recent speech urging canal construction "as a defense necessity on account of Nazi submarine sinkings," and called for the speedy passage of a bill to take advantage of what was turning out to be a propitious moment. In another letter penned just after the sinking of the *Gulfamerica*, Coachman concluded that "the U-boats, the increasing petroleum shortage and other transportational [*sic*] factors are in our favor." John Perry, owner of the *Jacksonville Journal* and the *Ocala Star-Banner*, seconded Coachman's beliefs. "Every ship plying from the Gulf to the Atlantic," he announced at a Chamber of Commerce luncheon, "which has been sent to the bottom is a monument to the stupidity of congress in refusing to go on with this great project. . . . Today it is a national defense project, pure and simple." Building on these arguments, the Canal Committee of the Marion County Chamber of Commerce put it succinctly in a telegram to Congressman Joe Hendricks—"Nows [*sic*] the time for action."[33]

The canal's opposition quickly prepared a formidable argument against those who maintained it would alleviate the growing gasoline shortage in the nation's East Coast. Once again, Arthur Vandenberg took up the cause. "I seem to have collided with the 'Florida Canal' again," he confided to Harold Ickes in early 1942. "This time the project is wrapped up in 'red, white, and blue' and offers itself as a palliative for our 'gas pains.' . . . Of course it would take two or three years to build the 'barge canal'—and by *that* time it seems to me that we certainly shall have conclusively mastered our enemies." Citrus growers, those other longtime canal opponents, raised similar concerns about the waterway's true benefits to the war effort. In April, the Lakeland-based Florida Citrus Producers Trade Association issued a news bulletin whose headline blared, "Help Fight the Canal and You Will Help Lick the Axis!" The growers called on Floridians to tell "Claude Pepper, Lex Green and Joe Hendricks that the water supplies of Florida fruit growers are more important than gasoline for joy-riding Yankees!" By asserting that the canal is "hardly likely to win the war or put more gas in auto tanks," the lobbying group played on the wartime patriotism of Americans. The bulletin concluded that the canal would cost approximately $25 million, funds that would be better spent to buy "either 1,250,000,000 .45 cartridges, 5,000,000 steel helmets, 4,166,000 anti-tank shells, 136,000 submachine guns, 4,630 barrage balloons, or 182 fighting pursuit planes."[34]

The *Miami Herald* expanded on those sentiments a month later. In a May 29 editorial entitled "Screen for the Big Ditch," the anticanal paper fervently attacked the project as a threat to "the social, economic, industrial, agricultural, tourist and business life of the most populous section of Florida." Calling the canal "an additional burden on a people staggering under the onerous load of war taxes," the *Herald* excoriated the canal as "a deception, which, if advanced as a war need, is hypocrisy of the first order." The paper even brought up the specter of a ship canal once again, suggesting that barge canal proponents had only "one purpose—construct a little ditch now under the guise of war effort and it will be a pushover to expand it to the big canal when the battling is over."[35]

Congressional supporters and opponents spent the first six months of 1942 arguing in increasingly convoluted fashion over the canal and its relationship to national defense. As in previous debates, the project became entangled with other large public-works expenditures that benefitted other regions. In the 1930s, the canal was tied to Maine's Passamaquoddy Bay Tidal Power Project. By 1942, it had been connected by both supporters and opponents to the massive St. Lawrence Seaway initiative. The Seaway, a pet project of FDR since his days as New York governor, had as checkered a career as the Florida Ship Canal itself. Designed to be built at a cost of somewhere between $550 million and $1.3 billion, the huge project envisioned a series of locks and dams on the international waterway, allowing ships of 24-foot draft to sail from the Atlantic all the way to Duluth, Minnesota, in nine days. Rejected by Congress in 1934, it came up again for congressional authorization in 1942 as part of an omnibus Rivers and Harbors Act that also included authorization for a barge canal across Florida following Route 13-B. The strategy of tying these projects together seemed obvious. It allowed representatives from disparate regions of the nation to support the waterways authorization, regardless of how they felt about the individual projects themselves. Yet, supporters of the Florida canal thought it might be better for their project if it stood on its own merits as a defense measure, rather than being tied to other projects like the expensive and unpopular St. Lawrence Seaway. Republican congressman William Pittenger, representing Minnesota's Duluth region and a strong supporter of the seaway, recognized the danger of this strategy to his project. "The best way to defeat the St. Lawrence Seaway Project," he wrote to FDR in April, "is to follow this procedure of dismemberment of H.R. 5593 [the omnibus bill]." The political calculations of canal supporters, however, overcame any thoughts about the seaway and its relation to either the Florida project or national defense. Building on a report issued by the chief of the Army Corps of Engineers, Eugene Reybold, stating that the value of a barge canal would "in time of war . . . be sufficient to warrant its construction,"

Texas congressman Joseph Mansfield crafted a bill specifically authorizing canal construction. Introduced in late April 1942, the bill, titled H.R. 6999, explicitly tied the Florida barge canal to "promot[otion of] the national defense." It called for the construction of a "barge canal twelve feet deep and one hundred and twenty-five feet wide . . . across Florida" at a cost of $144 million. Claude Pepper, with strong support from Charles Andrews, introduced a similar bill in the Senate, adding the proviso that the barge canal could be deepened and widened at a later date to serve as a ship canal. With these bills, authorization of the barge canal would succeed or fail on its own merits.[36]

Mansfield's Committee on Rivers and Harbors quickly held hearings and opened debate on the issue. The Texas congressman started the testimony by submitting a March 25 letter from the president that seemed to support the project. "I am hopeful," Roosevelt wrote, in his usual elliptical fashion, "that something constructive may be worked out without undue delay." As Mansfield was ill for much of the spring, Lex Green presided over many of the hearings, ensuring the canal would be placed in the best possible light. To expedite swift passage, Green tied the project to a pipeline designed to move oil and gasoline quickly from western refineries to eastern cities. He asserted that these two companion projects, costing $144 million, would alleviate the eastern oil shortage, estimated at 800,000 barrels a day. Green used all his political skills to get the bill a positive committee reading. "We did a lot of log-rolling and swapping on it," he remembered later, "to get it out of the committee." Through Green's legwork, the bill quickly received committee approval but ran into immediate snags on the House floor. In spite of the strong support of powerful House Speaker Sam Rayburn of Texas, the bill did not achieve the two-thirds majority necessary under special House rules to pass the legislation, and it died on June 2. From the other side of Capitol Hill, Arthur Vandenberg reveled in the House victory. "The Barge Canal," he wrote to a New York canal supporter, "could not possibly contribute to the Eastern oil shortage until 1944 and probably not until 1945. Therefore, the Barge Canal feature of the Mansfield Bill has no . . . bearing on the Eastern oil shortage." The usual lineup of opponents toasted their success in letters to such stalwart anticanal advocates as "Pete" Peterson. The widow of Frank Anderson, the recently deceased author of the leading missive against canal construction, wrote, "I wish Frank could have been here to enjoy that news." Randall Chase, a leading Sanford grower, telegraphed his "heartiest congratulations and genuine appreciation for . . . defeating the Mansfield barge and pipeline bill." Finally, the Miami Beach Chamber of Commerce congratulated the Lakeland congressman by telling him that "we are just as proud of you as can be."[37]

The joy expressed by canal opponents proved to be short-lived, however. The next week, Charles Summerall mustered all his political acumen in writing Rayburn, pleading for passage of the bill. "The barge channel across northern Florida . . . is vital to the defense of the United States," he wrote. "The debate on suspension of the rules served to cloud the real issue involved, i.e., the absolute necessity of this channel for war transport." Building on Summerall's assessment, Representative Joe Hendricks predicted that "the bill has a good chance of passing if it comes up under regular House rules." The next day, fellow Florida legislator Lex Green wrote a generic letter to all House members asking for the bill's passage. Reminding them that they had already voted "for guns, battleships, airplanes, bombers, barges, [and] tanks," Green asserted that a vote for the canal was a vote for "a war prosecution measure" and one "endorsed and requested by the war administration officials." Two days later, on June 17, Hendricks's prognostication came true as the House reversed itself and passed, by voice vote under regular House rules, the bill authorizing a high-level barge canal with locks as well as two oil pipelines. The *New York Times* reported that the "House debated sharply for more than five hours before they finally sent the legislation over to the Senate." There was "almost unanimous support" for the pipelines but significant rancor over the barge canal, "which, some members contended, was unnecessary and others maintained was a political trick on the part of interested groups to attempt to enact as a war measure the controversial Florida ship canal project." Total costs for the canal and pipelines were estimated at $93 million.[38]

While opponents angrily expressed their frustration over the House vote, supporters rallied around Claude Pepper's companion bill in the Senate. The day after the vote, the secretary of the Tampa Chamber of Commerce wrote to "Pete" Peterson decrying the hypocrisy of canal supporters. "This proposal by Mansfield and his cohorts is worse than rotten, worse than stinking," he railed, "and borders . . . close to treason." He continued that the canal involved "the squandering of millions of

dollars on useless and uneconomic ventures by narrow-minded, unin-
formed and pork-hungry demagogues" and even suggested that it might
be "better to point some of our guns on the firing line backward instead
of forward." Peterson sadly responded that he was "sorry we could not
strike the item out," but at least "we were able to get locks written into
the canal which will save the water supply." Since he was on the winning
side, Hendricks did not have to resort to such inflammatory rhetoric.
He did, however, give Pepper advice on the best means to shepherd the
canal bill through the Senate over the opposition of Arthur Vandenberg
and "the Republican side." Suggesting that Pepper play down any intention of turning the barge canal into a ship canal, Hendricks suggested
that Pepper announce that "the canal could not be dug one foot beyond
the present limitations of the bill without the approval of Congress." He
also tried to refute the charge that the canal would never be finished before the war was over. "No one knows when this war will end," he wrote,
"and if we are going to limit our plans to the needs which we can see at
the present time, we should not continue to build tanks and planes and
guns. . . . This canal is future planning." Hendricks concluded by stressing
the toughness of canal opponents. "The opposition," he warned Pepper,
"is not asleep and they are freely predicting that it will never be reported.
It is going to take a great deal of work to get favorable action on it."[39]

Initiated in the Senate Commerce Committee, the bill passed rather
handily by a vote of 11 to 3, in spite of the studious opposition of committee chair Josiah Bailey, the conservative Democrat from North Carolina.
Sent to the Senate floor, the bill ran into a curious argument presented
by none other than Arthur Vandenberg. Resorting to using the strategy
of saying the bill represented "unjustifiably Congressional interference
with presidential powers," Vandenberg hoped to convince the Senate to
defeat the measure. He took this stance on the assumption that if the
president alone was responsible for the allocation of wartime resources,
Petroleum Coordinator for War Harold Ickes, well known for his opposition to the canal, would simply kill the project. Vandenberg also attacked
the project on the usual grounds that it was "long-range wishful thinking
that totally fails to deliver the bill of goods." The measure passed the Senate by the barest of margins. For a moment it even appeared the measure
would fail, as the votes lined up 30 to 29 against the bill's passage. The
timely arrival of Idaho senator David Clark on the Senate floor saved the

day for canal supporters as his vote for the project created a tie. Acting in his role as presiding officer of the Senate, Vice President Henry Wallace broke the deadlock as he cast the deciding vote, assuring that the canal would finally be authorized again. Five days later, Franklin D. Roosevelt signed the "bitterly-disputed" bill under the guise of helping to "relieve the shortage of transportation for moving oil and gasoline to the eastern seaboard." The proposed $93 million authorization included $44 million for the barge canal, with the remainder of the funds earmarked for pipeline construction and deepening the intercoastal waterway along the Gulf coast.[40]

On July 18, Claude Pepper issued a statement expressing his gratitude for congressional support of a project "which in time of war will make an invaluable contribution to the winning of the war and in peace a great contribution to the recovery and the prosperity of the country and particularly the South." That same day, Joe Hendricks telegraphed one supporter revealing the pork-barrel nature of the project. "This means $43 million for my district," he proudly noted. The next day the president of the Marion County Chamber of Commerce wired supporters: "Following yesterday's approval, national defense will be strengthened, human lives saved, and the economic structure of our country fortified." Once again, it appeared that the canal's future was assured.[41]

It was at best a pyrrhic victory, for the bill simply authorized congressional approval. It did not, however, allocate any funding for actual construction. Since the project dealt with a wartime measure, Roosevelt could have provided those monies without congressional approval. Reflecting his longtime ambivalence, the president informed his director of the budget in August that he "doubt[ed] whether we should undertake its construction at this time." That stance referred funding back to a Congress that did not seem ready to provide those monies. By October, a joint House-Senate committee voted to remove money for construction from a general waterways appropriation bill. The vehemently anticanal *Tampa Tribune* editorialized that with this vote "the canal boondoggle is now dead," but warned that the fight was not over. "Whether this 'death' is permanent, or just another case of suspended animation, remains to be seen." The paper pleaded with senators Pepper and Andrews to give up their support. "Be yourselves," the editorial concluded. "Speak the truth to the Ocanala lobbyists and tell them to be gone." But Pepper and

Andrews, as well as Lex Green and Joe Hendricks in the House, were too deeply involved to give up now. They would spend the next three years desperately fighting for congressional funding.[42]

Those years saw proponents pushing Congress to appropriate construction funding. They again couched their arguments in terms of national defense, focusing on the canal's role in alleviating the East Coast oil and gas shortage. In January 1943, Summerall wrote a blanket letter to every eastern governor that asserted, "it is possible to complete this connecting channel and to initiate barge traffic on a scale adequate to supply the entire petroleum needs of the eastern states (without rationing) within ten months after the work is begun." Walter Coachman, in his position as executive secretary of the Canal Authority, made the letter public to build support for construction. With this initiative, Pepper and Andrews met with federal bureaucrats "to pry off dead center the pending barge-waterway program for relieving the fuel oil shortage." Pepper decried that "too much talk and too little action already has occurred." In his diary, Pepper expressed frustration. "Surely want to get this thing disposed of," he wrote dejectedly. "It has been an incalculable drain on my time." Yet he and others soldiered on. John Perry, the owner of the *Jacksonville Journal* and other papers nationwide, even took out a full-page ad in the February 24, 1943, *Washington Post* to extol the benefits of the canal. Citing opposition as "pure selfishness" on the part of railroad interests, the ad concluded that only the canal, the "small gap" in a great waterways system, could supply the East with the oil necessary to avoid disastrous shortages "without interference from the enemy." Conversely, opponents maintained the canal would not be finished in time to help ease the wartime shortages and, more importantly, construction would take valuable men and material from more pressing war-related projects. They cited the April 1943 opinion of Assistant Secretary of War Robert Patterson, who declared, "it would be inadvisable at this time to commit to the project the manpower, critical materials, and equipment necessary to complete the barge canal." That logic carried the day as Hendricks's attempt in early 1943 to allocate $44 million to fund canal construction failed to pass through the House Appropriations Committee by the slim margin of 21 to 19. The committee reported that the canal "cannot be looked to for any alleviation of the fuel oil or gasoline shortage in the Eastern seaboard for many months to come." *Time* asserted that the vote came as a result of "last-minute testimony from George A. Wilson, assistant to Petroleum Administrator Harold Ickes." Five weeks later, a similar close vote (16 to 14) in the Senate Appropriations Committee spelled defeat once again.[43]

Joe Hendricks placed the blame for the defeat squarely on Harold Ickes and his Petroleum Administration for War. In March, Ickes's deputy wrote to George Dondero, a Michigan Republican serving on the House Rivers and Harbors Committee. "We cannot support its [the barge canal's] construction," he wrote in bold letters, "since our conclusion is that it can have little or no utility in meeting the present emergency in the Eastern Seaboard states." Two months later, Hendricks wrote a stinging letter to Ickes. Maintaining that "the improper testimony given by your office to this committee was a deciding factor" in the House vote, Hendricks accused the presidential aide of duplicitous behavior. "I want to assure you that we shall not let this issue die," he warned. Continuing the war of words, Ickes wrote back in early June, defending his anticanal stance. Admonishing Hendricks that "the implications of your letter are unjustified," Ickes concluded that "the position which has been taken by this Office regarding the utility of the proposed Trans-Florida barge canal for the transportation of petroleum has been based solely upon the comparative merits of the proposed canal." On July 2, Hendricks wrote directly to Roosevelt, accusing Ickes of having "direct responsibility" for the "death toll of merchant crews and ships destroyed in attempting to do by sea what could have been safely done by barges . . . through the Florida Canal." Ickes's blind opposition to the project "raised the question as to the usefulness of his continued participation in the Government." Roosevelt responded to Hendricks later in the month by defending his confidant and advisor. "I cannot concur," he wrote, "that selfish interests of the Secretary or his staff have in any way influenced his judgment." He concluded by chiding the Florida legislator that "a more complete examination into the facts of the situation . . . will result in a changed attitude on your part."[44]

Roosevelt did not achieve his well-deserved reputation as a consummate politician by alienating supporters like Hendricks. While taking the Florida congressman to the woodshed for attacking Ickes, he simultaneously wrote a letter to Congressman Mansfield announcing his support for developing "detailed plans and specifications for the Florida Barge

Canal" so that "its construction can proceed without delay." Roosevelt's apparent go-ahead allowed the Army Corps of Engineers to return to Ocala and begin formulating plans for a barge canal. In spite of Hendricks's inability to convince the House to appropriate funds for the canal (another attempt failed in October 1943), the Corps' projections and reports gave supporters hope once again. In December 1943, the Jacksonville Office of the Corps issued a report on the canal entitled the *Definite Project Report*. This thirty-seven-page document, packed with voluminous and detailed scientific appendices and supplementary data, set the parameters for the high-level lock-and-barge canal that would eventually be built in the 1960s. Addressing everything from the construction of the locks and dams to the geology and hydrology of the areas surrounding the canal, the report estimated the cost of the project at $76,556,000. Supporters now had a detailed blueprint for all phases of canal construction. In spite of these latest plans, Roosevelt's genius lay in his ability to convince all sides that he favored their position. The stridently anticanal *Tampa Tribune* took a decidedly different message from the president's alleged support for canal construction. In a July 12 editorial that engaged in a detailed parsing of Roosevelt's letter to Mansfield, the paper concluded that "the canal couldn't have been built before now, it can't be built now, and it can't be built until the shortages in man power, material, and equipment have been overcome. . . . There isn't anything in that carefully guarded and evidently carefully conditioned statement to warrant the renewed beating of the canal tom-toms." Once again, the canal remained in a state of suspended animation, giving both sides just enough hope that it might either finally be built or killed.[45]

The remaining war years saw more of the same. Supporters like Buckman condemned Ickes for his obstructionism. In early 1945, he wrote to Hendricks that "we have not got the canal today because the PAW [Petroleum Administration for War] threw sand in the eyes of Congress by false estimates and promises." Northeastern maritime interests similarly complained of fuel shortages in their region that could have been prevented if the canal had been built. In spite of these arguments, Congress defeated repeated attempts by Hendricks to secure funding.[46]

With the war's end, proponents faced a new challenge. No longer could they couch arguments strictly in terms of national defense and regional petroleum shortages. A chaotic postwar economy, however, gave them an opportunity to resurrect earlier rationales and push for the canal as a public-works project. Within a month after the atomic bomb fell on Hiroshima, the commissioners of both Marion and Putnam Counties issued resolutions supporting the canal because "this work, if begun immediately, would furnish hundreds of jobs to relieve postwar unemployment." That same month, John Perry wrote to the editor of his *Jacksonville Journal*, sending copies as well to all members of the state Canal Authority. "Here we are facing real conversion," he wrote; "we are facing a labor surplus instead of a labor shortage. We are looking for useful jobs for released men. Why not make a concerted drive to get the money appropriated without further delay to carry out the law which has already authorized construction of the Florida State Canal?" Perry also saw the change in administration as a plus. "I understand Harry Truman is a friend of the canal," he wrote. In addition, "Secretary Ickes will probably be out in the next few months, and he was the greatest obstacle in the way of the canal. In other words, it strikes me that the road is clear, if we make a determined effort to get the money to build this canal without further delay."[47]

After ten years of Franklin Roosevelt's vacillating and often ambiguous commitment, the presidency of Harry Truman brought hope and optimism. But Florida's political guard was changing. On Capitol Hill, hard-liner Lex Green abandoned his House seat for an unsuccessful run for governor in 1944. Four years later, Joe Hendricks left office to pursue a business career. Senator Charles Andrews died in 1946 and was succeeded by Bartow resident Spessard Holland. While Holland would soon become a fierce canal supporter, some initially suspected that he would instinctively represent south Florida's interests. Claude Pepper remained standing, but that would only last until 1950, when George Smathers defeated him in the contentious "Red Pepper" campaign, in which the south Florida contender played up Pepper's ostensible Communist connections. A new generation of Florida's legislators—Smathers, Charles Bennett, Sydney Herlong, and Donald Ray "Billy" Matthews—became, in time, adept in pushing canal legislation forward. But in the late 1940s and early 1950s, the project's future languished. On Capitol Hill, appropriations faltered. Even with Arthur Vandenberg becoming increasingly preoccupied with the Cold War, Claude Pepper still found an unreceptive Senate.

For decades, the Santos bridge stanchions remained as mute reminders of the failed Ship Canal project. Courtesy of Florida Defenders of the Environment (FDE), Gainesville.

Pepper spent most of his energy courting the new president. Knowing Truman's support for the Ship Canal dated back to the 1930s, Pepper, along with Spessard Holland, visited the White House in October 1947. A week after the meeting, the new senator's assistant characterized the president's reaction as "favorable." In spite of this positive assessment and Truman's reputation for plain speaking, the president remained just as equivocal as his predecessor. In responding to Charles Andrews's request for assistance for the project, Truman warily noted, "I sincerely hope that eventually we can get it through." A year later, in writing to Maine congresswoman Margaret Chase Smith, he asserted that "neither of them [the Passamaquoddy nor the Florida canal] had yet been realized but may come about sometime in the near future." Four years later, Truman's ambivalence had not abated. In a May 1951 memo to his deputy secretary of defense, he conceded: "I've always been of the opinion that this Canal is essential to the welfare of the country. . . . Although my judgment is not final on it." Once again, canal supporters learned they could not rely on having a true friend in the White House.[48]

The Cold War offered a resurgence of hope as proponents found another chance to seek the president's endorsement of the project. More than anyone, Henry Buckman—now living in Washington's suburbs and employed as a consulting engineer to lobby for the canal—appreciated the power of the executive branch. In early 1951, he wrote a confidential letter to John Perry, asserting that "no project of this kind has a ghost of a chance of getting started unless it is *certified by the President* as being necessary in the national defense" (emphasis in original). On May 16, newly elected Jacksonville congressman Charles E. Bennett visited the White House to bring Harry Truman around to his way of thinking. Truman seemed interested enough to request, only days later, yet another survey of the canal's value as a defense project. This time, the wheels of the military bureaucracy turned surprisingly swiftly as Deputy Secretary of Defense Robert Lovett submitted a short summary report by the end of the month. Suggesting that the idea of a ship canal remained a possibility, Lovett estimated that its costs now would be $507 million. Conversely, a barge canal would be significantly cheaper, yet its three-year construction price had risen to a little over $108 million, considerably more than the $82 million projected only five years earlier. Lovett relied on the expertise of both the Corps of Engineers and the Joint Chiefs of Staff for their judgment. The Corps presented an essentially positive view of the project, saying it afforded "a short, economical, protected route across Florida for the movement of a large amount of long-haul traffic by barge." Conversely, the Joint Chiefs, while finding the canal would "reduce exposure of shipping to submarine attack," concluded that "the military aspects of the proposed problem are so limited that they should not be used as the primary basis for decision on this matter." As usual, bureaucratic indecision and ambivalence ruled the day.[49]

Back in Florida, hopes for canal construction continued to fade. In January 1951, *Jacksonville Journal* editor Robert Dow Jr. wrote Bert Dosh and observed that "the organization has pretty much fallen apart." A primary reason was the ill-health of Walter Coachman. As the managing director of the state's Ship Canal Authority since its inception, Coachman was old and frail and could no longer take an active role in coordinating support. By July, he was forced to resign. Within a year, he would be dead, "leaving a void in the Ship Canal Authority and the Navigation District that cannot be filled." Coachman would be replaced by another board member, Malcolm Fortson, a former naval officer and longtime canal supporter. To fill Fortson's shoes, the board nominated Bert Dosh. By

1953, the Ocala newspaperman became board chairman. Dosh's friendship with local attorney and rising political star Farris Bryant proved crucial in the eventual construction of the project. Henry Buckman saw their job as tough since they would "have to buck some of the ablest and most powerful groups in this country." In short, the lobbyist concluded, "this is no job for a boy."[50]

Canal boosters may have found a few good men, but they remained profoundly ambivalent on a construction strategy. Following the 1951 report from the Joint Chiefs, supporters could no longer overlook the negative assessments of the military's highest authority. Moreover, by 1953, Republican Dwight David Eisenhower occupied the White House. Compared to Roosevelt and Truman, who were willing to at least listen to canal boosters, the fiscally conservative former military commander showed no interest at all in the project. With a military strategy predicated on nuclear weaponry as the primary source of American military strength, a barge canal seemed positively archaic. Though supporters would take out advertisements in the *Washington Post* emphasizing the military rationale, they garnered little support, or even interest, among either the executive or legislative branches of government. The fact that the ads ended with the line, "We are already two wars late with its completion," indicated supporters were clinging to the past rather than looking to the future.[51]

Though Charles Bennett continued to advance military arguments through the 1950s, most proponents turned to the economic benefits to both Florida and the United States. By the early 1950s, inland waterways transportation had undergone a quiet revolution. Though barge traffic seemed antiquated in comparison to rail, truck, and, increasingly, air transport, the percentage of tonnage per mile carried by barges had almost doubled during the 1940s. While that percentage stood at a rather minuscule 5.1 percent by 1951, waterways advocates saw the numbers of both total tonnage and freight capacity increasing in the future. *Business Week* reported by late 1953 that "the inland waterway transportation industry has worked its way into the middle of a healthy boom." Southerners in particular recognized the importance of increased barge traffic. A 1956 *Time* article concluded that the Gulf intercoastal waterway saved Texas shippers $83 million annually. According to one Texas businessman, that waterway represented "a shining strand linking together the

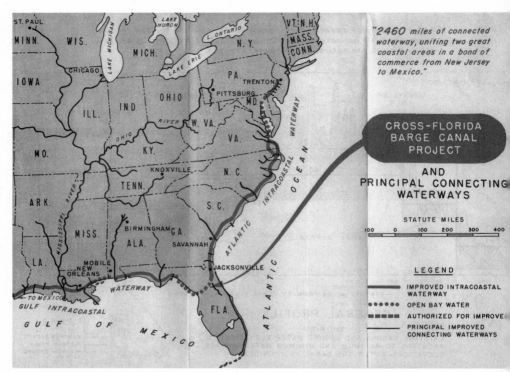

"*2460 miles of connected waterway, uniting two great coastal areas in a bond of commerce from New Jersey to Mexico.*"

CROSS-FLORIDA BARGE CANAL PROJECT

AND

PRINCIPAL CONNECTING WATERWAYS

STATUTE MILES

100   0   100   200   300   400

LEGEND

IMPROVED INTRACOASTAL WATERWAY
OPEN BAY WATER
AUTHORIZED FOR IMPROVE
PRINCIPAL IMPROVED CONNECTING WATERWAYS

Supporters viewed the canal as part of a national integrated system of inland water-ways. Courtesy of Florida Defenders of the Environment (FDE), Gainesville.

jewels of progress into a fabulous necklace along the curving bosom of the Gulf." All that was missing was "to cut a 160 mile cross-Florida canal to connect the Intracoastal system with the Atlantic Inland Waterway, thus creating a sheltered, 2,500-mile passage from Mexico to Massachu-setts."[52]

Florida's Canal Authority revived the idea of the canal as the "missing link" in a nationally integrated waterways system. To convince Congress to allocate funds for the moribund project, the Jacksonville Chamber of Commerce and the Canal Authority contracted with both a private consulting firm and the Corps to report on the economic impact of the canal. In August 1954, Jacksonville's Chamber presented a brief to the Corps explaining why an economic study would be beneficial. Chamber president Frank Norris wrote: "we are convinced that a thorough restudy of the canal's waterborne cargo potential will show it to be highly feasible on the basis of a benefit-cost ratio. It is a project which will aid virtually every large industry along the Atlantic and Gulf coasts." A month later, the Corps responded affirmatively, outlining a $25,000 plan for a nine-month restudy. Congress approved the measure, but allocated only $11,000. This restudy would drag on for years.[53]

In the meantime, the Canal Authority contracted with Gee and Jensen, a West Palm Beach engineering firm, for a similar economic analysis. Issued in November 1956, the fifty-eight-page report examined the canal's potential impact on a wide variety of agricultural commodities, extractive products like lime rock and coal, fertilizers and chemicals, and finished industrial products. Arguing that the canal would significantly reduce transportation rates, it asserted that the "estimate of tonnage for the Cross-State Canal would soon prove to be inadequate." In the midst of Florida's postwar boom, the engineers assumed economic growth would continue unabated. The report concluded that "industrial growth can be expected to follow the important 'missing link' of the Cross-State Florida Canal, perhaps even with a greater impetus because it represents the middle of a chain" of trade stretching from Texas to New England.[54]

The Gee and Jensen report provided much-needed ammunition. More came in January 1958 with the completion of the long-awaited Corps restudy. Brimming with such arcane data as the movement of Chicago sludge (concluding the canal would save $6.26 per ton on its transfer from Chicago to Jacksonville), the report contained the ultimate weapon for canal supporters—a positive benefit-cost ratio. This marked a sharp departure from analyses of the 1940s, when wartime consideration precluded any evaluation of commercial traffic. Indeed, during the war, the benefit-ratio for the canal was a paltry nineteen cents for every dollar spent. The new report assigned a positive ratio of 1.05 to 1.0. Engineers arrived at the figure by calculating estimated annual charges at $7,365,000 and estimated annual benefits at $7,758,000, of which more than $7 million were savings costs in transportation. Based on this economic analysis, the Corps produced a simple phrase that sent canal officials jumping for joy: "The project is therefore economically justified. The District Engineer recommends that the authorized Cross-Florida Barge Canal be placed in the *active* category of projects" (emphasis in original). There was a problem, however. In spite of the positive assessment, the cost of canal construction skyrocketed to a whopping $170.1 million, as compared to the 1943 projected cost of $76.5 million.[55]

Boosters conveniently ignored the dramatically inflated construction costs and quickly sought to capitalize on the positive recommendation. Barely a month after the report, four north Florida representatives— Charles Bennett, Sydney Herlong, "Billy" Matthews, and Bob Sikes— requested a modest congressional appropriation to start the canal. In April 1958, Governor LeRoy Collins met with the Canal Authority and other supporters and came away as "the first major Florida political figure to break the wall of silence which had surrounded the canal project for many years." For the next year, the governor and other state officials rallied supporters to lobby Congress. By 1959, Senator Holland, following the governor's lead, organized a large delegation of Floridians to visit Washington. Representing interests ranging from business enterprises to labor unions, and municipal leaders from Pensacola to Miami, canal boosters worked both the halls of Congress and the offices of the executive branch. Holland even managed to bring on board George Smathers, Florida's junior senator from Miami. In an October 1959 interview with the *Jacksonville Seafarer*, Smathers frankly admitted that he was opposed to the canal early in his senatorial career. "I've been educated and informed since then," he admitted, "and I am fully in accord that this is a national project of major scope which I fully support." The statewide effort made little headway, however. Holland could not even convince Allen Ellender of Louisiana, his personal friend and chairman of the powerful Senate Subcommittee on Public Works, to vote on a humble $160,000 in canal appropriations. "I find that the Cross Florida Barge Canal would cost a total of $165 million," Ellender explained, "and that it has a benefit-to-cost ratio of only 1.05 to 1.0, which is a very slim margin. . . . I regret that I will not be able to support your request for funds for this project in fiscal year 1960." If congressional support seemed at best lukewarm, the response from the White House was positively icy. With his penchant for fiscal conservatism and balanced budgets, there was no room at all in Eisenhower's 1960 budget proposal for any new waterways projects. As so often was the case, the canal seemed dead yet again.[56]

In spite of the lack of progress, Jacksonville's Charles Bennett tried to maintain a positive public face in what was obviously a frustrating situation. For him, support for the canal seemed unquestionable. In a time of increasing Cold War tensions, he could not understand how a former military man like Eisenhower could neglect the call for construction. Unable to convince the sitting president to support the project, Bennett had to turn to former President Truman. Now free from balancing competing interests, Harry Truman wrote, "in my opinion it is necessary for the whole country." Releasing Truman's remarks to the general public, Bennett expressed unwarranted optimism. He commended the state government and its citizens for providing "stronger and more universal" support than in the past. More importantly, the Jacksonville Democrat recognized that, in spite of what appeared to be a dire situation, conditions for construction were actually favorable. "This is the first time when the project has not only had authorization based upon its defense values but also economic justification recently ascertained by the Corps of Engineers, U.S. Army," he announced. "Very few projects ever enjoy this dual approval." Additionally, Bennett saw the canal as the crucial link in an integrated state and national transportation system. "The Cross-Florida Barge Canal is the key to important waterway developments in Florida [and] . . . would enhance materially the chances of other waterway projects." Spoken at the close of a decade characterized by delay and inactivity, Bennett's comments seemed prescient.[57]

The election year of 1960 created a new political dynamic, both nationally and statewide, that reenergized the canal debate. President John Kennedy's election restored to the White House a Democrat willing to devote federal dollars to public works. He also seemed determined to meet a Communist threat that, after the Cuban revolution, was just ninety miles from Florida's shore. In Tallahassee, the electoral success of Ocala's Farris Bryant as governor, Jacksonville's Tom Adams as secretary of state, Cedar Key's Randolph Hodges as Senate president, and Ocala's Bill Chappell as Speaker of the House assured the top layers of state government would be solidly behind canal construction. Working with an experienced congressional delegation in important committee positions, these north Florida politicians were positioned to secure federal funds for the canal. The project's now-positive benefit-cost ratio solidified the argument for its construction. And with the new design of a less intrusive shallow-draft barge canal, boosters felt the intrastate wrangling of the 1930s and 1940s was over.

Florida's Democratic establishment had pushed heavily for Kennedy's election. By 1960, Republicans were making inroads in the traditionally solid Democratic South. Issues of federal power and racial integration captured the southern imagination, and brought concerns that a Massachusetts liberal might be soft on the race question. North Florida canal supporters avoided Kennedy's stance on integration and instead emphasized his unyielding support for the canal. In late October, the *Ocala Star-Banner* trumpeted in bold headlines, "Kennedy Backs Florida Canal." It quoted a letter from the candidate to Charles Bennett: "I know of your persistent work for the . . . cross-Florida barge canal. If I am elected President I will be glad to cooperate with you in making this project a reality. I regard it not only as important to Florida, but to the economy of our entire country, which must fully utilize all our national resources if we are to achieve necessary economic expansion."[58]

Although Kennedy won the White House in an extraordinarily close election, he did not carry the Sunshine State. Losing Florida by only forty-six thousand votes, Kennedy could take comfort in the fact that north Florida solidly supported him. Boosters like Bennett and Dosh viewed the canal as the crucial issue in that support. Over the next three years, they would constantly remind Kennedy of both his pledge to build the canal and their hard work on his behalf.

Kennedy remained true to his word. In May 1961, he made a $195,000 request for final planning and initial construction. The optimism in Tallahassee seemed palpable as Secretary of State Tom Adams confidently stated, "if this project isn't sold now, it can never be sold." A month later, he and Governor Bryant, with other boosters in tow, descended upon Congress to appeal for federal funds for a host of Florida water projects, including the Cross Florida Barge Canal. To emphasize the importance of the canal to what the governor called "Florida's integrated inland waterways system," fully one-fourth of his congressional statement was devoted to the canal. Bryant asserted that both Florida and the nation would "immediately profit from the industrial development which would undoubtedly result from the completion of this vital canal link." Adams built upon Bryant's argument by claiming that the project would provide "a new economic vitality for the state which will be far more depression proof than tourism, citrus, and agriculture." He foresaw the canal transporting a variety of agricultural and industrial commodities, including rocket parts headed for Cape Canaveral. Indeed, he predicted that "within four years of its completion, the Cross Florida Canal will carry as much tonnage as is now carried by the Suez Canal."[59]

In addition to the economic benefits, the pressing need for flood control became another rationale for the project. Just as a previous generation had seized upon the grounding of the *Dixie* to garner support for the canal, Bryant raised the specter of the devastating floods of 1958, 1959, and 1960 in the Ocklawaha and Withlacoochee river basins. In the spring of 1960, more than 216 square miles of the Ocklawaha River valley had flooded, with an estimated loss of $3 million in homes, property, and agricultural production. To avoid such catastrophes, the canal offered additional discharges into the Gulf and the St. Johns River. Moreover, the large reservoirs necessary to the lock-and-dam system would allow for water storage during periodic droughts.

These arguments stood on their own, but boosters also reinforced their case with the pressing need for national defense. While the United States had dispatched the threat of German U-boats during World War II, the thought of Soviet submarines prowling the Atlantic coastline frightened Americans throughout the late 1950s and 1960s. As early as 1958, Louisiana congressman Overton Brooks announced that "the new Russian menace makes a Florida canal a vital link in the national transportation system." As one official from a waterways interest group explained, "with the submarine menace in the next [war] much more serious than that in the last, let us not be the victims again of too little, too late." A few years later, the threat seemed even more ominous following the successful Communist revolution in Cuba. "With the dominant Soviet influence in Cuba and the continuing threat of communist infiltration," the *Waterways Journal*, a national trade weekly, editorialized, "it is essential that this country have a safe inland route to the Atlantic coast."[60]

The Cold War, in short, created a more favorable climate for the national defense argument. Ironically, according to boosters like Bennett, the demands of modern warfare, with large-scale missiles and rocket boosters, required an archaic method to transport weaponry from the point of production to delivery. The barge canal would provide a safer, shorter route for transporting giant rockets to their launch site at Cape Canaveral. In light of the nuclear showdown over missiles in Cuba in the fall of 1962, boosters could easily maintain that "the urgent need for a protected Gulf-Atlantic waterway has been brought home to every thinking Floridian by the Cuban crisis. . . . Besides exposing these sensitive multi-million dollar missiles to the vagaries of open-water navigation and the lurching stresses of ocean travel, the growing Soviet submarine menace must be considered." Building on this, a 1963 House document concluded that "the Cross Florida channel would afford a safe inland route for these engines which can not be transported by rail or highway." However, the canal's stripped-down construction plan called for locks too small to accommodate the barges necessary to carry the giant Saturn rocket boosters. Even if completed, the Cross Florida Barge Canal would not have fulfilled the promise of national defense.[61]

With their rationales fully articulated, boosters pushed to reorganize the now outdated state Ship Canal Authority. In 1961, the Florida legislature dissolved the old board and established a new agency called the "Canal Authority of the State of Florida." Recognizing the importance of the integrated waterways strategy, this new bureaucracy claimed responsibility for all inland waterways programs throughout the state. The barge canal would be connected literally and figuratively to other projects, especially the proposed Sanford-Titusville canal, necessary for the movement of rockets and missiles to Cape Canaveral. Much like those of its predecessor, the Canal Authority's responsibilities lay not only in the purchase of lands and rights-of-way, but in promotional activities designed to convince Floridians of the need for a canal. The authority's new five-person board of directors had only Malcolm Fortson as a holdover, and was comprised of members from all over the state, rather than simply north central Florida. Cementing the importance of statewide membership, officials appointed William McCree of Orlando as chairman. President of the Central Florida Development Committee, the polished, forty-eight-year-old civil engineer became the face of the canal movement for the crucial next phase. By the end of 1961, Fortson was replaced as managing director by Giles Evans, a former Corps official who became the most strident public proponent for the canal. While Bert Dosh left the board in the bureaucratic reshuffling, he hardly remained on the sidelines. State officials appointed him the chairman of the Board of Commissioners of the Cross Florida Canal Navigation District, a newly established sister agency to the Canal Authority charged with securing state funding necessary to purchase canal lands.

The State of Florida spent the next two years encouraging federal officials to embrace the construction project. Evans, McCree, Dosh, and Buckman, now hired as engineering council to the Canal Authority, racked up thousands of miles in their travels between Tallahassee and Washington. Moreover they criss-crossed the state appealing to service clubs, chambers of commerce, and innumerable civic groups to raise the project's profile. The hard work, however, remained in Washington. Without congressional funding, the project was going nowhere. Kennedy had requested the $195,000 in 1961, but subcommittee votes and the final reconciliation process attached so many strings that the funding could apply only to the economic restudy, and not preconstruction planning. That restudy paid off, however, as a June 1962 report issued a new benefit-cost ratio of 1.17 to 1.0, higher than the previous figure of 1.05 to 1.0, as engineers added in benefits for flood control and land enhancement. Though the figure appears like the most marginal of bumps, it was critical in that it did not even consider national defense.

With an initial allocation in hand, boosters prepared for another round of budgetary battles in 1962. Buoyed by the support of the Senate and the executive branch, Florida's congressmen asked for another modest appropriation of $205,000 for preconstruction planning. However, they ran into the formidable Clarence Cannon of Missouri, the curmudgeonly chairman of the powerful House Appropriations Committee. At eighty-three, Cannon had served in the House for almost forty years. Noted both for his fiscal conservatism and his encyclopedic knowledge of arcane parliamentary procedure, Cannon seemed almost unbeatable. It would take the wily political skills of representatives Bob Sikes, Sydney Herlong, and Charles Bennett to shepherd the appropriation through the House in October 1962. Playing heavily on the defense angle, Sikes maintained the canal was needed to move missiles to Cape Canaveral and barges from the Gulf to the Atlantic. "The Caribbean is no longer a placid American lake," he warned. "The Panama Canal is no longer a certain, safe artery for American and world commerce. Castro and Khrushchev have changed all this." Recalling the threat of German U-boats off the coast of Florida during World War II, Sikes announced that "Khrushchev now has a naval base in Cuba . . . and can now create havoc at any time." To ensure success, Sikes attached the funding to a public works bill allocating funds to projects in Oregon, Texas, Washington, and Illinois. Bennett personally lobbied Majority Leader Carl Albert, asking for his help in passing a bill that "means more to me than any other thing." Upon the appropriation's passage, the *Orlando Sentinel* offered nothing but the highest praise for the congressmen and their struggle "against great odds." Though the appropriation was meager, the ability of the Florida delegation to outwit their formidable rival impressed many observers across the state, and indeed the nation. The *Sentinel's* editorial board expressed confidence "that the men who have already done that which seemed virtually impossible can now get the construction money." Cannon, however, did not seem so sanguine about the delegation's success. "No bigger bunch of pirates ever sailed the Spanish Main," he complained. "All the money that Captain Kidd and Long John Silver stole is infinitesimal compared to this raid on the federal treasury."[62]

The thrill of victory did not last long. Supporters had every reason for optimism until Kennedy mysteriously submitted a budget in January 1963 that did not include a provision for the canal. To make matters worse, Georgia officials resurrected a forty-year-old plan for a cross-Georgia canal. Georgians couched their proposal in terms of "bringing Cape Canaveral into Georgia—a cross-state canal viewed as space age boon." Floridians' response was both swift and shrill. On March 1, 1963, Henry Buckman fired a letter to Bert Dosh demanding that "it is absolutely essential that we advise the President . . . that his chances of winning Florida will in all probability vanish if he fails to provide for the canal this time." Dosh concurred and immediately wrote John Bailey, chairman of the Democratic National Committee in Washington. Stressing his credentials as a lifelong Democrat, Dosh reiterated Buckman's observation. He warned that "unless the President renews his support of the canal . . . Florida may be found in the Republican column when the returns of the 1964 election are tabulated." Dosh joined Governor Bryant, Bill McCree, and Giles Evans to meet face-to-face with Bailey in Washington at the end of April. McCree reported that "everyone laid it on the line that if Democrats expect to carry Florida in the 1964 presidential election, the canal must be built." In the months following, newspapers throughout the state editorialized on the need for the project. Even the *Miami Herald* muted its traditional opposition and joined in the fray. In late June, Kennedy bowed to the pressure, which included personal meetings with Bryant, Holland, and Smathers, and revised his budget to include $1 million for canal construction.[63]

By mid-November, the bill for construction passed its major hurdle— Cannon's House Appropriations Committee. On November 20, 1963, the full House overwhelming passed the measure, 359 to 27. With a successful Senate vote in early December, the *Orlando Sentinel* could ecstatically crow, "Dreams Do Come True." The culmination of so many years of hard work seemed bittersweet, however, as only two days later a nation grieved over the assassination of John Kennedy. His successor, Lyndon Johnson, would be another in a line of New Deal liberals, committed to large public-works programs. One of his first official acts as president was the December 31, 1963, signing of the massive $4.4 billion public-works bill that included the $1 million necessary to begin canal construction. For Bert Dosh, the year ended on a happy note. It appeared the initial allocation was but the first installment of full canal funding. In writing once more to John Bailey, Dosh said, "we have received . . . authoritative information that President Lyndon Johnson will recommend a budget item of $4 million to the Congress to continue construction on the barge canal." Finally, there would be another groundbreaking.[64]

Groundbreaking, Breaking Ground

February 27, 1964, dawned cold and gray, an inauspicious start for the "most significant day in Palatka's history." Public officials from around the state gathered at the Rodeheaver Boys Ranch, an eight-hundred-acre facility for troubled boys founded by Homer Rodeheaver, the former musical director for the famous evangelical preacher Billy Sunday. Located about ten miles southwest of Palatka, the facility marked the official groundbreaking of what the Corps of Engineers called the "century-old dream of a waterway across the upper neck of Florida." Excited by the prospect of President Lyndon Johnson's appearance, between ten and fifteen thousand Floridians assembled at the rural setting to mark the momentous occasion.[1]

Ordinary Floridians braved the cold, rainy weather and more than an hour of ceremonial speeches for an opportunity to see President Johnson celebrate a moment that would put north Florida on the map. The *Palatka Daily News* marveled at the commitment of men who stood "ten to fourteen hours in front of burning barbecue pits to serve as many as 1500 or 1700 without batting an eye, except to rest their eyes smarting and swollen." Younger observers, enthused by an unscheduled day off from school, compared the event to the new musical phenomenon of the Beatles and hailed the moment as more important than the mop-topped quartet's loud music.[2]

Following the lunch and an hour of speeches, Johnson stood in the pouring rain and gave birth to the dream of so many for so long. Bert Dosh, Claude Pepper, Henry Buckman, and the widow of Walter Coachman represented the old guard of canal supporters on the platform. Buckman was so frail he had to be lifted onto the dais in his wheelchair. The new generation of canal boosters was present as well, with Charles Bennett, Bob Sikes, Tom Adams, and Farris Bryant among others proudly joining the president. "God was good to this country," Johnson proclaimed. "He gave us great estuaries, natural locales for harbors, but he left it to us to dredge them out for use by modern ships. He gave us shallow waters along most of our coast lines, which formed natural routes for protected coastal waterways. But he left it to us to carve out the channels to make them usable. Today we accept another challenge—we make use of another natural resource. He gave us great rivers, but let them run wild and flood, but sometimes to go dry in drought." Then he added laughingly, "and sometimes to rain when we have a celebration." Johnson went on to praise the guests on stage, "for the work they did in making this barge canal possible, and more importantly, for making the American apparatus of freedom go forward to new dimensions and to new boundaries." Concluding on another laughing note, he said: "every time I left the House, I was button-holed by Senator George Smathers or one of the Florida Congressmen, and I thought Senator Spessard Holland was going to have a heart attack the last time he talked to me about the canal."[3]

With that impromptu ending, Johnson pulled the red switch and set off 150 pounds of dynamite. Though the explosion provided a dramatic touch, it was little more than theatrical illusion. According to the *Ocala Star-Banner*, "the blast sent a geyser of peat moss, powdered charcoal, and black oil (soaked into the ground) [sic]." These were added for effect and created a spectacular sight. "Ordinary soil," the Army Corps pointed out, "makes a poor showing when subjected to an explosion." While enhanced to captivate the audience, the blast was still strong enough to throw a mounted patrolman off his horse and into the mud. Nonplussed, Canal Authority chairman William McCree praised those government officials and citizens who spent the day in the rain celebrating the start of "this important public works project." To canal supporters, the bad weather symbolized the obstacles that had to be overcome for the canal's construction. In a March 1964 letter to Congressman Sydney Herlong, a platform guest at the groundbreaking, McCree expressed the can-do spirit that permeated the February festivities. "I honestly believe that hardship, such as the continuous rain," he wrote, "made [the crowd] appreciate the meaning of the historic occasion more deeply."

Planning and construction closely followed the *Definite Project Report* submitted by the Corps more than twenty years earlier. That December 1943 document proposed a "waterway 12 feet deep, 150 feet wide, and

President Lyndon Johnson presides over the groundbreaking ceremonies for the Cross Florida Barge Canal, just south of Palatka, February 1964. Courtesy of Ocala/ Marion County Chamber of Commerce, Ocala.

181.96 miles long, following generally the alinement previously selected for the Atlantic-Gulf Ship Canal, designated Route 13-B." It would have five locks, each 75 feet wide and 600 feet long, which would raise and lower the water level of the canal according to the Florida terrain. Plans called for two locks on the Withlacoochee (at Inglis and Dunnellon) and two on the Ocklawaha side (at Rodman—the so-called St. Johns Lock— and Eureka). The key structure, however, would be the Silver Springs Lock, which would tie the Ocklawaha part of the canal to a thirty-mile "summit" path cut across the heart of central Florida "through rolling sand hills underlaid by Ocala limestone" and eventually connect the canal to the Withlacoochee. Water for this summit cut "will be supplied by the Oklawaha River and Silver Springs Run, and will be pumped up into that section from the Eureka Pool." To help provide the steady supply of water necessary for this summit area, plans called for the development of three impoundment dams, two on the Ocklawaha River on the canal's eastern side and one on the Withlacoochee, near the canal's western terminus. The Ocklawaha dams, at Rodman and Eureka, were to be entirely new earthen and concrete structures, finished with concrete spillways. On the other hand, the Withlacoochee dam, located near Inglis, would expand upon the existing Florida Power Corporation structure in place since 1908. As that dam had deteriorated significantly since its initial construction, much work was needed to bring it up to the standards necessary for use in the canal system.[5]

Under the supervision of Corps engineers, private contractors immediately began excavating a six-mile stretch of the canal west of the St. Johns River to the proposed site of the St. Johns Lock. By early 1965, the Corps completed this section and began construction on the lock, as well as continuing the canal's path toward the anticipated Rodman Pool, an artificial lake on the Ocklawaha that was a necessary element of the lock and dam system. Simultaneously on the western end of the project, dredging and dragline operations cut a straight line from the Gulf of Mexico to the Inglis Lock, where construction started in April 1965. In addition to the canal work, contractors built the first of three mammoth bridges designed to span the width of the ditch. To accommodate traffic along U.S. Highway 19, the major north-south artery along Florida's west coast, Corps staff designed a 65-foot overpass more than 150 feet long. By the end of 1965, the Corps proudly announced that construction was 4 percent complete, with an additional 8 percent under contract. With more than $16 million expended from groundbreaking to the end of 1965, it appeared the canal was well on its way to completion.

Despite the Corps' optimistic assessment, a mere four-and-a-half years later, the U.S. Department of Interior concluded that, "at the present rate of appropriation and construction it would take approximately thirty years to complete the project." The difference between these vastly divergent evaluations lies in the impact of an emerging environmental movement exemplified by Marjorie Harris Carr. Previous opposition to the canal either centered on its wasting of taxpayer money or its assumed destruction of the Floridan Aquifer by allowing salt water to intrude into the freshwater supply so necessary for both human consumption and agriculture. Corps engineers figured they had both of those problems solved by the late 1950s. The 1958 and 1962 economic restudies showed conclusively, at least to the Corps and canal supporters, that the project was a worthwhile economic enterprise that would benefit the nation more than it cost the taxpayer. Furthermore, the design changes, from a ship canal to a high-level lock barge canal, mitigated against the problems related to the aquifer. In 1959, Robert Vernon, the Florida state geologist, testified before the federal Bureau of the Budget that "the proposed barge canal construction of locks would maintain the groundwater level of the area adjacent to the canal . . . [and] has none of the inherent dangers present in the formerly proposed sea level ship canal." Vernon even went so far as to claim the canal would help Florida's "underground freshwater resources by serving as a conservation and flood-control measure." With those water problems ostensibly solved, who could argue with the new canal plan?[6]

Canal supporters did not figure, however, on a new calculation in the building process—the value of nature itself as an irreplaceable entity. While opponents railed about other issues, particularly the canal's relationship to the state's water supply, little had been said in the 1930s, 1940s, and 1950s about the canal's effect on the beauty of natural Florida, especially the relatively pristine Ocklawaha River. Only Theodore Hahn, a DeLand doctor and avocational boater, brought up this concern. His many boat trips down the Ocklawaha inspired him to begin a book-length manuscript about the beauties of the crooked stream. In early 1949, he wrote Carl Carmer, editor of the popular and critically acclaimed Rivers of America series, a collection that included Marjory Stoneman Douglas's masterpiece *The Everglades: River of Grass*. Hahn inquired whether Carmer would be interested in a book about the Ocklawaha, a stream important for "its wild beauty." We do not know if Carmer ever replied, but Hahn's uncompleted manuscript was never published. Writing to the editor of *Sailing South* magazine in June 1950, Hahn explicitly addressed the Corps' plans. "I am writing a book on this Oklawaha River," he penned, "which I hope to get published some day in order to stimulate interest in the river and also to keep perhaps the despoilers of nature from trying to make a ditch out of it for a cross state canal." Nothing came of either Hahn's book or his desire to stop those "despoilers of nature." But Hahn's feelings about nature and the river itself would appear again in the 1960s, this time much more publically, through the person of Marjorie Harris Carr.[7]

Marjorie Carr embodied the economic, environmental, and scientific opposition that would eventually prove the canal's undoing. Representing a new environmental ethos, Carr would fuse sentimental attachment to the preservation of wild land with a scientific understanding of the fragile nature of ecological systems. With a talent for grassroots organization and a sophisticated understanding of publicity, Carr pushed the canal from a local to a national issue and thus made it a part of a broader environmental movement. Born in Boston in 1915 and coming of age in the frontier area of southwest Florida in the late 1920s, Marjorie Harris combined a New Englander's love for learning with a Florida Cracker's love for the land. "We lived out in the woods," she later remembered, "and I had parents who could answer your questions about what is that bird, or what is that snake." Harris entered Florida State College for Women in Tallahassee in 1932 and majored in zoology. After graduation, she attended graduate school at the University of Florida in Gainesville and was consigned to a laboratory because of unwritten rules regarding a woman's place in academia. It was there she would meet Archie Carr, a newly minted professor of zoology. After a whirlwind courtship, they were married in January 1937. Archie Carr proceeded to have a illustrious career as a wildlife biologist at the University of Florida, becoming nationally known through such works as *The Windward Road*, a classic study of Caribbean sea turtles. Though Marjorie gave up her formal career as a scientific researcher to raise her family, she remained a scientist at heart. Their relationship depended heavily on their mutual scientific concerns. "Our whole interest in conservation evolved together, batting the ball back and forth," she remembered in 1984. "We were both concerned. We both defined the problems as we saw them and what could be done about them." In a later time, Marjorie Harris Carr would have had an esteemed career as a published author and a university professor. In the mid-1960s, however, she found a higher calling in the defense of the Ocklawaha.[8]

More than a year before the groundbreaking, as plans for the canal were being hammered out in Congress, Carr and other north Florida residents started to weigh the consequences of construction. While organizing a series of public forums on statewide conservation problems, Carr happened upon the canal issue when a Jacksonville Garden Club member notified her of the Corps' intentions. As a founding member of the Alachua Audubon Society, she, along with David Anthony, a University of Florida biochemistry professor, organized a special program in November 1962 to raise the question, "Is the cross-state barge canal worthwhile?" Meeting in a Gainesville high school auditorium, the event featured two state bureaucrats who gave presentations on "the effects of the proposed cross-state canal on wildlife and wilderness areas" as well as "the economic possibilities of such a canal." In spite of their arguments, accompanied by "impressive charts, statistics, and figures," most of the nearly two hundred audience members remained skeptical. They felt "in evaluating this project, it is not easy to introduce statistical evidence or show effects on charts. It is not easy and perhaps not very sensible to weigh real estate values against the pleasure people get from unspoiled wilderness." Listeners concluded that "their uncertainty [about the wisdom of canal construction] is based to a great estent [sic] on the fear that the canal will irrevocably change this part of the State and destroy many of our loveliest wilderness areas." The question-and-answer session after the official presentation quickly turned confrontational as audience members drilled the befuddled state officials on their rationale for the project. Indeed, the session turned so vitriolic that David Anthony later recalled that he had to apologize to the speakers for the audience's rather intemperate remarks.[9]

Energized by the level of discussion, Carr came away feeling that she had to do something. Combining an almost mystical love of wilderness with a scientist's understanding of ecology, Carr had found her calling. "The first time I went up the Ocklawaha," she later remembered, "I thought it was dreamlike. It was a canopy river. It was spring-fed and swift. I was concerned about the environment worldwide. What could I do about the African plains? What could I do about India? How could I affect things in Alaska or the Grand Canyon? But here, by God, was a piece of Florida. A lovely natural area, right in my backyard, that was being threatened for no good reason." While Carr expressed her opposition to the canal in poetic language that recalled Sidney Lanier, she would use science as the rationale to stop construction. Carr concluded: "Gainesville is a university town, and many of our Audubon members are professors who have a habit of questioning and testing statements. A blizzard of questions followed the presentation. Questions about the economics of the project, about the effects of construction on the geology, hydrology and ecology of the canal project area. These were questions for which the government speakers had no satisfactory answers."[10]

Carr's connections with the University of Florida proved crucial in her crusade to save the Ocklawaha. This was ironic since, just a few years earlier, university professors had played a significant role in promoting the canal. In March 1960, Ralph Eastwood, an economist in the department of agricultural economics, organized a three-day symposium on the campus dealing with the canal. With the help of the State Ship Canal Authority, he obtained such influential speakers as Robert Thomas, president of the Florida Waterways Committee, and State Geologist Robert Vernon to address the gathering. The conference emphasized the economic importance of the project to the state as well as the extolling the virtues of the new design of a lock-type barge canal that minimized damage to the Floridan Aquifer. Unlike the November 1962 meeting that so galvanized Carr, this academic seminar saw only positive feedback. Only days after the conference's end, Canal Authority chairman Malcolm Fortson wrote to the dean of the university's College of Agriculture. "Efforts to promote the canal in particular, and the interests of the State of Florida in general," he wrote, "will gain dignity and stature as a result of these seminars being sponsored by the University." Dean Willard Fifield wrote back to Fortson that "the Cross-Florida Barge Canal project is a tremendous undertaking. . . . I do hope our participation will yield worthwhile results."[11]

Marjorie Carr did not share those feelings about the canal and its effects on the Ocklawaha. Dissatisfied with the government's plans for a canal so close to her backyard, Carr became driven to do something, and she was not alone. By 1963, a few hardy activists from Marion and Alachua Counties had established an organization with the unwieldy title of Citizens for the Conservation of Florida's Natural and Economic Resources (often shortened to the "Citizens for Conservation") with the goal of saving the Ocklawaha River from the canal. While Carr was involved from the beginning as a noted conservationist, she was not the group's principal leader. That mantle fell to fellow Alachua Audubon member F. W. "Wally" Hodge, a retired army colonel and a recent migrant to Gainesville. Another major player was John Couse, a prominent Palm Beach County businessman and a member of the "Old Florida" elite. In 1913, Couse's father had built one of the first homes on the shores of Lake Worth, located along Florida's then almost undeveloped southeast coast. Couse made a living in the air-conditioning and refrigeration business before he moved northward to retire in the early 1960s, finding south Florida increasingly overdeveloped. As he later put it: "I just got sick of all the destruction. Nothing was sacred anymore; nothing was pure." After searching across the state for "a piece of Florida as God had created it," Couse settled in a cabin along the Ocklawaha for an ostensible peaceful retirement with his wife. He soon discovered, however, that his "natural paradise" was threatened by the bulldozers and draglines of canal construction. With the help of his daughter, Margie Bielling, a high-school science teacher, Couse organized anticanal forces to preserve the river and "repay Florida for all the wonderful things it has done for me." Bielling proved a crucial link in the organizational strategies designed to stop the canal. Living along the edge of the Ocklawaha Valley with her husband, Paul, a U.S. Forest Service biologist, Bielling, like Marjorie Carr, was a university-trained biologist. A generation younger than Carr, she had the opportunity to study at the now coeducational University of Florida. Driving back and forth between her home in Fort McCoy, close to the Ocklawaha, and Gainesville, Bielling became an important catalyst in the early stages of the opposition movement. Her hostility to the project made for a lonely life in booster-dominated Marion County. Seeking to bring more Gainesville residents into the fight, Bielling found kindred spirits among the academics and environmentalists of Alachua County.[12]

Anticanal forces, led at this time by Hodge, Couse, and Bielling, confronted a daunting task. Following the drama of a presidential groundbreaking, the steady roar of machinery soon created a momentum of its own. Seeing their dream at last started, the Corps and its local supporters confidently presumed the project was now on course for rapid completion. With each day of construction, decades of anticipation literally became concrete reality. Facing such a formidable foe, Citizens for Conservation quietly marshaled its arguments to challenge canal proponents. As articulated in two pamphlets produced in a six-month period from late 1964 to early 1965, the group opposed the project in terms of conserving natural resources and reining in wasteful government spending. Showing that a full environmental ethos had not yet emerged, arguments on this front centered on the destruction of hunting and fishing habitat rather than the protection of a pristine river. One pamphlet stated that the canal construction would "flood 27,350 acres of productive game habitat," causing "the drastic reduction of the wild turkey." Instead of relying on sentimental language that would later be associated with 1960s environmentalism, the group drew upon the hard data of the U.S. Fish and Wildlife Service. Describing the canal's impact in terms of lost man-days for hunting and commercial fishing, the group concluded the project would have deleterious results on these important outdoor activities. Reviving claims from the 1930s, Citizens for Conservation also claimed that the western reaches of the canal endangered the Withlacoochee River with saltwater intrusion. This time, however, the threat had less to do with the aquifer than its destruction of freshwater fisheries. At bottom, the organization cautiously addressed environmental issues in terms of traditional conservation rather than a belief in the preservation of wilderness as an end in itself. Yet there was a harbinger of the future. Denouncing the Corps' myopic measures of monetary values, the group asserted that in none of the engineering reports was there an "attempt made . . . to evaluate the losses from an esthetic viewpoint."[13]

In March 1965, Citizens for Conservation introduced its second point of contention: the canal was not economically feasible. Producing a pamphlet on the financial problems of the project, the group asserted that "the tragedy [of canal construction] takes on the dimension of folly when this unique and irreplaceable natural asset is destroyed for the sake of an industrial canal that (1) is not economically sound and will not benefit the general taxpayer, (2) has negligible value for national defense, and (3) will encourage the development of industries not in keeping with the widespread desire for Florida to remain a pleasant environment for both tourists and residents." They challenged the Corps' benefit-cost ratio projection of $1.17 returned for every dollar expended. Bolstering their charges with references to a 1963 *Life Magazine* feature entitled "Pork Barrel Outrage," the group concluded that the ratio "is far below the minimum 2 to 1" necessary for government projects to be economically viable. The pamphlet went on to claim that the Corps' figures were misleading as they were based on the unrealistic annual interest rate of 2.625 percent and a completion time of 6.5 years. By 1965, construction time estimates had already been extended to eleven years, raising project costs from roughly $158 million to over $171 million.[14]

Far from simply extolling the virtues of wilderness preservation, Citizens for Conservation demonstrated that they too could use statistical data and attack the Corps on its own terms. In some respects, this reflected a growing national strategy to confront the Corps' rationale for massive water projects. As a February 1965 editorial from the *Stuart News* suggested, "the best approach on these deals is to challenge the economic justification put forward by the Corps." Wading through reams of technically obscure engineering reports proved important, but in the end, the arguments of Citizens for Conservation hinged on a simple, but important, question about Florida's economic future. To their mind, canal boosters were encouraging "the development of industries not in keeping with the widespread desire for Florida to remain a pleasant environment for tourist and resident." Fearful that the canal would attract only "dirty" industries, leading to the creation of a "second Ruhr Valley" in which "petroleum, fertilizer, wood products and other bulk materials" would constantly traverse and eventually destroy the Sunshine State, they suggested Florida should instead attract "'polite' industries [with] no smoke, no dust, and a high quality of professional and technical personnel." The group ominously warned that "no one wants to turn Orlando, St. Petersburg, and Tallahassee into another Newark, Jersey City, or Trenton."[15]

Throughout the spring and summer of 1965, while opponents developed their arguments against construction, those in favor of the canal pushed forward. In early May, they saw visible proof of their hard work as ground was broken for the Inglis Lock and the U.S. Highway 19 bridge on the western end of the canal. The occasion was celebrated by a special supplement to the *Citrus County Chronicle* and the *Dunnellon Press*, in which editors waxed poetic over the canal and its benefits. "As for the event this edition commemorates, its naked hugeness could never be adequately covered . . . . The Cross Florida Barge canal, our barge canal, is as American as a Florida vacation." The ceremony was filled with all the hoopla associated with a patriotic holiday. Though not as grandiose as the festivities at Palatka fifteen months earlier, the Inglis affair still showcased the canal and its boosters. Backed by three high-school bands, master of ceremonies Congressman "Billy" Matthews presided over a crowd approximated at three thousand people. While local supporters used three thousand pounds of fried mullet and seven hundred pounds of cole slaw to feed the crowd, the highlight of the event was

Local towns clamored with excitement about the canal's promise for economic progress and prosperity. Crystal River, near the canal's western terminus, 1966. Courtesy of State of Florida Photographic Archives, Tallahassee.

Governor Hayden Burns, the former longtime mayor of Jacksonville and ardent canal supporter, manning a bulldozer to symbolically move the first load of dirt. Corps regional official General A. C. Welling addressed the assembled supporters and thanked Florida's congressional delegation for their "well-coordinated and well-determined organization that runs wonderful interference . . . on the Hill." He also complimented his Corps workers on the project as well as "local and state governing officials—a superb state organization working on projects for development of water resources." Officials anticipated that these projects on the canal's western end would take two years to complete and cost approximately $7 million.[16]

Supporters were genuinely delighted with the tangible results of their efforts on both the western and eastern ends of the canal. Yet any sense of satisfaction was diminished by the disquieting problem of a growing anticanal movement. Grappling for a way to cope with their critics, proponents vacillated between overconfidently downplaying—or even ignoring—the opposition and developing a public relations effort to vigorously promote the project throughout the state. An alliance of Corps officials and boosters from Ocala, Palatka, and Jacksonville, as well as state officials and Florida's congressional delegation, supplied a steady stream of publicity that they hoped would drown out the opposition. In many respects, the canal advocates had every advantage. The weight of time, precedent, and sheer bureaucratic inertia worked in their favor. With persistence and tenacity, they could easily muscle the project through to completion.

With these attitudes in mind, the thought that anyone would challenge the canal invited ridicule, at best, and, at worst, outright hostility. As Jim Hughes, manager of the fervently procanal Ocala/Marion County Chamber of Commerce, put it in January 1965, "I doubt you will be impressed by the Citizens . . . in their opposition to construction of the Cross-Florida Barge Canal." A week later, Congressman Sydney Herlong concurred with Hughes, adding that "what this corporation in Gainesville does is of no concern to me. The Canal is on its way to being built." Even more vociferous in his support, and dismissive of the opposition, was Florida Secretary of State Tom Adams. Among the most powerful politicians in the state, Adams was born into a prominent Jacksonville family in 1917. After attending the University of Michigan, he returned to Florida to become a successful dairy farmer and real-estate investor. Adams, known for his engaging public speaking and his rough-and-tumble style of politics, became the point man for canal supporters within the state. In May, buoyed by the Inglis groundbreaking, he opined that "when the Cross-Florida Barge Canal is completed and tied together with the Gulf and Atlantic Intracoastal Waterways, Florida will indeed find a wealth more valuable than gold."[17]

While Adams made the case for the canal's benefits, Canal Authority chairman William McCree gleefully mocked and belittled canal critics in a presentation before the Gainesville Kiwanis Club. Calling Citizens for Conservation "a so-called conservative group from your fair city," McCree charged that their "false" accusations about the canal's impact impeded the progress of construction. "I don't know much about this organization, but they have never come to their greatest source of information—the Canal Authority." Sydney Herlong also dismissed their influence, especially after another round of congressional appropriations in the summer of 1965. Another $10 million for the continuation of construction, he remarked, signified "the dying gasp of opposition from people who object to the project." While Herlong, who represented Leesburg and Lake County, remained optimistic about the outcome of the dispute, Gainesville congressman "Billy" Matthews was beginning to feel the political heat of canal opponents. Confiding in the manager of the Ocala/Marion Chamber of Commerce in late July, Matthews conceded that although he was "100% for the Canal," he was "taking a whipping at home in Gainesville . . . and every day I am getting threatening letters and telegrams."[18]

Much of that heat stemmed from the growing strength of statewide and, indeed, national opposition to the project. In April 1965, the Florida Audubon Society finally took a stand opposing canal construction. Protesting "the loss of this irreplaceable natural asset—the Oklawaha wilderness area, the Oklawaha Valley, and the 45 mile stretch of the Oklawaha River—which will result from the construction of the Cross-Florida Barge Canal," the organization encouraged members to lobby state and federal officials and called for a public hearing on the canal issue. This was a sharp departure from their previous hesitancy on the matter. Marjorie Carr expressed her frustration that "there appears to be a lack of understanding of the basic responsibility and obligation of the Society." Having secured the support of Florida's oldest environmental organization, local members of Alachua Audubon pushed the issue even harder. While canal proponents celebrated increased appropriations and continuing construction, anticanal forces were emboldened to challenge officials publicly. In May, Wally Hodge, by now president of both Alachua Audubon and Citizens for Conservation, engaged Canal Authority chairman McCree in a debate sponsored by a Gainesville civic club. Echoing previous concerns, Hodge announced that construction would "drive our wildlife such as turkey to higher, drier ground." Remarks like this showed that broader concerns for wilderness preservation for its own sake were still in the future. Even so, Alachua Audubon concluded their campaign with a mailer to nearly four hundred conservation groups throughout

Florida, alerting them to "the extent and nature of the destruction to the Oklawaha River Wilderness, which will brought about if the present construction plans for the Cross-Florida Barge Canal are carried out."[19]

The efforts to turn the canal controversy into a national issue succeeded when the *New York Times* issued a critical editorial in July 1965. The *Times* argued that "it is time to stop the work and weigh the total public interest in the Oklawaha before irreparable tragic damage is done to this stream for benefit of a project of what is at best dubious merit." It also decried the lack of public hearings on the project. Response from canal boosters was swift and vociferous. The *Orlando Sentinel*, playing on time-worn opposition to northern interlopers, proclaimed that "presumably for lack of any problems up that way to stick its nose into," the *New York Times* "is demanding editorially that work be stopped on the Florida Cross State Barge Canal." After blasting the *Times* for inaccuracies and half-truths, the *Sentinel* concluded that the river will be "even more valuable because more will be able to enjoy" it. Proponents continued this tack, accusing Citizens for Conservation of feeding the *Times* misinformation. The *Ocala Star-Banner* suggested that they did not "question the sincerity of some of those who lent their names to the letter writing, but sponsors of this campaign have violated every rule of fairness and objectivity in the literature which they have prepared. . . . We can conclude that this letter-writing campaign has been inspired and master-minded by traditional opponents of waterways improvement in a last-ditch effort to inject confusion and misdirection into the path of progress."[20]

Other proponents confronted their adversaries behind the scenes. Speaking for the Florida Board of Conservation, Director Randolph Hodges responded directly to the *Times* editorial page editor John Oakes in an August letter. He announced that "this editorial must have been based on misinformation, and certainly a lack of understanding of the urgent need for the economic development of the north-central area of the state of Florida." Hodges forcefully made the case for canal completion, by asserting: "To deny economic opportunity to north-central Florida for no other reason than to preserve forty-five miles of an admittedly beautiful wild river, would be a crime that no responsible state official ever could commit." Attacking opponents of the project as "an almost infinitesimal minority of our people," Hodges concluded that the state's "support of the project is reflected in the virtually unanimous backing of the project by Florida's major newspapers." Writing to one Corps engineer, Bert Dosh similarly accused "that Gainesville group of issuing much propaganda, much of it false and misleading."[21]

In spite of the increased opposition, proponents had good reason for optimism in the summer of 1965. Congress approved $10 million for the continuation of the project with "a thoughtful disregard of certain minute but anguished protests on the Cross-Florida Barge Canal." With this authorization, Canal Authority chairman McCree proudly announced that "no longer is the Cross-Florida Canal a dream. . . . We are sitting on top of the most dynamic inland navigation project in the nation." With construction continuing at both Rodman and Eureka on the east and Inglis on the west, Corps district engineer Colonel R. P. Tabb confidently reported that "the barge canal job has been described as the fastest moving civil works project by the Corps of Engineers in six Southeastern states."[22]

To build on this momentum, the Canal Authority and the Corps put much of their effort into managing public support for the canal. Cognizant of the defeat of the Ship Canal thirty years earlier, they could not afford to remain passive. As early as January 1965, in response to the publication of the first Citizens for Conservation pamphlet, canal supporters systematically kept tabs on the activities and publications of their critics. From the beginning, Canal Authority officials understood that "the perpetrators of the [anticanal] folder" had "no tolerant attitude" in reconciling the demands of progress and conservation. Indeed, "they want to kill the canal." A month later, they expressed their concerns in even starker terms. The "Chairman considers a concentrated local public relations effort now mandatory in order to ensure continued success of the project. A local 'conservationists' group has recently started attacking the project and apparently are [sic] financially well supported. The effort over the next few months must be political as well as strictly public information in nature." For much of 1965, canal supporters followed this advice. Exemplifying this approach, the head of the Corps' Planning and Reports Branch suggested there should be no "specific rebuttal to the present folder. To do so would tend to reinforce and bolster what appears to be a rather small voice of discontentment." Congressman Herlong concurred that it would be best to "give the opposition to the canal the silent treatment."[23]

With much vested in the project, the Corps, however, was not simply content to ignore or even monitor their adversaries. Taking the initiative to eliminate the perceived threat, officials emphasized the need for "publicity efforts [that] take the constructive form of public information and education." Corps public relations officer Gene Brown fulfilled that mission by steadily "providing stories to papers in each county citing [the] effect of works on canal." Accompanying a flood of press releases and news bulletins, Corps officials relied on both the editorial pages of supportive state newspapers and direct public promotion. State Canal Authority manager Giles Evans worked hand in hand with Corps officials to disseminate "weekly news clippings germane to the Canal project." The authority's public information staff distributed pamphlets and brochures, organized displays in public buildings, lobbied congressmen and other public and private officials, and presented slide shows to interested service clubs and chambers of commerce throughout Florida.[24]

The public relations strategy of both the federal Corps of Engineers and the state's Canal Authority showed dividends in August as newspapers around the state rallied to support the canal. Within a two-week period, papers from Panama City to Orlando editorialized in favor of the project. Even the *Gainesville Sun*, hometown paper of anticanal activists, expressed the beliefs of project supporters when it announced that "since nature didn't provide the state with a cross-peninsula waterway, we applaud the ingenuity, the intelligence, and the perseverance that is at last bending nature to the service of human needs and the welfare of the state." Coming from the home of Hodge, Carr, and Anthony, the *Sun*'s position especially pleased supporters. A Canal Authority consultant noted that in spite of having to "make certain concessions to their local people," the *Sun*'s editorial board vigorously argued "that the Canal must be completed."[25]

Though boosters had good reason to remain positive about the ultimate fate of the canal, they also had cause for concern with increasing national criticism of the project. The *New York Times* was not alone in its questioning. In a June letter to Carr and Anthony, National Audubon Society president Carl Buchheister expressed his concerns about the canal, building on local claims about the destruction of "some of the largest concentrations of wild turkeys in the state." The Citizens Committee on

Natural Resources, a national conservation organization, similarly supported withholding the $10 million 1965 appropriation "in view of the serious damages to the natural environs with the resulting uneconomic project." In September, the Sierra Club issued a policy statement out of its national office that "favors preserving the Oklawaha River of Florida as a Wild River." Such national support strengthened the claims of the Federated Conservation Council, an umbrella group of sixteen Florida environmental organizations that included garden clubs, women's clubs, and Audubon Society chapters.[26]

While money for the project was secured for the moment, this opposition ensured that future reauthorizations would be difficult. As national groups began to express doubts about the canal, local activists voiced their concerns to a broader political audience. Writing to White House counsel Lee White in June, Marjorie Carr implored that "we desperately need your help . . . only the immediate intervention by President Johnson can prevent the tragic loss, this summer, of this valuable natural resource." Deputy Budget Director Elmer Staats responded for the White House, blandly commenting that "while we truly appreciate your views, there appears to be no reasonable alternative but to follow through with the construction of the canal because of the work already done and existing contract commitments." Receiving little help from the executive branch, Carr, this time writing on behalf of Alachua Audubon, contacted the Florida congressional delegation a month later. "A mistake—a terrible mistake—has been made in planning the route of the Cross-Florida Barge Canal," she wrote. The canal would lead to the "obliteration of the Oklawaha River," as the Corps remained outside public scrutiny with "its contempt for public opinion." Carr listed a wide-ranging number of state and national conservation groups that "have all gone on record vigorously protesting the projected loss of the Oklawaha River and its magnificent forested valley" and asked the congressmen to support these organizations and stop (or reroute) the project. By September, she pleaded with vocally procanal congressman Sydney Herlong for "your support—your active support—in this critical situation." Herlong responded that though he both "admire[d] and respecte[d] you and Archie," he didn't "plan to do anything to interfere with the orderly construction of the Cross-Florida Barge Canal." In spite of being rebuffed once again, Carr's

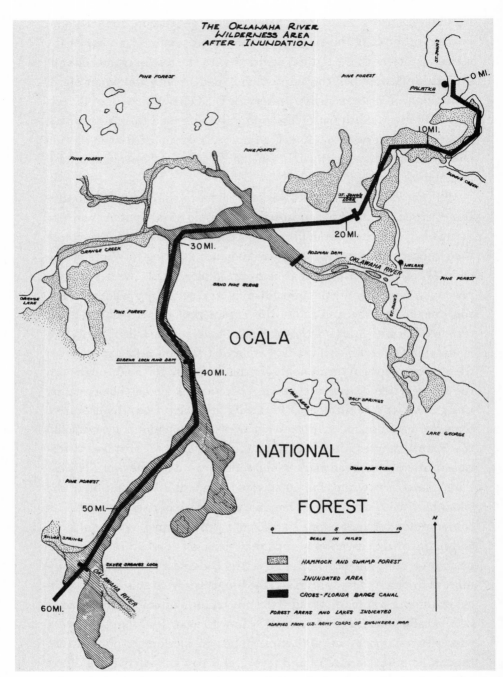

Proposed canal route and its impact on the Lower Ocklawaha, bypassing the river's natural outlet into the St. Johns River. Courtesy of Florida Defenders of the Environment (FDE), Gainesville.

message was getting out—the Cross Florida Barge Canal would not be completed through the Ocklawaha Valley without a struggle.[27]

The summer of 1965 marked two milestones in the fight to stop construction and save the wild Ocklawaha River. One was tactical and offered what seemed to opponents as a legitimate alternative to the canal's path through the river valley itself. Anticanal forces announced that if the project could be rerouted away from the Ocklawaha, they could support completion. Plans were drawn showing a canal cutting southward, hooking up with the St. Johns at Lake George. The alternative design would allow the river to remain in its primitive state. In July 1965, Carr wrote to the governor "request[ing] plans for a new route for the canal from the Silver Springs area to the St. Johns, which would avoid the Oklawaha River Wilderness Area." These possibilities showed that Citizens for Conservation and other opponents saw themselves as not simply obstructionists to progress and opposed to any development. They could support both the canal, albeit a rerouted one, and protection of the Ocklawaha River. Canal advocates did not consider these arguments in this light however. Instead, they viewed them as specious and disingenuous, contending that the increased costs and time needed to reroute the canal would doom the whole project. Boosters saw the stopping of the canal, not the rerouting of it, as the real goal of Carr, Couse, Bielling, and their supporters. These technical concerns over differing routes proved the major issue of mid-1965 and obscured the second, and ultimately more important, issue.[28]

Anticanal forces, especially Marjorie Carr, were gradually moving toward a broader notion of wilderness and its relationship to the canal itself. Earlier environmental opposition had centered on the "destruction of natural resources that will result from the construction of the Cross-Florida Barge Canal." Conservation-oriented, this approach depended on statistics of man-days of hunting and fishing as well as loss of habitat acreage as reasons for stopping the canal. The image on the late 1964 brochure opposing the canal on natural, rather than economic, grounds, featured a wild turkey symbolizing the loss of valuable habitat as "the turkey population would be drastically reduced." By July 1965, Carr had all but abandoned the turkey-conservation strategy. Building on an environmental ethos stressing the preservation of wilderness, rather than simply the conservation of natural resources, she forged an opposition to the canal based on the ideas of Aldo Leopold, Charles Torrey Simpson, and Rachel Carson.[29]

Aldo Leopold was a product of Yale University's forestry program, graduating as a professional forester in 1909. Developing a philosophy of wilderness preservation while a U.S. Forest Service employee in New Mexico, Leopold honed these ideas in the 1930s as he helped establish the Wilderness Society. Taking a position at the University of Wisconsin, Leopold increasingly saw nature in scientific, ecological terms rather than sentimental, aesthetic ones. His seminal work, *A Sand County Almanac*, published posthumously in 1949, broke new ground in its delineation of the importance of wilderness. "We abuse land," he wrote, "because we regard it as a commodity belonging to us. . . . It is inconceivable to me that an ethical relationship to land can exist without love, respect, and admiration for land, and a high regard for its value. By value, I of course mean something far broader than mere economic value; I mean value in the philosophical sense." This vision of the inherent importance of wilderness for its own sake was becoming increasingly popular by the early 1960s and would become a foundation for Marjorie Carr's evolving notion of why saving the Ocklawaha was necessary.[30]

Charles Torrey Simpson came to south Florida in 1902 and spent the last thirty years of his life writing about the Everglades as a natural system worth saving. At a time when even Marjory Stoneman Douglas, the "Mother of the Everglades" herself, spoke positively about the attempts to drain the Everglades, Simpson railed like an Old Testament prophet about the necessity of preserving the natural systems of south Florida. As early as 1920, he conceded bitterly that "soon this vast, lonely, beautiful waste will be reclaimed and tamed; soon it will be furrowed by canals and highways and spanned by steel rails. . . . We shall proudly point someday to the Everglade country and say: Only a few years ago this was a worthless swamp; today it is an empire. But I wonder quite seriously if the world is any better off because we have destroyed the wilds and filled the land with countless human beings." Though few heard and even fewer listened to Simpson's concerns during his lifetime, his ideas about the uniqueness of the Florida environment and the importance of preserving it gained new life in the 1960s through the work of Art Marshall, Marjory Stoneman Douglas, and Marjorie Carr herself.[31]

Carr's intellectual debt to Rachel Carson was both more personal and less specific than that to either Leopold or Simpson. Carson's *Silent Spring*, published in 1962, was an environmental jeremiad, raising the specter of an American landscape ravaged by chemical pesticides. With the eye of a trained scientist, Carson analyzed the devastating effects of DDT on birds and other wildlife. While Leopold and Simpson spoke to a small but growing constituency, *Silent Spring* became a national best-seller and the bible of an emerging environmental movement. In hind-sight, its publication was "effective and timely for an American public that had blindly accepted the comforts technology provided." Carson's work was much more than a book about chemicals and birds. To Carson, the key concept was not wilderness absent of human contact, but ecol-ogy, where people were an integral part of the natural world. Though more comfortable with Leopold's vision of nature and Simpson's view of the Everglades than Carson's more jaundiced views, Carr also embraced Carson's gendered notions of feminine nurturing and motherhood. It is easy to imagine Carr nodding in agreement with Carson's assertion to a national sorority of women journalists in 1963: "I believe it is important for women to realize that the world of today threatens to destroy much of that beauty that has immense power to bring us a healing release from tension. Women have a greater intuitive understanding of such things." Representing a turning point in the women's movement, Carson, and by extension, Carr, combined traditional female concerns with the mind-set of a scientist. Though opponents would often denigrate Carr and Carson as mere female interlopers in a masculine world of business and science, their very gender aided their cause by appealing to moral notions of a mother's concern for her family.[32]

Carr's genius, besides her incredible organizational skills, lay in her ability to take Leopold's abstract musings, Simpson's understanding of Florida's ecology, and Carson's gendered environmental vision and turn them into concrete actions to save the Ocklawaha. Capitalizing on the momentum of the just-passed 1964 Wilderness Act, Carr increasingly saw opposition to the canal in terms of preservation of wild places. Drawn to the late nineteenth-century writings of Sidney Lanier on the pristine nature of this beautiful area, Carr dropped (at least for the moment) sci-entific rationales in favor of more sentimental ones. Her seminal March 1965 piece "The Oklawaha River Wilderness" (widely disseminated in the August 1965 issue of Florida Audubon's publication the *Florida Natural-*

ist), graphically showed her intellectual evolution. Starting with Lanier's lyric lines, "the Oklawaha, the sweetest water lane in the world," Carr penned a paean to the virtues of wilderness. She ended by announcing that "this valuable complex of wilderness which, mostly by plain good luck, remains as a sample of the original Florida, if saved and cared for could forever serve the enjoyment, inspiration and education of man." Building on this vision, Carr, writing for Alachua Audubon, asked Governor Burns in July for his "support and assistance to save the Oklawaha." The rationale for this help was the maintenance of wilderness. Carr reminded Burns "that the intangible, but nonetheless, real values of this unique wild river area did not receive consideration by those responsible for planning the canal route." By September, while writing to Congressman Herlong, Carr attached a new piece of anticanal documentation. This time it was not the dry economic statistics, or even the hunting man-days lost, brought up in the earlier pamphlets of Citizens for Conservation. A single photograph of the pristine Ocklawaha wilderness, accompanied by lengthy Sidney Lanier quotations on the beauty of the river, marked the new strategy of what was developing into a broader environmental movement.[33]

By early 1967, Carr revealed just how far she had come in her thoughts concerning the inherent value of wilderness. In a paper delivered at the Orlando meeting of the Southeastern Section of the Society of American Foresters, Carr contended that "once wilderness is gone, it is gone forever." Generalizing about the problem, Carr continued, "even if I think my descendant [*sic*] could be content in a world without wilderness I am sure I don't want to be responsible for making him live in such a world." Carr saw "a lag between public reaction and official recognition" of an emerging pro-wilderness constituency. Nationwide, she maintained there has been "a serious under-assessment of the size of the contingent of citizens who place high value on the intangible qualities of our original landscape, and for a gross underestimate of public support for the preservation of our remaining samples of wilderness." Moving to the specifics of the Ocklawaha controversy, Carr underscored her new line of thinking. "While the Oklawaha is indeed a superb river for fishing and boating," she concluded," these were not the primary reasons why all of these people have spent their own money and taken the time to make the long trip to Tallahassee on that cold January day. These people

Marjorie Harris Carr, "Our Lady of the River," at the Ocklawaha River, 1966. Courtesy of Florida Defenders of the Environment (FDE), Gainesville.

wanted to save the Oklawaha simply because it is one of the most beautiful rivers in America."[34]

While Carr's vision of wilderness provided the impetus for future opposition, strategy for the duration of 1965 centered on developing an alternative canal route—one that would completely bypass the river. Audubon activists concluded that "the Oklawaha River Valley can be preserved if the section of the route of the Cross-Florida Barge Canal from Silver Springs to the St. Johns River is realigned." Suggesting a route "so that the canal will run due east from the Silver Springs Lock," they concluded that "realignment of this section of the canal *would not preclude* the realization of the alleged benefits of the canal project" (emphasis in original). With the development of this strategy, activists had to find a way to convince the Corps to adopt their plan. They settled on the need for a public meeting in north central Florida "for the purpose of exploring the advisability of realigning the section of the canal. . . . We think it would be most desirable—and just—to have such a public hearing held before the Corps of Engineers and/or the Florida Canal Authority." Carr announced that the "hearing must be held before January 15, 1966," because "the Corps will let the contracts for the construction of the Rodman and Eureka Dams in February, 1966." This would allow conservationists "time for adequate preparation of statements and testimony." Opponents also maintained that this would be the first opportunity since 1940 for the public to debate the merits of the project.[35]

Supporters saw no need for such a meeting. As one booster put it, "those who would at this late date come up and castigate a project of much importance to the nation, and the state, and the national defense, and . . . make erroneous and false statements about the canal, none of their comments should be given any weight." Supporters especially took issue with the charge that the project lacked public input. In an editorial attacking Citizens for Conservation, Bert Dosh dismissed as "incorrect" their claim "that the canal project has never received a public hearing in the state of Florida." In fact, Dosh listed three previous public hearings on the canal held in Jacksonville and Ocala in 1935, 1940, and 1959. The irony that there had been only one opportunity for public input in twenty-five years seemed lost on the semi-retired Ocala editor.[36]

Frustrated with the Corps' intransigence, Citizens for Conservation attempted to turn a routine October 1965 meeting held in Ocala into a debate on the broader issue of the need to reroute the canal. As Canal Authority manager Giles Evans put it, "opponents unsuccessfully attempted to disrupt the proceedings into a sounding board for their feelings on the canal." During the meeting, Authority chairman L. C. Ringhaver dismissed all critics as out of order, concluding that "this meeting was arranged to discuss right-of-ways only." Blindsided by the unexpected vociferousness of the opposition, Ringhaver mentioned the possibility of future hearings. William Partington, a West Palm Beach environmental educator who would become another key figure in the anticanal struggle, pressed Ringhaver on the issue. "I am writing to you personally," he wrote, "to request that you arrange a public hearing on this controversial project . . . in suitably large quarters." By early December, however, the Canal Authority chairman still had not come up with a date. "We still haven't come up with a hearing," he reported; "December is pretty busy with the holidays and all." Critics were also rebuffed for hearings at a national level when Senator George Smathers concluded that "no useful purpose could be served by additional public hearings." Frank Philpott, the *Gainesville Sun*'s outdoor editor and a vocal opponent of the canal, observed in early December that "the strategy of the Authority thus far has been to dodge, table, and postpone" the establishment of a meeting. Undaunted, Carr and her allies applied pressure in the public arena. By the first week of December, headlines from across the state addressed the controversy and, in spite of their continued editorial backing of the canal, supported the need for a meeting. A *St. Petersburg Times* headline screamed "Oklawaha Use for Canal Protested." Papers from as far as Miami, Fort Pierce, and Lake Wales chimed in to alert Floridians about the potential loss of the river. Carr even resorted to plaintively plead to longtime canal supporter Claude Pepper, "please, will you help us secure a public hearing?"[37]

The efforts of Carr and fellow conservationists paid off when Governor Hayden Burns announced on December 9 that state officials would hold a public hearing on the canal at the Tallahassee meeting of the Board of Conservation regarding water-resources development. Proposed for January 25, 1966, the meeting, according to Burns, would be "the proper time and place for any citizen to be heard on matters relative to water development." Though the meeting was not on Carr's terms as to either the time or place, anticanal forces had gained a significant victory. The Corps and the Canal Authority yielded grudgingly to their demands. Both sides viewed the upcoming meeting in confrontational terms. With the meeting little more than a month away, they would have time to galvanize forces and meet their adversaries in what two north central Florida newspapers called "the last battle" over the project.[38]

Amid the growing debate, canal construction lumbered on. Indeed, by year's end, supporters could boast of numerous accomplishments. Though proponents always clamored for increased federal funding, Congress' $10 million annual appropriation was more than double the total monies expended in the two previous years. Having conquered an eight-mile section of the project in the east, army engineers proudly announced "the canal is complete from Palatka to Rodman Pool." To the west, draglines and dredges continued to plow through land and sea from the Inglis Lock to the Gulf. Moreover, lock construction at both

This Corps publicity shot proudly shows off the first completed mile of canal near Palatka, August 1964. Courtesy of Ocala/Marion County Chamber of Commerce, Ocala.

ends was occurring on an immense scale. At the St. Johns Lock alone, more than "600 tons of crushed rock, 300 tons of sand, the equivalent of 1800 bags of cement, and 600 bags of fly ash [were] being used daily at the special concrete mixing plant set up at the lock site in Putnam County." By December, more than half of the St. Johns and a third of the Inglis locks were completed. The Corps was making progress on ancillary structures as well. In addition to the completion of a $45,000 project office building at Silver Springs, construction of the western Highway 19 bridge proceeded at a steady pace. Bolstered by such significant gains, the Corps confidently looked toward the immediate future, planning to begin construction on the Rodman Dam and spillway and the Eureka Lock and Dam by the end of April 1966.[39]

As the Corps and its contractors engaged in the more dramatic aspects of canal construction, the Canal Authority and the Navigation District quietly continued to purchase land necessary for future phases of the project. While the work was considerably less glamorous than that of their uniformed federal counterparts, state authority officials reminded observers that the "construction progress can proceed no faster than right-of-way acquisition regardless of federal construction appropriations." For 1965, they had gained considerable ground, obtaining the Rodman Dam site, most of the Eureka Lock and Dam site, and all rights-of-way on the western side of the canal. Hardly deterred by the growing opposition, they spent the latter half of the year relentlessly acquiring land for the Rodman Pool. Nothing would distract them from their mission. The same week Governor Burns ceded to demands for a public hearing, authority manager Giles Evans blandly issued a monthly report emphasizing an order for the taking of the Rodman Dam area and a condemnation suit "against [the] Chilean Corporation for Levy County portions of their property." In the report, Evans made no mention of the January meeting.[40]

Like their adversaries, anticanal forces also had much to be proud of in 1965. Considering its modest origins from a local Audubon Society meeting in November 1962, the movement started in Marion County by Margie Bielling and John Couse, and in Gainesville by Wally Hodge, Marjorie Carr, and David Anthony, had coalesced into a potent political force. Like a "terrier on the end of a trouser leg," these activists had pushed the canal issue to the front pages and editorial columns of newspapers across the state. Moreover, they assiduously forged alliances with a wide variety of nascent and well-established environmental groups on state and national levels. Intent on organizing a broad-based coalition, and perhaps in recognition of future political battles, Citizens for Conservation had by year's end allied itself with the Florida Bi-Partisans, an environmental group organized by the Lake Wales activists Ken and Helen Morrison. Together these groups would provide "an amalgamation of organizations fighting to save the Ocklawaha."[41]

All of these accomplishments pointed to the emergence of an increasingly influential environmental movement, yet canal boosters contemptuously dismissed their opponents as mere "bird watchers and do-gooders." In an otherwise favorable article in early December, the *Gainesville Sun* recognized the difficulties facing anticanal forces. Characterizing their struggle as "a mismatch of David and Goliath proportions," the *Sun* noted that "an intrepid band of nature lovers and outdoorsmen" were "trying to save a river from execution." To the paper, the contrasting positions could not be sharper. On one side were the defenders of "45 miles of one of the loveliest rivers in America." On the other were "the invisible hands of economics with its commercial interests clamoring for a water route across the midriff of Florida."[42]

Canal opponents had to sadly concur with the newspaper's discouraging assessment. In a last-ditch effort in the fall of 1965, Carr and Anthony wrote to Lady Bird Johnson, who had taken the lead in fostering environmental awareness with the Highway Beautification movement earlier that year. With remarkable bluntness, they admitted that "from the standpoint of support, encouragement, and endorsement, we have been astonishingly successful—but we are going to lose the Oklawaha River." For more than a year, they had pursued an exhaustive campaign that had "explored every possible path recommended by several respected and knowledgeable conservationists." And yet public apathy remained. To Carr and Anthony, there was "a bitter similarity between the case of the doomed Oklawaha and the case of the young woman, Miss Catherine Genovese, whose prolonged attack and murder in full view of 38 citizens of New York, on March 13, 1964, so shocked the nation. . . . Miss Genovese could easily have been saved. Any one of the 38 silent citizens could have called the police in time to prevent the third and final murderous attack. It would have taken very little effort and hardly any expense. The Oklawaha River can easily be saved." Carrying the Genovese analogy for three angst-ridden pages, Carr and Anthony played upon the subsequent guilt that was associated with the inaction of innocent bystanders in that infamous event. They concluded this overblown portion of the letter with a plea to the public conscience, "We are sure that our descendants will bitterly deplore this incredible act of folly of our generation." Carr and Anthony plaintively ended their appeal by asking the First Lady to "please help us save the Oklawaha. We have delayed writing to you before now in the hope that a solution would be reached at other levels. We have failed—and time for the Oklawaha is running out." Johnson's noncommittal reply a month later underscored a sense of hopelessness and an understanding of the importance of the January meeting.[43]

While working from a position of relative weakness, opponents still made some gains by year's end. In addition to securing the upcoming public hearing, they were forcing the *Gainesville Sun* to reconsider its position on the canal. In an editorial entitled "Pirates at Work?" published just days after the announcement of the public hearing, the paper asserted, "no one can deny the conservationists have raised serious questions about the cross state barge canal." Agreeing with David Anthony that the canal was "a primer of bad government," the *Sun* called for a reappraisal of the project. This sudden change of position seriously rattled boosters, as a local paper that once "strongly urged proceeding all out" in favor of the canal was now assuming the tone of south Florida publications that consistently attacked the "calamitous canal"[44]

Girding for the upcoming battle, canal supporters both extolled the benefits of the canal and downplayed the concerns of naysayers. Using the power of his office, Tom Adams dedicated the December 1965 issue of the Florida Department of State's *In the Capitol* to the positive impact of the canal. Lauding the project as "an asset to all," Adams asserted its completion would place the state "at the apex of the greatest trade potential in the nation." Not content to remain in Tallahassee, Adams crisscrossed the state, from Jacksonville to Englewood, extolling the virtues of the project. As a "stimulus and . . . lifeblood of modern economic development," he announced, the canal would be part of a steady procession of progress. Its completion would lead to "the biggest economic boost to the state since the first railroad lines tied wilderness Florida to the nation." While Adams remained the positive public face of the project, Canal Authority manager Giles Evans shored up congressional support by attacking canal opponents. On January 3, 1966, Evans mailed a vitriolic seven-page letter to Congressman "Billy" Matthews. Discrediting both organized opposition groups and the *Gainesville Sun* for reversing its editorial policy, he concluded that there was "no worthwhile purpose in entering a debate with these late-comer opponents." Calling anticanal activists "obstructionists, "hysteria-ridden," and "well-intentioned but ill-informed," Evans maintained that their use of "innuendo" and "ridiculous extremes" obscured the obvious benefits of the project to "the consuming public." Evans viewed the upcoming public hearing as a waste of the Canal Authority's time and an impediment to canal completion.[45]

Facing an increasingly strident opposition, canal boosters counted on the press to publicize the project's value. Newspapers representing communities along the route continued to extol the benefits of canal completion. The December 23 headline in the *Dunnellon Press* vividly showed the ongoing optimism among supporters. "Many Express Faith in Canal Future," its headline exclaimed. These papers, however, saw storm clouds on the horizon. Rather than discussing the recent construction successes, the *Ocala Star-Banner* published an article entitled "McCree Defends the Canal." Even though the *Dunnellon Press* could say "Barge Canal Forges Ahead," it needed to qualify that by noting "notwithstanding opposition." By year's end, the *Citrus County Chronicle* ominously proclaimed, "there's trouble in the wind in '66 for the Cross Florida Barge Canal." Determined to confront the challenge ahead, the paper cautioned readers that the "unrest can be soothed provided those who believe in the canal expend the energy necessary to thwart the efforts of those who have recently organized to stop it." Marshaling the troops for the upcoming showdown, the *Chronicle* proclaimed: "We must raise our voices once again and be heard."[46]

With the start of the new year, all eyes now focused on Tallahassee and the January 25 public hearing. Though excited to discuss the matter, anticanal forces approached the meeting with caution. The long-sought hearing, wrote a south Florida conservationist, "has several unfortunate aspects. The Time, which is close to when the contracts for the locks affecting the Oklawaha River will be let out, Place, which is far from the Canal route and appears discriminatory by being so inconvenient for many honest and knowledgeable critics to attend, and Circumstances, as there will be an annual meeting of water resources developers in Tallahassee at that time, are all factors obviously and intentionally 'stacking the deck' against the Canal critics."[47]

Nevertheless, opponents organized for what they viewed as their last chance to protect the Ocklawaha. Gathering on the banks of the river at Orange Springs two days before the scheduled Tallahassee meeting, members of various environmental and conservation organizations participated in an emotional assemblage that played upon sentimental and indeed spiritual notions of nature. Considering it was Sunday, Reverend Ulysses S. Gordon, longtime pastor of Gainesville's First Presbyterian Church, set the mood with a prayer that—unlike Lyndon Johnson's 1964 vision of creation, where there was always room for human improvement—established a direct relationship between God and nature. "Almighty God," he intoned, "who has revealed Thy splendor and bounty in this created earth, we praise Thee for the order and beauty of nature in

Completed canal portion on the western end at U.S. 19. Courtesy of State of Florida Photographic Archives, Tallahassee.

sea and cloud, in field and forest, in the blue lakes and the swiftly flowing stream." Moving from the general to the specific, Gordon thanked the Creator for "this shining stream, fed by deep, clear springs, winding its course in banks of peace and solace for men in the push and noise of these hurrying days." As both a warning and plea for forgiveness, he reminded listeners "that we have squandered our resources, that we have polluted our streams, that we have substituted ugliness for beauty and desecrated with man's clumsy efforts the handwork of God." He ended the homily by appealing to the Almighty to "bless the efforts of this group dedicated to saving the remnants of a vanishing Eden, awaken our law makers and all citizens of this commonwealth to the sin of waste and to the immediate importance of saving for our children and for all posterity the Oklawaha and its surrounding wilderness." For true believers, the rhetoric was enlivening, and it no doubt encouraged supporters to descend upon the Capitol confident in their ability to change the course of both history and the canal.[48]

In addition to relying on the power of prayer, Carr and her supporters grew increasingly sophisticated in drawing attention to their struggle. In the weeks before the Tallahassee Showdown, they repeatedly called for a massive boat-a-cade to attract attention to their cause. At the same time, they held "Save the Oklawaha" fishing contests that granted prizes to those who not only caught the largest fish but also those who traveled farthest to participate. Most importantly, they invited Holly Fisher, a young, idealistic New York filmmaker, to the January 23 event on the banks of the Ocklawaha. Fisher, the daughter of Richard Fisher, a wealthy boat manufacturer and early convert to the anticanal cause, intended to make a film of both the Sunday riverside revival and the public meeting in Tallahassee. Her finished product, a twenty-seven-minute documentary entitled *Progress, Pork-Barrel, and Pheasant Feathers,* produced in conjunction with New York independent filmmaker Romas Slezas, carried the message of the anticanal activists to a wider audience. The film, though overly melodramatic and preachy at times, tied anticanal forces to the emerging environmental movement nationwide and became an important element in the overall strategy to discredit the Corps and save the river. According to *Newsweek,* the film became a "classic example of government indifference to public outcry [which] whipped up a storm strong enough to sustain the battle."[49]

Undaunted by the cold and rainy weather that seemed to define the major events of the canal controversy, hundreds of conservationists, representing all facets of the movement, crowded the Florida House chamber on January 25. Following a routine morning session devoted to statewide water-resource issues in which local officials and government bureaucrats backing other projects also "expressed their support of the canal and its present alignment," opponents finally had their chance. Held before the Florida Board of Conservation, the public hearing lasted from 2:30 to 9:30 p.m. Despite a packed chamber, board members—the entire state cabinet—did not think the issue merited their attention, or even attendance. With the exception of board director Randolph Hodges and vociferous canal advocate Tom Adams, all the other members left the room after the pro forma morning session. Speaker after speaker, including Bielling, Carr, and Anthony, addressed Hodges and Adams, imploring them to change the route of the canal. Frank Philpott, outdoor editor of the *Gainesville Sun,* joined the fray by casting himself as a representative of "many alternative route supporters." Castigating the board for refusing to hold the meeting "in the vicinity of the Oklawaha River 'where the action is,'" Philpott demanded—in the name of fishermen, tourists, and "the ordinary Florida citizen who has never seen the Oklawaha"—that "the eastern section of the barge canal be relocated in order to preserve this beautiful river for wildlife and man." Dozens of others representing themselves and conservation organizations from across the state spoke passionately in favor of rerouting the project.[50]

Canal supporters marshaled all their considerable power to rebut the opposition's testimony. While Carr and her allies fully expected resistance, they naively assumed government officials would provide at least an opportunity for a balanced hearing. Instead, they faced the hostility and contempt of Tom Adams, who "took it upon himself in the earlier parts of the afternoon to be the voice speaking for the canal." Virginia Forrest, director of Florida Audubon, complained of Adams's "disgraceful treatment of citizens invited to air their views." The secretary of state "endeavored to intimidate and belittle the statements of value made by renowned and respected citizens." Moreover, "Mr. Adams repeatedly challenged and badgered conservationists but never challenged the canal supporters." Another activist characterized Adams as a "prosecuting attorney" who bullied his opponents with ruthless cross-examination. Even more disturbing than Adams's surly behavior was the suspicion that the hearing was "most amazingly rigged." The president of Florida's Izaak Walton League concluded wearily, "we don't feel that we got an adequate hearing. . . . I would say the decision was already made. They've committed themselves on it and they are going along as planned." Other speakers concurred that the meeting seemed designed to "let the conservation groups blow off steam." Indeed, one speaker directly challenged Adams and demanded "an immediate answer" to prove "that this hearing was not just 'sound and fury signifying nothing.'" Adams held the line and summarily dismissed his opposition, claiming "this canal has been underway a number of years and this is the first time we have heard these protests." It would be irresponsible, he added, to stop "the orderly process of a public works project just because of the protests."[51]

For weeks afterward, Adams became a symbol of the canal controversy. To boosters in such places as Ocala, the secretary of state was a hero who single-handedly faced down "the well-meaning bird watchers." Charging that the rerouting argument was nothing more than a ruse, the *Ocala Star-Banner* hammered canal opponents for their repetitious "phoniness which cannot be ignored." Mimicking Adams's confrontational tone, the paper snidely questioned "how much they know about the situation" when many opponents "not only have never visited the river but can't even pronounce the name." Calls for a public hearing by Carr and others were simply an attempt "to obtain a stage on which [they could] produce a documentary on video film and in writing to put into the hands of congressmen in Washington" who might intercede on behalf of the anticanal forces. Thanks to Adams's strident performance, "canal support remains deep and dedicated. The foundations of faith have not been shaken."[52]

It is hard to imagine that canal opponents watched the same event as project boosters. Undaunted by Adams's "shabby treatment," Carr and Anthony "felt a moral responsibility to do what we could to prevent an irrevocable mistake of enormous proportions." Fisher's filmed portrayal of the event deftly played on Adams's performance by juxtaposing images of the state official, looking imperious and condescending, with wholesome shots of Margie Bielling testifying on behalf of a doomed river. Carr and her allies were now even more determined to save the Ocklawaha. And for the time being, they continued to stick to the rerouting argument. As Carr put it in July 1966, "altering the route of a section of the Cross Florida Barge Canal will preserve the Oklawaha and will cause no more interference with the canal project than if the Corps of Engineers decided to change brands of cement—maybe less."[53]

Lessons learned from the January showdown were as different as the positions of the two adversaries. For canal proponents, Tallahassee was but a minor impediment on the road to completion. Critics had their chance to blow off steam, but the project would go on. As the *Sarasota Herald-Tribune* sardonically editorialized: "All that the objectors could possibly have hoped to achieve by insisting on this week's hearing was to put the conservationist's viewpoint once again on the record—in the hope that the next river or forest or bay or reef may be saved. The Oklawaha, friends, already lies in the path of Progress Inexorable which can easily be heard hooting in the near distance." Calling the meeting "the sideshow in Tallahassee," the editorial concluded, "the Army engineers have already let contracts for considerably more canal work than Congress has as yet given them the millions for." Though admittedly battered by the confrontation, canal critics were of a different mind. Having faced an adversary that one newspaper characterized as "a rolling 'dozer . . . one on which all the major figureheads of the political power structure of the state are riding," anticanal activists felt they had held their own. Indeed, they did more than that. As one conservationist presciently noted before the meeting, "if this 'Public Hearing' turns against the wishes of these public spirited citizens, who are as concerned with the moral image of Florida as they are with the idiotic destruction of the Oklawaha, then you may be assured that there will be a redoubling of efforts to demand investigations into the actions of the Canal proponents." Years later David Anthony recalled how people came to Tallahassee "hat in hand," and "left angry and galvanized." Margie Bielling similarly remembered leaving the Capitol far from dejected, but "stimulated to see people from all over the state backing us." Far from being a final confrontation, the ground was now broken for a "fledgling environmental movement."[54]

New Visions, New Strategies

The Tallahassee Showdown of January 25, 1966, settled little. Boosters seemed more determined than ever to see the canal finished as quickly as possible. For the moment, opponents quietly retreated and reorganized their efforts to protect the river and stop the canal. But underneath this extremely public dispute, the ground was shifting. Florida was changing—irrevocably, fundamentally, and permanently. The struggle over the canal was a part of that change but a broader postwar economic and demographic revolution contributed mightily to the shift. By the mid-1960s, Florida emerged as a leader in a phenomenon political scientists would later deem the "Sunbelt." Massive federal spending during World War II sparked a boom that, until recently, has hardly slowed. An astonishing period of growth followed as an unprecedented postwar economy delivered the promise of middle-class prosperity. Aided by the relentless boosterism of state agencies and local chambers of commerce, few could resist the siren call of Florida. Drawn by cheap and bountiful land, low taxes, and the lure of continuous sunshine, new arrivals from the Northeast and Midwest crossed the state line at a rate of one thousand people a week by the mid-1950s. Modern transportation—jet travel and especially the newly developed interstate highway system—drove the process. Technological change also helped. Air-conditioning and pesticides blunted brutal tropical conditions and made the summers bearable. Now more than a temporary winter retreat, Florida living symbolized a perpetual vacation. With climate-controlled housing amid the sun and surf, retirees and northern transplants gladly abandoned the harsh winters of the urban industrial North.

In addition to new people, the region was attracting new industries. Previously grounded in agricultural production and extractive industries such as lumbering and phosphate mining, postwar Florida embraced a modern economy of high-tech manufacturing. Closely tied to Cold War military spending, new industries such as Martin-Marietta, Harris Corporation, Pratt and Whitney, Honeywell, and General Electric gravitated toward an emerging Space Coast and its metropolitan neighbors. Simultaneously, Florida became even more dependent on tourism and the service industries that accompanied it. Indeed, tourism became the primary engine of economic growth. The new tourism would have only a casual relationship with the river boats of the nineteenth century. Instead it would be organized around the imaginary world of Walt Disney and his plans for a self-contained vacation destination in central Florida, a destination that would be in Florida but not of Florida.

Such widespread changes also made their mark on Florida's political landscape. The steady influx of snowbirds seeking refuge in the cities and suburbs of central and south Florida raised important political questions. Florida had long been one of the most malapportioned states in the nation, with less than 14 percent of the northern rural population determining more than half of the state's senators and only 18 percent electing a majority in the House. Challenged by the growing influence of south Florida, a tight-knit clique of north Florida politicians, nicknamed the "Pork Chop Gang," became even more resolute to maintain their legislative grip on Tallahassee. Throughout the 1950s and 1960s, politics divided the state as the legislature struggled with issues associated with economic development and population growth—especially highway and school construction. The most vociferous struggles hinged on demands for reapportionment, which would alter the legislative balance of power southward, into the hands of transplants with little interest in north Florida's traditional rural values. Part of that tradition was segregation, and the politics of race and the emerging civil rights movement also exacerbated regional tensions. Facing federal demands to desegregate public schools following the *Brown* decision of 1954, northern Pork Choppers stood firmly in defense of Jim Crow. On the other side were Governor LeRoy Collins and south Floridians who, while quietly emphasizing a commitment to southern segregation, sought a more conciliatory resolution in the name of stability and racial moderation that would enable the Sunshine State to continue to attract newcomers and diversify its economy. Such a position aroused the ire of north Florida politicians, who viewed newcomers with suspicion. Wedded to ideas of white supremacy and one-party politics, the Pork Choppers seemed increasingly anachronistic amid the world that was taking shape in the 1960s. Yet, they would continue to dominate state politics until the United States Supreme Court finally intervened and declared Florida's reapportionment law unconstitutional in 1967.

Much like the state's political traditions, the Cross Florida Barge Canal was becoming a relic as well. In a world of jetports, interstate highways, and a burgeoning trucking industry, a cross-state canal seemed incapable of meeting the demands of a modern economy. With Florida's growth predicated increasingly on tourism and services, there was little use for a canal and the industrial infrastructure associated with it. Images of lumbering barges and grimy industrial parks violated the vision of a subtropical paradise imagined by northern migrants. And though Tom Adams seemed pleased with himself for shouting down Marjorie Carr and her allies, pictures of a crew-cut Pork Chopper—complete with a seersucker suit and white socks—berating ordinary citizens did not play well to the state's new constituency. Indeed, within a year Florida politics took a dramatic turn when Claude Kirk, championing modernization and democratic reform with a new constitution, became the Sunshine State's first Republican governor since Reconstruction. Finding themselves in a new and different political world, canal boosters would have to alter their strategies.

The gubernatorial campaign of 1966 revealed just how much the political ground had shifted. On the surface, the canal controversy disappeared from the headlines as the campaign between Republican Claude Kirk and Democrat Robert King High centered on broad generalities regarding Florida's future. High, the young dynamic mayor of Miami, had made a name for himself by defeating the sitting governor, Hayden Burns, in a closely contested Democratic primary. Anticanal forces saw themselves as major players in the upset. They enthusiastically embraced the south Florida politician because he opposed the canal. In making her case to a sympathetic reporter, one conservationist explained, "it was not the political machine in Florida or Washington, nor High's political machine that defeated Mr. Burns. . . . It was the behind the scenes efforts of conservationist[s] that did the job for him." Marjorie Carr saw High's victory in even more specific terms. In a May 30 handwritten note to Congressman Sydney Herlong, Carr asserted that "we are confident the recent victory of Mayor High reflects the growing unhappiness of the Florida public over the *needless* waste of Florida's natural assets in general and the Oklawaha River in particular" (emphasis in original). Herlong, like most Florida politicians, remained unconvinced. Responding to Carr, he declared, "I can't find any connection between the nomination of Mayor High and the defeat of Governor Burns with the Cross State Barge Canal." As evidence, he cited the support of longtime canal backer Congressman Claude Pepper for High's candidacy.[1]

Though Herlong understood Florida's political culture far better than single-issue activists like Carr and her allies, conservationists felt empowered by High's primary victory. Heady with optimism, canal opponent Richard Fisher wrote the gubernatorial candidate to congratulate him for his "recent victory and the unseating of the Pork Chop Gang." For Fisher, High represented a rare opportunity to fundamentally change the political landscape in favor of conservation interests. "We plan to dramatize you and your victory perhaps even a bit out of proportion," he asserted, "to arouse courage in other states for individuals and organizations to get on their feet and go after other 'bad guys.'" Unable to contain his enthusiasm, Fisher ended his appeal with a rousing postscript that saw the canal controversy as a catalyst for not only statewide but national change. "The prospect of the State of Florida almost overnight being transferred from an iniquitous arena of cynical exploitation to a stronghold of enlightened husbanding of natural assets has a scintillating grandness that staggers the imagination and what's more important can, we are completely sure, galvanize to concert tremendous forces throughout the country. We hope that you leave no little thing undone which can help to realize this prospect."[2]

Within weeks the realities of statewide politics dashed such hopes. Recognizing the need for north Florida's support for his campaign victory, High publically changed his position on the canal, endorsing its completion in a speech at the Eureka Dam groundbreaking in June 1966. Decrying this apparent betrayal, one activist wrote to High immediately afterward. "You seem to have abandoned your stand on conservation," the woman proclaimed. Playing off of High's campaign slogan, "Integrity Is the Issue," she reminded him that "no man with integrity can support this pork-barrel project. Therefore we must conclude that you have sacrificed your integrity to political expediency. It follows that you are not deserving of support from any conservationist."[3]

High recognized the potential fallout of his decision. Writing to the president-elect of the Izaak Walton League, he pleaded for understanding: "Had I been in an authoritative position only a few years ago, I am sure that an alternate route not destroying the Ocklawaha could have been developed. . . . However, if our efforts in the field of conservation are to be meaningful they must be based on realities. I do feel that the Cross Florida Barge Canal is a valuable addition to Florida not only in terms of our economy, but also in terms of increased recreational facilities. . . . Let us hope adequate long and intermediate range planning can make possible both economic development and preservation of our national environment." High's call for understanding fell upon deaf ears. Embittered by his changing position, many canal opponents switched their support to the Republican Kirk, whose position on the canal remained steadfastly unknown. Kirk's platitudes seemed a better alternative than High's pandering.[4]

Kirk's unprecedented electoral success seemed to validate the political clout of anticanal forces. As John Couse put it in early 1967, "our Republican victories in the past election were the result of a revolt by the citizens." Though obviously overstating their role in Kirk's victory, conservationists took pride in the new governor's close relationship with one of their own, Hobe Sound's maverick Republican activist Nathaniel Reed. Young, wealthy, and good-looking, Reed represented a throwback to the Republican conservation ethos of Theodore Roosevelt. Kirk quickly appointed Reed his chief environmental advisor at the bargain-basement price of a dollar per year. Throughout the late 1960s and early 1970s, Reed would profoundly shape the environmental policies of not only state but national figures. Though losers in the battle with Tom Adams, anticanal forces saw the future optimistically, for they now seemed to have a voice in Tallahassee.[5]

In spite of the changing political winds, 1966 proved to be "another year of achievement" for the Canal Authority and the Army Corps of Engineers. As a Canal Authority report put it, the project, having overcome "major difficulties" caused by "virulent opponents from throughout the state and further afield," had "made excellent progress in 1966." In terms of promotion, land acquisition, and construction, canal supporters pressed onward with little concern for the arguments of their adversaries. Becoming much more assertive, the Canal Authority increased its "public information activities intended to educate the people of Florida, and the Congress, to . . . the importance of the Cross-Florida Barge Canal project." Authority representatives criss-crossed the state, setting up displays in schoolrooms and courthouses, and speaking to a wide variety of service clubs and interest groups. Pamphlets and brochures were distributed extolling the virtues of the canal's economic and recreational benefits to all Floridians. Recognizing the need to appeal to a wider audience, the authority celebrated the recreational possibilities of almost two hundred miles of newly created shoreline. Officials even hired a public relations consultant to provide "a vacation-type story on the canal" designed to answer "inquiries from people in distant places who want to spend their vacations on the barge canal." Additionally, the agency developed a new promotional movie entitled *Main Street USA*, which was not only distributed to civic clubs and chambers of commerce across the state, but broadcast on local television stations, reaching "almost 13 million television sets in 1966."[6]

While the Canal Authority promoted the project, its main function was securing the land necessary for construction. Though at first glance tedious and arcane, acquiring rights-of-way was the crucial first step toward the project's completion. However, nearly two years following the ceremonial groundbreaking, the Authority had not accomplished its goal. By November 1965, authority manager Giles Evans complained that "we have not made as fast progress as needed on acquiring [land]." Determined to pick up the pace, Evans recommended a "thorough streamlining and revision" of procedures "to ensure that we keep to schedules and priorities." At that point, the agency controlled only 30 percent of the project's right-of-way, and that was with the head start of several thousand acres acquired by the Ship Canal Authority of the 1930s. Through orders of taking, condemnation suits, easement arrangements, leasing agreements, and outright purchases, state officials aggressively accelerated the drive for land.[7]

By the end of 1966, the authority controlled thirty-three thousand acres, roughly 48 percent of the total right-of-way. A year later, working diligently, it had acquired another thirteen thousand acres. With little regard for the environmental concerns raised by Carr and her allies, much of that procurement centered on the Ocklawaha Valley, where almost twenty thousand acres were designated as sites for the massive

Rodman and Eureka Reservoirs. The scale of real-estate transactions was as dramatic as the construction that followed. Two years after Evans's admonition, the authority, spending nearly $4.5 million, obtained nearly twenty-three thousand acres from 335 large and small landholders. With another $8 million, the agency assured the public it could achieve its goals by decade's end.

Clearing tons of earth and constructing massive water control facilities at both ends of the project, the Army Corps moved quickly to complete its mission. In bland bureaucratic prose, it cataloged its impressive achievements. By the end of June 1965, the Corps had completed only 10 percent of the St. Johns Lock, and 5 percent of the Inglis Lock. Within three years, however, project engineers had made great gains with the construction of the St. Johns, Inglis, and Eureka locks and dams. At the same time, work continued on the digging and dredging of the canal itself and the clearing of reservoir pools to provide the water necessary to operate the locks. Moreover, the Corps was building four expansive bridges for the highways that traversed the canal's right-of-way. In addition to all of this, the project demanded a host of ancillary buildings, spoil banks, and spillways to complement the canal itself. Contracting with local and national construction firms, the Corps appeared well on the way to achieving the dreams of generations of Floridians.

The dry statistics of canal construction belied the reality of the environmental destruction taking place in the Ocklawaha Valley. Particularly invasive, and especially visible when compared to the traditional use of draglines, bulldozers, and cranes, was the specially designed crusher. Called the "Crusher Crawler" by engineers, "Big Charlie" or "Paul Bunyan's Bull" by its operators, and "The Tree Killer" by environmentalist critics, this 22-foot-high, 306-ton amphibious machine cleared the land for the ten-thousand-acre Rodman Reservoir located on the Ocklawaha River. Developed by the Leesburg engineering firm of Gregg, Gibson, and Gregg, and considered the pride and joy of firm executive F. Browne Gregg, the crusher provided a low-cost efficient mechanism to rapidly clear a relatively inaccessible area.

Built at the Jacksonville Shipyards at a cost of approximately $600,000, the crusher quickly became the symbol of both the canal's promise and its problems. Engineers and technocrats marveled at its massive power and huge dimensions. Operated by a two-man crew and powered by twin

Construction of St. Johns Lock (later renamed after longtime booster Henry H. Buckman), 1967. Courtesy of State of Florida Photographic Archives, Tallahassee.

270 horsepower diesel engines, the machine was designed to work in swampy terrain and could negotiate rivers and streams "without hesitation." Gregg considered the machine as "the greatest break-through in construction work in recent years." Summing up the consensus regarding the crusher's destructive power, an observer said, "I feel like I'm in a science fiction movie."[8]

Gregg and the Corps of Engineers proudly premiered the new invention to an obviously impressed group of newspaper reporters in April 1967. In effusive, flowery prose, local writers lauded the machine's overwhelming size and power. In a time when technological achievements were automatically associated with progress and modernity, the crusher symbolized man's conquest of nature, a positive assumption to be taken as a matter of fact. While journalists likened the machine to legendary giants of the past—comparing the crusher to a "rumbling behemoth," a "self-sufficient monster," and a "goliath"—they finally stood in awe,

The crusher. Courtesy of Florida Department of Environmental Protection, Office of Greenways and Trails, Tallahassee.

impressed by the "newly developed mechanical marvel." It was a "monster," but not "of flesh and blood," instead "comprised of metal, hydraulic systems, ballast pumps, and electronic gear." If the Corps felt their unveiling of the crusher would provide much positive publicity, they were mistaken, as images of the machine and its destructive capacity quickly galvanized opposition and raised questions about the legitimacy of the project.[9]

Moving through the heavily timbered lowlands of the Ocklawaha Valley, the crusher cleared terrain at the rate of one to two acres per hour, much faster than the conventional rate. Observers noted how the "newly developed mechanical marvel . . . mows down trees and pushed them underground as though they were matchsticks." With its concrete-filled reinforced steel push-bar, the machine could easily topple cypress trees six feet in diameter by simply "crush[ing] them into the earth or muck." Speed and efficiency were the watchwords of the crusher's operations, as Gregg developed the machine "to fill a need by engineers and industry for an economical method of clearing flood basins." Able to work on a round-the-clock basis, the crusher was expected to clear the area for the Rodman Pool in approximately four months, "as debris left in the monster's path will be left to become part of the bottom of the reservoir."[10]

F. Browne Gregg symbolized the engineering mentality that pushed for canal completion. "I think it has now come to be recognized that my tree crusher was not so bad a thing," he reminisced in a 1971 newspaper feature, "that going through a swamp and pushing trees over and embedding them in the mud and sand is the natural way, the way nature puts those trees down. It's just evolution in a sense." Echoing Tom Adams's dismissive condemnation of canal opponents, Gregg later compared Carr and her allies to Jacobite radicals: "I kind of feel like it is similar to the French Revolution. . . . We are having an ecology revolution," he concluded. "And with this ecology revolution, people who are ordinarily sane, well-judging individuals have apparently gone berzerk [sic] on the subject." A true believer in the Corps' notion of progress and improving the natural environment, Gregg refused to concede ground to his critics. Articulating the shared assumptions of the Corps and canal boosters, he asserted that "man's development of natural resources has either got to continue or we're going to find ourselves in the dark ages." With regard to the canal, "certainly we upset the environment in the Ocklawaha. But we did not destroy it, not by a long shot. And really, we improved it." For Gregg, there was no room for debate: "in my opinion the cross-state barge canal is worth every bit of ecology it upsets."[11]

Fresh from his perceived triumph at the January 1966 meeting, Tom Adams fleshed out the logic of Gregg's vision in a March 1966 speech. Adams proclaimed that "Florida will be at the apex of the greatest trade po-

The crusher in action, clearing the land for Rodman Reservoir, 1968. Courtesy of State of Florida Photographic Archives, Tallahassee.

tential on the continent . . . once the barrier of our peninsula is breached with the Cross-Florida Barge Canal." Continuing in this vein, Adams foresaw the canal as the engine of Florida's progress. Four years after the project's completion, "it will carry more tonnage than now passes through the Suez Canal . . . allowing the state's ports to become focal points for transshipment to the emerging markets of the world. . . . People are coming to Florida," he concluded, "industry is coming to Florida. Agriculture is growing in Florida. And we spend millions of dollars each year in advertising to encourage both people and industry to come here. Thus, it is imperative that we develop the capacity to support them." For Adams, as for Gregg and other supporters, that development meant the canal's completion.[12]

Canal supporters continued to see the project as only one part, albeit the crucial part, of a broader system of national waterways. As he did so often, Tom Adams best articulated that vision, this time in a 1967 speech. "This canal is a key to Florida's total waterway system. . . . Once that key is completed it will enhance the value of all other waterways currently in Florida's system and those planned to complete our system of commercial watercourses, such as the missing link between Tarpon Springs and St. Marks." Interest in this complementary project coincided with the canal's progress. Deemed economically unfeasible by the Corps in the late 1950s, the Gulf coast project gained greater support with the potential tie-in to the canal. Florida representative William C. Cramer proclaimed the waterway "one of the essential links in the Gulf, Atlantic and Mississippi waterways systems." By the fall of 1966, the Corps announced plans to develop the two-hundred-mile inshore Gulf waterway at a cost of $87 million. This, of course, was dependent on congressional authorization, but Florida's congressmen seemed optimistic. Tampa congressman Sam Gibbons crowed: "I am grinning all over. This is a necessary project to keep our area in competition with the rest of the United States as far as transportation is concerned, to make sure we can take advantage of new markets." As the *St. Petersburg Times* editorialized in October 1967, "the full usefulness of the cross-state canal depends on the creation of a sheltered barge route south to Tampa Bay and north to other major Gulf ports," for it would unite Florida's west coast with a national waterway system from Brownsville, Texas, to the coast of New England.[13]

Corps members and canal boosters seemed to live in their own world, where rhetoric so often became reality. Any sign of progress was superlative. Indeed, if the supporters had been playing college football, they would have been flagged for excessive celebration. As an Ocala Chamber of Commerce publication put it, "After forty years of talk . . . there is water in *one mile* of the Florida Cross State Barge Canal!" (emphasis in original). Other, more objective observers viewed the glass as half-empty. A 1967 feature article in the *St. Petersburg Times* Sunday magazine exemplified the growing skepticism: "From Inglis on the Gulf to Palatka on the St. Johns, the air reeks with the smell of boom. But . . . progress is slow." Despite all of the news articles and press releases touting the rapid rate of construction, "to the average sightseer driving along the routes that edge and careen along the canal's path, there is little to see. For so little has as yet been done. Of one hundred miles of canal to be excavated, only about twelve are built. . . . Of 87 million cubic yards of earth to be excavated, only about a twelfth has been dug. . . . Of five locks to be built, only three are under construction—contracts for the other two have not been awarded."[14]

Boosters, of course, took exception to the *Times*' judgment. As L. C. Ringhaver, chairman of the Canal Authority, complained in the authority's 1967 annual report, "few of our general public really appreciate just how far construction of the canal has progressed." From his perspective, much had been accomplished: "The St. Johns Lock is nearing completion, and Rodman Dam is ready for closing the Oklawaha as soon as the lock is operable—probably about March 1968. Inglis Lock is already operable; the 23 miles of channel excavation are complete; the U.S. 19 Highway Bridge was open nearly a year ago. S.R. 19 is now open. The U.S. Coast Guard is marking the Western end of the Canal from its terminus several miles out in the Gulf on past U.S. 19 to the junction with the Withlacoochee River near Inglis Dam." In spite of Ringhaver's rosy outlook, problems were emerging on the horizon. Amid the usually glowing year-end summary, Ringhaver expressed profound concerns about "disappointingly low federal construction appropriations." In fact, the vigorously procanal *Orlando Sentinel* announced, "Pessimistic Note Sounded in Canal Report" as it headlined Ringhaver's year-end assessment. In his report, Ringhaver noted "the Corps of Engineers had been able to award *no* new canal construction contracts during the entire calendar year of 1967" (emphasis in original). Such a halt in construction was inherently problematic to the chairman. Unless appropriations increased, Ringhaver saw a future of "deleterious delays in realizing the project's economic benefits; increased overall costs of the project; and deferred development of national defense aspects."[15]

Those national defense aspects continued providing a significant rationale for the canal's construction. In March 1966, Representative Charles Bennett, a strong canal proponent, made the case that "we need the canal today more than ever before. . . . The situation to the South of Florida—a threat only ninety miles from our state—demands it. The world situation in our perilous times demands it." Echoing Bennett's concerns a year later, Ringhaver reminded the public that "this project was authorized in 1942 as a National Defense measure, after German submarines rapidly began decimating our tanker fleet." For Ringhaver, time had not diminished the threat. He cited "reports of unfriendly submarines off our shores" as a continued justification for the project. Failure to complete the canal would be "a gamble," especially "when we recall the 1962 Russian missile crisis in Cuba . . . and the recent 'unconfirmed' newspaper stories about further Russian missiles in Cuba."[16]

Ironically, national defense spending was undercutting congressional funding for the project. Though Charles Bennett asserted the "Viet Nam War demands" the canal's completion, the conflict actually siphoned dollars away from the Army Corps of Engineers. In December 1967, the *St. Petersburg Times* boldly proclaimed, "War Slows Progress of Cross Florida Canal." The paper noted that "the snag is Vietnam, which is receiving federal funds which otherwise would be used to hasten completion of the barge canal, among other things." Included among those "other things" was the "Missing Link," which fell from the federal priority list. Having only received 55 percent of a requested $20 million for fiscal year 1968, the Corps found its hands tied even tighter with a presidential demand for an indefinite freeze on all public works contracts. As authority manager Giles Evans noted, "the curtailment of the canal funding was more drastic than for the other major Florida projects." When the *Times* concluded that "the Corps and the state must sit idly by and see the culmination of a seventy-year-old dream so near and yet so far," it was obvious that the canal would not be completed as scheduled. Originally slated for completion in 1970, the best case for completion was now early 1974; the worst pushed construction well into the 1980s.[17]

To Ringhaver, the realities of congressional budget cuts were both immediate and long-term. Complaining about the uncertainty of the budget freeze, he announced "we have no indication as to how long the freeze will continue, but while it lasts, no new Barge Canal contracts will take place." Though this seemed bad enough, the situation grew worse over time. With a benefit-cost ratio tied to quick canal completion, any delay lowered the economic feasibility of the project, and gave more ammunition to those critics concerned with the project's fiscal propriety. For Ringhaver, "Stringing-out the construction life tends to increase the sums accounted as 'interest during construction'—and these already amount to about 8 percent of the projected federal investment of $146,900,000. There also exists the continuing threat of inflation in actual construction costs year by year." In short, the situation was not as rosy as it appeared in the promotional literature.[18]

Despite the looming fiscal problems, supporters felt they could count on bureaucratic inertia as a loyal ally. As early as 1965, Representative Sydney Herlong sought to minimize fears that the project would not be completed. Writing to the Ocala Chamber of Commerce, he concluded that "where work has already started on a project such as this, and the amount of money that has already been appropriated and spent is as great as it is, it is hardly conceivable that the Congress is going to quit right in the middle of the stream." Based on years of experience with congressional wrangling, Herlong's insight seemed reassuring. But two years later, his correspondence seemed less sanguine. Writing to his friend Lance Ringhaver, Herlong admitted, "I would be less than frank if I told you that there was any possibility of upping this appropriation from 11.4 to $20 million in the House regardless of what we in the Delegation do." Herlong's rather dismal forecast reminded canal supporters of the precarious position of living by annual budgets. If the canal was to be completed, supporters would have to make their case to a broader constituency.[19]

Though the justification for canal construction was always expressed in economic terms, boosters recognized early on that other rationales could drum up support for the project. As usual, Tom Adams put it best, "Let it never be said, as some have suggested, that this project will provide only navigation benefits to be enjoyed by but a few." Building on the claims of earlier boosters, Adams cataloged a multitude of purposes. The canal would provide for flood control in "an area that has been plagued in the past by raging flood waters." It would also provide for water conservation as "water control structures will be placed in such a manner as to afford maximum storage for water which could then be used in times of need." While these considerations were important, the allure of recreational activities along the route soon became a paramount selling point for the canal.[20]

Well before the Palatka groundbreaking, proponents touted the potential for fishing, boating, camping, picnicking, and sightseeing that the canal would provide. A 1962 promotional pamphlet issued by the Florida Waterways Committee, a lobbying group working hand-in-hand with the Canal Authority, asked the rhetorical questions: "Have a boat . . . like to fish . . . want to go on a picnic?" The answer was obvious as "several hundred miles of waterfront property will be created by the Canal—thousands of acres of beautiful crystal clear lakes will surge into being." The Corps' 1964 master plan incorporated recreational ideas suggested by the Waterways Committee, with more than twenty public-use areas spanning the entire length of the canal. Never more than window-dressing for the canal's real purpose, the promise of recreational activities would nonetheless gain more significance as budgetary shortfalls and an increasingly vocal opposition challenged the project.[21]

Within a month after the January 1966 Tallahassee meeting, Florida's Board of Conservation adopted a resolution requesting "supplementary actions to ensure maximum recreational benefits" along the canal route. Later that year, Corps Colonel R. P. Tabb announced, "We want to show you that we do care about recreation, esthetics, birds, fish and wildlife, clean water, and the other matters in which your groups have an interest." Even the intensely procanal Ocala Chamber of Commerce saw the need for new rationales. Echoing Tabb, a Chamber spokesman declared, "we feel it is essential that areas be set aside to accommodate the various recreational activities." In addition to the traditional joys of boating and fishing, the Chamber saw potential in everything from water skiing to pleasure driving. To show how much ground Citizens for Conservation and their allies had gained, the Chamber also listed archaeological sites, nature study, and "sanctuary areas for selected species" as being of "primary interest to us in the enhancement of the natural resources" within the canal area. Even utilizing a major argument of canal opponents, the

Chamber concluded, "efforts should be made to preserve habitat for wild turkey which would ensure that this species would continue to be a part of the Oklawaha Valley."[22]

The Corps' newfound emphasis on the canal's recreational benefits was designed to build a larger constituency of supporters in the wake of increasing federal budget cuts. But the recreational activities envisioned by supporters were radically different from those advocated by Marjorie Carr. The roaring sounds of power boats and the laughter of children playing along newly developed shorelines would replace the treasured pristine wilderness of Carr and her conservationist allies. To canal supporters, the project would democratize recreation along the Ocklawaha, allowing anyone and everyone the chance to appreciate the river in its new and improved state. Unlike the existing waterway, which was a "swampy, shallow, and crooked boat trail," the developed canal would be "wide, deep, and inviting" enough to allow "hundreds of pleasure craft from skiffs to palatial yachts" to "cross Florida from one coast to the other on the same day." As the Florida Board of Conservation proclaimed, the "construction of the twelve-foot-deep and a 150-foot-wide canal will make the swampy forbidden area more accessible to a greater segment of the public . . . it will open new recreational areas which do not now exist—areas which now are almost inaccessible except to the hardiest sportsman." Canal boosters claimed the new recreational facilities would provide a "rare and welcome opportunity," even a "blessing," for a state poised on the cusp of exponential population growth. With the dramatic demographic boom of the 1950s and 1960s, and with anticipated projections of even larger increases in the late twentieth century, these recreational areas would "emerge as among the most beneficial assets" of the canal.[23]

Recreational benefits became the basis of a new publicity blitz throughout the latter half of the decade. Pamphlets and press releases resulted in dozens of newspaper articles extolling the virtues of the canal's "recreation potential." While acknowledging the incongruity of recreational facilities clustered along an industrial waterway, proponents saw no contradiction in multiple uses of the canal. It was all things to everyone. In a pamphlet extolling the "virtually limitless facilities for recreation and sports activities," the Canal Authority admitted the "actual canal will be comparatively unpretentious" and "typically commercial in use." Mari-

Digging the ditch on the Inglis section of the canal, 1968. Courtesy of State of Florida Photographic Archives, Tallahassee.

nas, industries, municipal waterfronts, and commercial ventures would no doubt line the project. Yet radiating from "the canal proper, there will be miles of shoreline encircling acres of reservoir waters which will be clear and refreshing, with sandy shores and beaches providing countless, unlimited natural swimming, picnic, and camping areas." Moreover, the creation of 254 miles of additional shoreline would result in "some of the finest fishing in the southeast." Promoters envisioned "spectacular fishing for trout, red bass, and other species.[24]

Canal critics were much less convinced. For Marjorie Carr, the river did not need improvement "because it is one of the most beautiful rivers in America." Recognizing the success of the Corps' publicity machine, Carr concluded "the press failed to get the message. With a few notable exceptions, articles . . . brushed aside the pleaders for the Oklawaha as a few impracticle [sic] visionaries who wanted to preserve the wilderness for birds instead of for man." Carr was taking the opposition to a new level. Conservation was no longer the major focus of canal criticism. Instead, Carr was moving to a more preservation-based ethos. To her, swampy, shallow, and crooked boat trails were exactly what made the Ocklawaha distinct, and thus worth saving. Nature was an end in itself, a place for the original meaning of "re-creation," where individuals could quietly reflect on nature's beauty. For Carr, nature was not something to be "as tame as a recreation park." Rather, her highest priority was to "preserve these samples of original wilderness."[25]

Though Carr bemoaned the press's support for the project, she was media-savvy enough to shape the ongoing debate. Ironically it was the success of Corps and the crusher's destructive power that gave Carr an opportunity. Though F. Browne Gregg could crow about "the positive publicity on the Crawler Crusher [that] permeated the business pages of newspapers throughout the world," the extraordinary devastation wrought by the machine provided critics with vivid visual evidence of the destructive path of progress. The crusher soon became a symbol for the canal itself. For men like Tom Adams, Giles Evans, and F. Browne Gregg, it was the fulfillment of an engineer's dream, of man's ability to shape the environment and create a better world. For their opponents, the machine embodied the exact opposite: the destructive capacity of humanity for little or no discernable end. Years later, Dave Bowman—a former Corps employee who later became employed with the Florida

Department of Environmental Protection and supervised canal property for nearly thirty years—underscored the crucial role of the crusher in shaping public opinion. "It's hard to get a groundswell of interest in [an environmental movement], and then something just comes along that's a real pivotal point, and I think [the crusher] was it." More than anything, he noted, "the pictures that FDE took and put in newspapers all around the state probably did more to mobilize people into saying 'what the hell are we doing with this project,'" especially after "they saw all those floating stumps" as a result of the impoundment.[26]

For those living in the vicinity of the project, the roar of heavy equipment served as a constant reminder of how the Corps had successfully steamrolled the opposition. Bemoaning the destructive force of the crusher, Jane Maiden, from the small community of Hog Valley along the eastern bank of the river, fired off an angry letter to the Corps, extending an invitation to "share with us several all night and all day sessions of having to listen to canal machinery in the Oklawaha river valley, the drone of which is relieved by the tremendous crashing of trees as they are destroyed." Mocking the agency's vision of progress, she added that such a visit would provide engineers with "an overwhelming feeling of real accomplishment." Enclosing a copy of her prescription for sleeping pills, she demanded reimbursement for her expenses, as she "did not ask for the canal nor the machinery" to operate twenty-four hours a day. Maiden closed by reminding the Corps that "the money for their projects and their salaries comes from the very people who are treated as though they didn't exist." Maiden's frustration indicated a growing perception that government was, at best, indifferent, and at worst, a malevolent force.[27]

Contested notions of progress were at the heart of the ongoing debate over the canal and Florida's future. This controversy reflected a broader struggle over the legacy of liberalism as it withstood assaults from both fiscal conservatives and left-wing activists who became increasingly radicalized during the 1960s. At first glance, monumental struggles over civil rights, free speech, and the war in Vietnam seem far removed from the fight to save the Ocklawaha. Yet, like much of the politics of the 1960s, the ground was shifting rapidly in the debate about the river. For true believers like Adams, Evans, and Gregg, the efficacy of the project was beyond question; for them the future belonged to the engineer, the expert who held the power to shape the world and therefore improve it. In many respects, their assumptions were grounded in a postwar liberal consensus in which most Americans had faith in the promise of technology and government. It was a world of unbridled optimism, of unprecedented prosperity, and ever-expanding abundance. Admittedly more complex, this world created new demands and new solutions, and all could be met by government and its growing reliance on an army of experts. In the early 1960s, politicians like John F. Kennedy and Lyndon Johnson—and the majority of Americans who elected them—believed that many domestic and foreign problems were merely technical in nature. And with government—by definition representing the best interests of the people—willing to tackle any challenge before it, all that was necessary to achieve more progress was more commitment, more money, and more action.

Reflecting this vision of democracy, the canal promised a wealth of benefits for Florida and the nation. Having addressed the relatively mundane problem of saltwater intrusion that dogged earlier plans, boosters could not imagine the emergence of a viable opposition to the project. The same could be said for Lyndon Johnson, who embodied the spirit of liberalism behind canal construction, a tradition in place since Franklin Roosevelt's New Deal. Confidently announcing during the 1964 groundbreaking that the canal was just "another challenge [to] make use of another natural resource," the president found himself making a similar speech, though in a far different context, a little over a year later. This liberal consensus would be put to the test in Johnson's approach to Vietnam, where it would also meet its failure. Frustrated with the stubborn resolve of the Communist resistance, Johnson, in his famous "peace without conquest" speech of April 1965, proposed that the best way to ameliorate tensions was to develop a vast flood-control project along the Mekong River "on a scale to dwarf even our own TVA." Once again, public-works spending would solve all problems. Presuming everyone—even Communists—could agree that "a dam built across a great river is impressive," Johnson thought the North Vietnamese would drop their weapons and jump at the chance to improve their lives by altering a river. He could not understand how Ho Chi Minh could summarily rebuff his proposal. Over time, canal proponents would become similarly frustrated with the rejection of the obvious progress—to their minds at least—associated with the Cross Florida Barge Canal.[28]

More than a failed construction project, the canal was part of the patchwork quilt of modern liberalism that began to fray in the 1960s. Created as a New Deal work project, and later resurrected amid talk of new frontiers, the barge canal also represented the expansive vision of Johnson's Great Society. As a tangible symbol of the promise of liberalism, the barge canal suffered from many of the inherent contradictions and challenges that plagued the Johnson administration through the mid-1960s. In just a matter of years following the 1964 groundbreaking, many of the project's underlying assumptions, much like the seemingly immutable liberal consensus, lay in tatters. Caught in the whirlwind of the 1960s, the fate of the canal was both reflected in and affected by broader social and political changes that disrupted the relative calm of an earlier postwar period. Most important was the dramatic polarization of politics as race riots, countercultural excess, and antiwar protests became the defining images of the decade. As the Left split into a variety of movements—coalescing around issues of race, class, gender, age, and sexuality—that reflected a new rights-consciousness and anti-establishment activism, an emerging conservative backlash also threatened to tear the country apart. Much of the division was exacerbated by the war in Vietnam, which magnified and distorted social and economic problems that had long been festering in postwar American society. Moreover, the prolonged conflict vividly revealed that even a nation as bountiful as the United States could not afford both guns and butter. In so doing, Vietnam not only raised doubts about the capacity and honesty of government, but the promise of progress as well.

The Vietnam War influenced the fight over the canal in both overt and subtle ways. With Americans fighting in the jungles of Southeast Asia, project supporters now had another rationale, and could push the national defense angle as part of a larger effort to contain communism and protect the United States. Yet, money earmarked for the canal was continually syphoned off to pay for the increasingly expensive conflict overseas. Canal supporters decried this steady loss of funding, and demanded their fair share of federal dollars. On a deeper level, many of the assumptions behind canal construction were parallel in some ways to

America's experience in Vietnam. Much like the course of the war, the best of intentions could lead to unintended consequences. In Vietnam, it reached the height of absurdity with the apocryphal dictum that "in order to save the village, we had to destroy it." By 1970, the rationale for canal completion harkened to that bizarre sensibility. Reflecting on the uselessness of a twisting, turning river surrounded by swampland, Corps spokesman Joe Livingston intoned that "the preservationists talk nonsense when they say they want to save the Oklawaha. The barge canal will save the Oklawaha from nature."[29]

Amidst all of the upheaval of the late 1960s, anticanal activism remained relatively quiet. Stung by their lack of success in the 1966 Tallahassee Showdown, opponents retreated homeward to regroup, unsure of their next step. The next three years marked a crucial transition, as Citizens for Conservation evolved into the Florida Defenders of the Environment (FDE) by the summer of 1969. Never bound entirely to an environmentalist position, canal opposition represented a variety of viewpoints that at times seemed contradictory. In what would eventually become a marriage of environmental liberalism and fiscal conservatism, anticanal activists drew from a broad constituency. In the end, the success of the anticanal movement rested upon a variety of elements: focused determination, political acumen, and public-relations savvy. But the ultimate success of the movement lay in its ability to bridge the ideological gap between antigovernment Goldwater conservatism and activist environmental liberalism.

While the public face of anticanal forces became more and more associated with Marjorie Carr, John Couse was in many ways crucial to the movement's success. Though Carr represented a strain of liberal activism that valued citizen involvement and scientific expertise, Couse brought a bedrock belief in the conservative nature of conservation. In many respects, he was the movement's Theodore Roosevelt, while Marjorie Carr played the role of John Muir. The success of the movement depended on the ability of both people to work together in spite of their differences and focus on a single goal: to stop the canal and save the Ocklawaha.

Never one to concern himself with personal aggrandizement, Couse remained quietly behind the scenes and allowed Carr and her cohort of university scientists to represent the movement. Years later, David Anthony characterized Couse as "the money man," a fund-raiser with countless political and financial connections across the state. But Couse was more than a mere financial conduit. As with Marjorie Carr, his vision of Florida was shaped by growing up in a relatively pristine frontier region. Having spent much of his life altering that environment as a successful businessman—making a living in, of all things, air-conditioning and refrigeration—he came to appreciate the natural world of his youth.

In addition to this very personal motivation, Couse's deep-seated conservatism drew him toward canal opposition. In November 1962, the same month that Carr and other Gainesville residents first confronted public officials about the canal, Couse wrote a confidential letter to Florida senator Spessard Holland. Expressing "deep and abiding conviction," he announced that he was "violently opposed to the newly proposed cross state barge canal." Concerned with preserving Florida's vanishing natural areas, Couse tied the failed ship canal of the 1930s to its newer incarnation. "During the many trips that I have made from Palm Beach County to the University of Florida to visit my children," he wrote, "I have driven by those large concrete structures, out there in the wilderness, which are a part of the preliminary construction of a cross state canal, in which I have always felt stand there was mute testimony of the folly of Claude Pepper." The association with Pepper was deliberate, for in the mind of many conservatives, Claude "Red" Pepper was the embodiment of New Deal excesses. Playing upon that theme, he continued, "My own experience in operating my business here in Palm Beach County . . . compels me to the view that the give-away programs of our Federal Government and the boondoggling in connection with unnecessary projects all over the nation cannot produce anything but future trouble, and I believe that if we continue with it indefinitely, we will end up in a completely socialized, if not a Communistic, state." For Couse, the canal represented those boondoggling projects, for it was "wholly unnecessary" and would "cost the taxpayers of our country incalculable sums of money."[30]

Far from a lone voice, Couse was part of a growing backlash against wasteful government spending. This critique harkened back to Arthur Vandenberg's attack on the ship canal of the 1930s, and played out in the 1960s with the rise of fiscal conservatives like Barry Goldwater and William Proxmire. Couse even wrote William F. Buckley both commending him for his "excellent campaign" in his unsuccessful New York City mayoral bid and asking for his help to stop the Cross Florida Barge Canal, which would tie into Buckley's "drive for conservative responsible government." Reaching mainstream America through the pages of such middle-brow publications as *Reader's Digest* and *Life Magazine*, Couse's vision struck home with Americans concerned with an ever-increasing tax burden. Terms such as "boondoggle," "pork-barrel," and "pet projects" were becoming increasingly commonplace as critics blasted massive public-works projects that appeared to have little or no benefit to the nation. They particularly singled out expensive water-resource projects, shepherded through Congress by Ohio congressman Mike Kirwan, derisively labeled "the undisputed Prince of Pork." In a June 1967 *Reader's Digest* article entitled "Mike Kirwan's Big Ditch," author William Schulz assailed on a national level what John Couse attacked locally. Calling Kirwan's plan to build a 120-mile canal from the Ohio River to Lake Erie "indefensible from any standpoint," Schulz concluded, "this billion-dollar boondoggle is one of the most brazen pork-barrel projects ever foisted on American taxpayers." Central to Schulz's argument was Kirwan's reputation for political horse-trading. Controlling the purse strings for federal public-works projects, Kirwan got support for his "ditch" by doling out funds for other projects, including the Cross Florida Barge Canal. Couse easily concurred with Schulz's assessment: "Such engineering projects as the Cross Florida Barge Canal are juicy plums. They are vigorously and lavishly promoted by those interests who expect to profit by them. They are hard things to buck. Nearly every item in the Federal budget can by some logic be associated with defense. Something must give."[31]

Couse's connections and fund-raising abilities proved just as important as his concerns over the canal's economic viability. Tied to a network of influential south Florida financiers and businessmen, he raised funds throughout the anticanal campaign. Despite Couse's obvious strengths and commitment, he never assumed a high-profile role within the movement. For one, his strident conservatism was out of step with many environmentalists dedicated to the fight. Also, his south Florida roots and his obvious personal interest in preserving the Ocklawaha as a result of his retirement home on the river made him suspect in the eyes of canal proponents and gave them an opportunity to question his motivations. Better to let scientists and concerned citizens take the lead.

Marjorie Carr filled both roles perfectly. As a scientist, she attacked the canal with objective authority. As a concerned citizen, she used both her interest and her gender to disarm her opponents and raise uncomfortable questions, all wrapped in the mantle of domestic feminism. Neither June Cleaver nor Betty Friedan, Carr consciously capitalized on the traditional role of protecting the household to include saving the Ocklawaha. Far from a radical, Carr saw herself as a defender of home and hearth, a sphere she extended to a broader natural world under attack by outside forces. Indeed, her status as a faculty wife enhanced her position by allowing her the time to pursue the issue. Simultaneously, Carr's brilliant leadership allowed her to create a scientific rationale for saving the river while allowing others to portray her as a "mere Micanopy housewife." In a time before computers, cell phones, and fax machines, Carr mobilized a loosely formed coalition and almost single-handedly turned it into a tightly focused organization designed to take on the government itself.

Using her kitchen as a war room, Carr marshaled her forces like a master strategist. More Dwight Eisenhower than George Patton, Carr's genius lay in her ability to crack the whip and keep unruly scientists and activists on task. She also had the unique talent to find common ground between disparate interests within the anticanal community. Combining John Couse's Goldwater Republicanism with the countercultural vision of left-leaning scientists took someone who could make both see the importance of the ultimate goal. Perhaps most remarkable was her ability to do all of this with so few resources. Armed with a primitive, and frequently broken, copier and an unpaid staff that often included her children, Carr proved indefatigable in managing reams of information efficiently. Her son David reminisced that the huge, loud Xerox machine originally placed in the kitchen, leveled the playing field between her and the Corps. The importance of copying memos, petitions, letters, and research studies is shown by her persistent letter writing to the Xerox Corporation for newer, faster, and better machines. Cataloging materials in an idiosyncratic manner that would have driven a librarian to distraction, she had immediate command of everything ranging from the addresses of potential donors to the latest hydrologic studies of the Ocklawaha. Commenting on her filing system in 1971, Carr stressed that information management was hardly a "piddling housekeeping detail," for it is "important to set up a system that makes it convenient for an ever-changing group of people both to assemble information and extract it."[32]

Carr continued her fight following the Tallahassee Showdown but was unsure of how to proceed. As she explained later: "We didn't give up. We waited . . . we just waited. . . . For the next two years a dozen or so of us did little more than telephone each other now and then to reassure ourselves that we had not quit." Having lost a direct political confrontation, she "realized then that our only recourse would be the courts. But we were unable to find lawyers interested in environmental problems of this kind." In the meantime, Carr and her Gainesville allies honed their organizational skills for another round of battle. More importantly, they increasingly focused on conducting scientific research that would build a case for stopping the canal. From his home along the Ocklawaha, John Couse simultaneously pursued what he did best: cultivating contacts and raising money. Couse's friendship with Fisher-Pierce Company president Richard Fisher, manufacturer of the Boston Whaler line of watercraft, led to a series of national advertisements in boating magazines opposing canal construction, and to the financing for the documentary film *Progress, Pork-Barrel, and Pheasant Feathers*. Produced and directed by Fisher's daughter, the film publicized the plight of the river by focusing on the Tallahassee Showdown. Embraced by anticanal activists, the movie verified their belief that the fight to save the river was a clear-cut confrontation between good and evil. It also served as a counterpoint to the equally heavy-handed promotional films produced by the Corps in the early and mid-1960s. Over time, it became a staple of FDE's publicity campaign against the canal.[33]

With Holly Fisher's movie centering on the political conflict in Tallahassee, it seemed only appropriate that the next confrontation would take place in the same venue. In January 1968, canal opponents descended on the State Capitol once again for another annual Water Resources Committee meeting, on what they called "the second anniversary of the infamous Oklawaha 'hearing.'" At this meeting, the Florida Board of Conservation turned its attention to the "Missing Link," the proposed Gulf Coast waterway from St. Marks to Tarpon Springs. Board director Randolph Hodges had earlier noted the importance of this project. Announcing "it represents a broken segment in the inland waterways system," he called for quick completion of the project. Summoning activists to protest "the Waterway Boys," environmental activists Ken and Helen Morrison issued a call to arms to avoid being "flattened by the canal steamroller again." Writing to Hodges, Marjorie Carr laid down a marker: "The lessons learned by Florida conservationists during the long and frustrating endeavor to save the Oklawaha River . . . have not been forgotten. Florida has lost the Oklawaha because appropriate studies in the broad field of biology were not made. . . . Let us not endanger the natural resources of the Gulf Coast of Florida from St. Marks to Tampa Bay by asking for Congressional authorization . . . until *after* appropriate studies in the areas of biology, hydrology, and geology have been completed, evaluated, and reported on to the Florida public" (emphasis in original).[34]

The meeting proved to be much less contentious than the one two years earlier. As if to assuage their critics, the Corps announced plans "to determine just which route would be most compatible with the important biological resources of the area traversed." The board concurred, setting aside $600,000 for "detailed biological, ecological, hydrological, and geological studies" before "a decision is made to proceed with the project." This time, even Tom Adams, the bête noir of this emerging environmental movement, appeared more conciliatory regarding the opposition's concerns: "I would like to assure you, that in this construction the very minimum of damage to the ecology of the area will be done." Forced to concede the Cross Florida Barge Canal damaged "to a certain extent the ecology of the Ocklawaha River area," Adams praised planners for minimizing the damage "so the least harm would be done to the natural ecology of the area." The board's recommendations were a world apart from the rebuff Morrison and Carr had received two years earlier. Whether this decision represented a legitimate shift in governmental attitude, a public-relations ploy to placate conservationists, or the fact that no money had yet been spent on construction, opponents saw a glimmer of hope.[35]

Though important on its own terms, the battle over the "Missing Link" remained closely tied to the larger effort to stop the Cross Florida Barge Canal. More than a fight to preserve the environment, the struggle against the canal raised important questions about irresponsible government spending. Canal opponents saw the project as the ultimate fiscal boondoggle. And any opportunity to show its economic useless-ness would be a powerful argument against completion. For years, pro-ponents argued the cut across Florida's peninsula was a vital part of an integrated waterways system, providing cheap efficient transportation from Texas to New York. A crucial component of this vision was the de-velopment of the "Missing Link," which connected the western terminus of the canal to a broader waterways system and provided for safe barge transportation along Florida's Gulf coast. Without it, the economic vi-ability of the canal would be seriously compromised. More than a side-show, the struggle over the "Missing Link," especially for conservative opponents like John Couse, became a crucial part of a broader antica-nal strategy throughout the late 1960s and early 1970s. These economic arguments were even imbedded in FDE's monumental environmental impact statement published in March 1970. Though focusing on biologi-cal and ecological arguments against the canal, there was an underlying theme of economic profligacy that fit well with conservative criticisms of government waste. Sounding more like *Reader's Digest* than *Audubon*, the statement concluded that "the need for, and cost of, this Missing Link waterway is not considered in calculating the costs of the proposed canal."[36]

The failure to build the "Missing Link" was no doubt a setback for canal proponents. Yet a larger problem loomed over the project with the course of the Vietnam War. Throughout early 1968, newspaper editors and canal officials decried budgetary cuts made necessary in light of the expanded war effort. A June 7 *Orlando Sentinel* headline said it best: "Viet to Sap Waterway Projects $." Though President Johnson's budget reduced federal funds across the board, Canal Authority manager Giles Evans felt no other project "has suffered such drastic deprivation in the budget request." Randolph Hodges concurred, noting dourly, "unless we can do something with our friends in the Congress, we will be in serious trouble." L. C. Ringhaver led the charge, announcing that the "budget re-quest of a mere $4.6 million to continue construction work on the Cross Florida Barge Canal indicates a wanton disregard for the national, re-gional, and local values of the project." Contending that these cuts would delay completion until the 1980s, he concluded that these "delays are particularly adverse to the over-all economy of the entire State." For the Canal Authority chairman, the cuts marked a direct attack on the canal itself, and required a significant response. "We seek all possible support for correcting this unrealistic budget figure," he proclaimed. "We intend to leave no stone unturned in this effort." Recognizing the importance of this struggle, the Canal Authority's Public Information Office concluded in March that Ringhaver's statements on the budget crisis were "more widely quoted than anything a canal authority official has said in quite some time."[37]

A sense of defensiveness now permeated canal proponents. Especially vehement in its opposition to the cuts was the *Ocala Star-Banner*, hometown paper of the city that stood to gain the most from the canal's completion. Conceding that the "Vietnam War, of course, is proving to be a massive load for the taxpayers to shoulder," an editorial writer appealed to President Johnson to restore "the $20 million sought by the U.S. Army Corps of Engineers and the Canal Authority," as the canal remained "an economically sound investment." In March 1968, Ocala Chamber of Commerce president C. C. Leiby wrote to Congressman Sydney Herlong to complain that the "canal item seems to be carrying a disproportionate share of the over-all cut in the civil works program," and pleaded for his "help in having the Congress increase the appropriation for the canal." Herlong's response was disappointing. Though one of the project's most vociferous advocates, the Leesburg congressman came down firmly on the side of budgetary restraint. Herlong explained that he had been "urging the President to cut down on federal spending," adding, "I cannot in good conscience go to the Appropriations Committee and ask that committee to appropriate more money for the Cross State Canal than is presently provided for in the budget."[38]

Mirroring the sense of futility among canal supporters during this budgetary battle, the death of Henry Buckman in March 1968 underscored the fact that the opportunity to complete the canal seemed to be slipping away. Apart from Bert Dosh, no other person had been more closely associated with the long course of canal development than "Harry" Buckman. As engineering counsel for the Ship Canal Authority of the 1930s, he was instrumental in securing funds for canal construction. An astute bureaucratic infighter throughout the 1940s and 1950s, "he did much to keep hopes alive for the canal back in those days before leaders in Washington recognized the need for the waterway." While president of the National Rivers and Harbors Congress (a position he held up to his death), Buckman tirelessly advocated for the importance of the canal in a national system of waterways. With his death, an era had passed. Yet, canal advocates capitalized on the event as another opportunity to reverse their flagging fortunes. Within two weeks of Buckman's passing, Jacksonville congressman Charles Bennett introduced a resolution on the House floor calling for the renaming of the St. Johns Lock in his honor. Extolling "his goals for manmade projects to help advance our society," Bennett called Buckman the "canal's father-confessor for four decades." Perhaps the dedication of the lock to Buckman's memory would jumpstart a renewed effort to complete the canal.[39]

In spite of the budgetary cuts that forced even the *Jacksonville Seafarer* to proclaim "canal progress prospect dim," the last three months of 1968 provided hope that Buckman's dream would become a reality. Key to this was the completion of Rodman Dam and the soon-to-be-renamed Buckman Lock. These massive complementary structures provided the mechanism to move barges through the eastern end of the waterway. Stretching 7,200 feet across the previously untamed lower stretches of the Ocklawaha, the $3 million dam siphoned water toward the newly finished lock. The completion of the 22-foot-high earthen dam and spillway also allowed for the flooding of the land behind it at a rate of five to six thousand acres within its first two weeks of operation. Eventually this flooded area would create a ten-thousand-acre lake with an operating depth of 18 to 20 feet above sea level, which would provide the necessary water for the operation of the accompanying lock. Called variously Rodman Pool, Rodman Reservoir, and Lake Ocklawaha, this impoundment would become the center of the canal controversy for decades to come.[40]

The completion of the dam would route most of the Ocklawaha through the canal cut leading to the St. Johns Lock and eventually the St. Johns River. The lower portion of the river would receive a significantly reduced flow through a four-gate spillway attached to the dam. The 600-foot-long and 84-foot-wide lock itself opened in October 1968, at a cost of more than $5 million. Engineers excavated more than 1 million cubic yards of earth in the building of the lock, which required 89,000 cubic yards of concrete to complete. These dry figures belie the sense of drama that accompanied the completion of the first two structures on the east side of the canal. With headlines that screamed "St. Johns Lock Is Born," the

Palatka Daily News took pride in announcing the first foot of water that appeared in the lock. For the local editorial board, it was "evidence that the canal no longer is a dream." Here was—quite literally—concrete evidence of success: "No longer can people speak of [the canal] in speculative terms." Recognizing what seemed to be the perfect bureaucratic fait accompli, the paper asserted "the sight of the lock filling also leads to the conclusion that appropriations will continue to be forthcoming from Congress to insure completion of the 107-mile waterway." Indeed, the flow of water into the lock carried new life into what seemed to be a flagging canal movement. As the paper concluded, "Yes, water in the St. Johns Lock is the signal that soon sportsmen will have new fields and that some day the Canal will provide a navigable water-way from the Atlantic Ocean to the Gulf of Mexico, bringing with it an economic boon to Putnam County."[41]

The completion of the eastern portion of the project seemed a perfect time to commemorate the year's achievements. On December 14, 1968, members of Florida's congressional delegation, as well as politicians from across the state, joined Corps and Canal Authority officials and local supporters in a gala celebration at St. Johns Lock. Two thousand dignitaries and interested onlookers participated in what began as a rain-soaked ribbon-cutting ceremony that included the dedication of the lock, the $7 million Rodman Reservoir, seven miles of canal excavation, and a $2 million bridge that spanned the waterway near Palatka. Feted with an obligatory dinner of fried mullet and hush puppies prepared by the residents of the Rodeheaver Boys Ranch, guests listened to Massachusetts congressman Edward Boland's keynote address that extolled the virtues of canal completion. As a stand-in for his Ohio counterpart Mike Kirwan, the powerful chairman of the House Appropriations Subcommittee on Public Works, Boland dedicated the lock "with hope that it will add to the greatness and goodness of our land." Reminding everyone that the occasion was merely a respite amid a long struggle, Boland proclaimed: "the task now is to be sure that the project goes on. Hopefully ample funds can be budgeted to bring it to an early completion—by 1973." North Florida congressman Don Fuqua seconded the notion, adding that "this canal is more important today than when Congress authorized it in 1962. Particularly with these troubled times it will offer great potential for defense as well as transportation." For Lieutenant General William

F. Cassidy, chief of the Army Corps of Engineers, everything was falling into place: "At last we are approaching the fulfillment of a 142-year-old dream of the people of Florida." All the decades of surveys, studies, and contentious political debates were finally paying off. "Several times the project reached the verge of construction," he observed, "only to be halted at the last moment by a war or some other crisis. It has remained for us in our generation to keep faith with past generations by making their dreams come true."[42]

Many notable canal supporters sat on the dais to watch Tom Adams cut the ceremonial ribbon with a giant pair of scissors. Many of these same guests had attended a similar event four years earlier when Lyndon Johnson had inaugurated the project. Though the location, the menu, and even the weather of both ceremonies were similar, much had changed since February 1964. While Johnson had once stood in a muddy field where people could only envision a future canal, Ed Boland and Tom Adams were now surrounded by tangible results of progress. Not only were the structures completed, they were used for the first time as assorted watercraft came from as far as Jacksonville to pass through the lock. Even more propitiously, the sun came out this time. As Adams cut the ribbon, a cacophony of air horns and boat whistles from pleasure craft, barges, and a Coast Guard cutter signaled realization of a dream. In spite of all the problems of the past few years, the brightness of that December day seemed to guarantee canal completion. And then the logs popped up.

Floating Logs, Dying Trees, and Clogging Weeds

During his legendary campaign for the United States Senate in 1970, Lawton Chiles walked the length of the state to take the pulse of his fellow Floridians. By June, on the nineteenth day of his journey, Chiles reached Silver Springs, feeling "pretty good, realizing that instead of pounding the pavement . . . I would be sitting and riding on the water." While traveling along the Silver and Ocklawaha rivers, Chiles felt pleasantly surprised "that the damage further up the river was not really as great as I had imagined it would be." He quickly reversed his judgment, however, upon reaching Rodman Reservoir. There he not only encountered the crusher, but the devastation it wrought. Calling the machine a "colossal failure," Chiles saw the residue of trees supposedly pressed "into the muck to get rid of them." Instead, he noted, "for the last year and a half they've been popping up like corks and the Corps is now having to spend tremendous sums of money keeping a dredge out picking up the logs, piling them on the banks for burning." For Chiles, the situation was fraught with mistakes, representing "somebody's grandiose idea [that] just didn't pan out, and the taxpayer pays."[1]

Nearly everyone who encountered the Rodman Reservoir in the late 1960s commented on the environmental destruction wrought by the crusher. While observers acknowledged a wide range of problems associated with canal construction, they consistently focused on the logs that continually popped to the surface of the reservoir. This suggested that the crusher was not only inherently destructive, but inefficient as well. Built specifically for the project, the crusher was designed to smash trees "into the relatively soft ground so that even the biggest trees will not float when the reservoir fills." Yet it dramatically failed to deliver that promise. As Florida Audubon's Bill Partington remarked, "it would take up to three years to clear the log jam mess they have created at the so-called recreation pools." Standing as an unsightly monument to government waste, Rodman Pool did contain a silver lining for environmentalists, for it convinced many neutral observers to oppose the project. According to Partington, "several told me that although they had originally backed the canal, in view of the . . . general mess of the remaining waterway itself (logs and other debris which make motor boating hazardous) . . . if they had to do it all over again, they would oppose the project with everything they had."[2]

The literal logjam provided opponents a significant opportunity. The persistent problem of logs breaking the placid surface of the reservoir created a moment that not only verified earlier scientific concerns but captured considerable public interest. In December 1968, critics appealed to the press, especially *NBC News*, to publicize the story. Partington boasted that "we encouraged them" to arrive in Silver Springs and fly over the canal: "Bob Lissit of NBC had said he wanted shots of someone in a boat talking about the natural river. They used Dr. David Anthony of [the] University of Florida and myself for that part. . . . Hopefully this will get on TV." Their gambit worked, as the issue reached a national audience with a news segment on the popular *Huntley-Brinkley Report* in January 1969. Addressing the canal's potential for environmental damage, the report featured interviews with key critics like Anthony and Partington. In addition to complaining about the wasteful clearing methods, they raised fundamental questions about the impoundment itself. More than just an ugly mess, the artificial lake encouraged the growth of invasive aquatic plants like water hyacinths and hydrilla. Anthony charged that in exchange for a free-flowing, shade-covered wild river, the Corps had created a stagnant pool filled with blankets of floating vegetation that would choke off recreational use, interfere with commercial navigation, and greatly accelerate the accumulation of rotting organic material. Indeed, "following the initial flooding of Rodman Pond in September 1968, during the roughly two-and-half months growing season remaining in the year, an estimated 300,000 tons of water hyacinths grew."[3]

For canal proponents, meddlesome environmentalists and a press increasingly sympathetic to their cause provided problems. Supporters quickly denounced the *Huntley-Brinkley* news clip as one-sided. The *Waterways Journal*, the weekly voice of national barge transportation inter-

Piles of logs from the ground clearing of Rodman Reservoir testified to the waste and destruction of canal construction. Courtesy of Florida Defenders of the Environment (FDE), Gainesville.

ests, condemned the program as "particularly critical" and complained that "only the conservationist side of the story was heard." Such defensiveness mirrored the increasing testiness of the various interests and government agencies involved in the project. As late as March, the Florida Board of Conservation felt it necessary to "clarify misinformation which has been recently publicized by news media such as the January 1969 *Huntley-Brinkley* television program."[4]

The Board of Conservation considered the problems of logs and weeds as little more than a public-relations issue. "Some complaints have arisen as a result of floating logs and snags associated with the method of reservoir clearance," the board admitted. However, the agency continued, "this is a temporary condition which was expected after impoundment began . . . [and] will soon be resolved." Government officials readily "admitted that the impoundment plan will produce some major changes for parts of the Oklawaha valley." For them, such alterations were a mere byproduct of improvement. The board asserted it was "inaccurate to think of the river as being destroyed or despoiled. Instead a different set of wildlife and aesthetic values will emerge."[5]

Though ostensibly an argument over floating logs, the Rodman controversy reflected deeper concerns over the nature of conservation and preservation. The dire predictions of canal critics were all too soon real-ized, but boosters continued to dismiss their evidence of failure as only momentary impediments. In a rare moment of self-reflection, the board grudgingly admitted "the river in its original form is . . . a stream of great beauty, but its retention in its original state would become a preserva-tionist ideal involving enjoyment by a comparatively small group of elite purists rather than fuller use and greater enjoyment by a broad segment of the people. The economic benefits that would be foregone by failure to complete the canal would place an extraordinarily high premium and economic burden on a less elite but overwhelming majority."[6]

Against those arguments, Marjorie Carr and her allies needed to persuade the public that the Ocklawaha was inherently valuable in its natural state. They would also have to demonstrate that the boosters' assumptions about the canal's economic benefits were seriously flawed. Only then would their arguments resonate with a broader constituency who remained at best disinterested about the fate of the river. As oppo-nents labored to find a way to defeat the canal, they searched for an issue that would vividly illustrate both the project's economic failures and its environmental destructiveness.

By the beginning of 1969, with apparent momentum on their side, a confluence of events enabled critics to develop a four-part strategy—never cogently organized, more improvisational than an actual blue-print—for saving the Ocklawaha. Foremost was the public-relations coup that fell into their laps as the Corps encountered a myriad of problems with Rodman Pool. The unsightly logjams, obviously caused by the ac-tivities of the crusher, and the explosive growth of water hyacinths and hydrilla verified much of what Carr, Couse, and Anthony had been saying over the last three years. The defensiveness of canal builders and sup-porters regarding this issue appeared to give even more credence to the concerns of opponents. Second, building upon what they saw as the self-evident truths of their concerns, anticanal activists reinforced their case with the power of science. Tapping into the talent of the University of Florida's academic community, Carr and Anthony quietly gathered reams of irrefutable data to overwhelm their adversaries. Next, opportunities for political action seemed more open than before. By the decade's end, a new generation of lawmakers, both Democratic and Republican, gained power and diminished the influence of north Florida's Pork Chop Gang.

Burning logs to clear the land for Rodman Reservoir. Courtesy of State of Florida Photographic Archives, Tallahassee.

Nationally, the election of Republican Richard Nixon held out hope that the commitment to a leaner federal government would lead to a reconsideration of the canal's questionable economic benefits. Finally, while politics remained important, it was the legal system that provided the best avenue to challenge the project. And strangely, a serendipitous reading of *Sports Illustrated* provided the inspiration, as well as the road map, for the legal assault.

With their coverage of the New York Jets' stunning victory in Super Bowl III out of the way, the editors of *Sports Illustrated* desperately searched for material to fill the pages of their February 1969 issues. With environmentalism becoming an increasingly visible topic, they sent one of their top reporters, Gilbert Rogin, to Long Island to interview Victor Yannacone. Not a sports figure but a young, idealistic attorney, Yannacone was a trustee of the recently established Environmental Defense Fund (EDF), which operated out of Brookhaven, New York. Rogin's article, "All He Wants to Do Is Save the World," was published in the February 3 issue. The piece would spark a legal challenge to canal construction and become part of the institutional memory of what would soon become the Florida Defenders of the Environment (FDE).[7]

Established in 1967, the Environmental Defense Fund quickly made a name for itself by combining scientific expertise with legal advocacy. Forged from a local struggle against the use of DDT for mosquito control, EDF quickly moved from the politics of Suffolk County, New York, to a more national arena. Yannacone's landmark 1966 suit against county mosquito spraying galvanized the scientific community of the State University of New York (SUNY) at Stony Brook and the Brookhaven National Laboratory. Charles Wurster, an assistant professor at SUNY at Stony Brook and chair of EDF's Scientific Advisory Committee, was at the forefront of research showing the effects of DDT upon bird populations. As he later recalled: "A lawyer came into our midst. He had already filed a suit to stop the mosquito commission from using DDT, although he did not have much scientific support behind him. In joining with him, we brought the science and the law together. . . . We were electrified. We had known nothing about getting action through the courts. That was the catalyst that got us started." Full of boundless energy, the small group of scientists, allied with a single attorney, constantly struggled for both publicity and financial support. Long after they were well established, they jokingly looked back on their origins and referred to themselves as the "Fundless Environmental Defenders."[8]

With the knowledge of Wurster and his scientific associates, and the legal skills of Yannacone, EDF pursued a strategy similar to that of civil rights activists twenty years beforehand. The idea was to place issues before the court and use expert testimony to convince judges of the validity of the litigation. For the NAACP Legal Defense Fund, the testimony of psychologist Kenneth Clark was crucial in adding scientific legitimacy to their arguments: for EDF that role would fall to biologists like Wurster. However, the organization's energy rested on the provocative rhetoric of Victor Yannacone. Addressing the 1967 National Audubon Society Convention, he illustrated the importance of legal action to social change. In many respects, a nascent environmental movement was following paths blazed by earlier movements for social justice. "At this time in American history," he argued, "litigation seems to be the only way to focus the attention of our legislators on the basic problems of human existence short of bloody revolution. Conservationists cannot riot in the Everglades, for who would notice but the few remaining great denizens of the swamp? Look at the 50 year history of the human rights struggle in the American courts. Look at the success of the American labor movement. . . . All of the major social changes which have made America a finer place to live have their basis in fundamental constitutional litigation. Somebody had to sue somebody before the legislature produced long overdue action."[9]

Though EDF's successes must have been known to Florida activists like Carr, the *Sports Illustrated* article raised the group's profile significantly. Gilbert Rogin portrayed Yannacone as the quintessential crusader: "a fiery attorney . . . for the nation's most militant conservation group" who "hauls into court those who wantonly defile our habitat." He synthesized Yannacone's credo to a three-word battle cry, "Sue the Bastards!" Yannacone dismissed out of hand one of the favorite tactics of Carr and her allies: "Don't write letters . . . do something! Yannacone's Law: when someone shoves, shove back. Don't write sterile prose. A letter to the editor? Bleahhh!" For him, the only solution to public problems was legal action: "Litigation frames the issues as no other procedure short of physical combat can. I practice law with the philosophy that if there is something that morally must be done, there is a legal way to do it. Of course, you've also got to make the thing swing. If it doesn't swing, you're going to get yawned out of court."[10]

Combined with this element of showmanship was Wurster's commitment to science and radical social action. "We're hawks," Wurster noted, "but we operate entirely within the socio-legal structure. We don't block traffic. We don't sit in. We don't riot. EDF isn't content to do things in the usual slow way with limited accomplishment." He went on to say, "We want to do more faster, even if we have to crack a few skulls." To the Stonybrook scientist, this is what separated EDF from existing mainstream conservation groups like the National Audubon Society and even the Sierra Club. He saw conservationists as "a nice, placid, quiet, law-abiding group of citizens, some of the best we've got, but they have a way of talking to themselves in a closed eco-system. They are legally weak, scientifically naive and politically impotent. They lack an offense." Even for 1969, this was heady stuff, especially for biologists. For Carr and other canal opponents, Wurster and Yannacone offered both the necessary tools and energy to tackle their adversaries.[11]

When Alachua Audubon member Lee Ogden read the *Sports Illustrated* article, he became so excited that he telephoned Yannacone personally and tied up the line for more than two hours as he recounted the story of the canal. Here was the answer to the opponents' prayers: a strategy that combined their own scientific expertise with the new model of legal activism. Within weeks, Marjorie Carr established formal contact with EDF concerning the organization's possible role in stopping the canal. The New York group seemed interested, yet hesitant. EDF board member Joseph Hassett responded to Carr cautiously, by writing "there is another matter which must be clarified before EDF would decide to go ahead with legal action. It is the matter of financial arrangements." Consistently broke, EDF demanded $250,000 to take the case. Complaining of their compromised financial situation, Hassett lamented, "EDF cannot risk initiating legal action which it cannot finish, if protracted, due to lack of funds." Putting it on the line, Hassett concluded: "Your committee may feel that $250,000 could not be raised. This to us would indicate either that there is not sufficient public support for your committee and its objectives or that the public is interested but not willing to pay a minimal amount for the preservation of a quality environment." The price seemed exorbitant, yet Hassett justified it on the basis of needing to appropriately compensate Yannacone for his work. Already overextended, the attorney found that "his work for EDF has been so time-consuming that he cannot maintain his normal legal practice and there is no other present source of income for his family." Thus his fee was set at ten thousand dollars a month. EDF also requested that the unorganized Florida group incorporate as an organization designed specifically to stop the canal and save the river. When those conditions were met, "EDF will send a team to Florida to prepare the legal action."[12]

Jumping on the positive response, Carr and her allies spent the next four months securing the support necessary to bring EDF on board. A new organization would have to be developed, one designed specifically to not only raise funds and public awareness for saving the Ocklawaha but to challenge the Corps in the courts. It had to be divorced from the more general conservation movements like the National Audubon Society as EDF's confrontational posture had the potential to alienate more traditional conservationists. Indeed, according to David Anthony: "National Audubon Society got nervous about our activism. The U.S. government, at the instigation of Nixon, removed tax exemption from the

Sierra Clubs. This made Audubon twitchy." Advocates also seemed eager to avoid repeating the mistakes of the past by giving the organization such an unwieldy name as "Citizens for the Conservation of Florida's Natural and Economic Resources." Instead they would aim for simplicity with a clever twist on EDF's moniker. Established in July 1969, Florida Defenders of the Environment (FDE) emerged as the preeminent voice of anticanal protest.[13]

It was a propitious moment. Meeting in an Orlando area motel, environmental activists established not only FDE but Conservation '70s, an umbrella organization designed as a conservation lobby. The C-70s, as the new organization became quickly known, stood in sharp contrast to the confrontational EDF. Tallahassee insiders like Nathaniel Reed filled the roster of the organization's board of trustees. Fully half of the board was comprised of state senators and representatives. Their mission was to provide a voice on conservation issues in the clubby network of Tallahassee politics. Other board members included officials from mainstream conservation groups like Florida Audubon, which in many respects served as a bridge between the C-70s and the more action-oriented FDE. The EDF scientist Charles Wurster attended the birth of both organizations and seemed pleased with the fruits of a growing environmental awareness.

While Conservation '70s centered on providing a place for conservation within government, FDE focused on creating an independent scientific voice for environmental protection. Overtly replicating EDF's emphasis on scientific expertise, FDE established a twenty-nine-member board of trustees headed by David Anthony. Twenty of the members held Ph.D.s, most of them in the biological sciences. Though the majority of members were affiliated with the University of Florida, other academics came from the University of Miami, University of South Florida, and Florida Presbyterian College. A key member was the University of Florida economist Paul Roberts, who proved important in challenging the Corps' claims of the project's economic viability. The only nonscientific member—with the exception of lawyer Harvey Klein—was John Couse, still responsible for raising funds necessary for the legal challenges ahead. Marjorie Carr served as a board member and vice president of the organization. Carr's husband, Archie—by this time a nationally recognized graduate research professor—also lent his name and credibility to the group. The board selected William Partington as its inaugural president, a wise move that brought significant experience and institutional cachet to the struggle. In assuming the position, Partington was forced to take an eighteen-month leave of absence from his post as assistant director of Florida Audubon. Lacking an income, he would have to rely on fellow board member John Couse, who covered his salary during his tenure in office.

Though organized officially in the summer of 1969, FDE did not formally incorporate until ten months later. The articles of incorporation provide a revealing glimpse at FDE's mission. Nowhere in the document, however, is the primary rationalization for FDE's existence—to save the Ocklawaha and stop the Florida barge canal. Instead, the organization's charter focused on the general themes of protecting Florida's environment. In keeping with EDF's request for funds up front to move the lawsuit along, FDE elaborated its first purpose as "to receive gifts and grants of money and property of every kind, and to administer the same exclusively for educational and scientific purposes." Additional duties included coordinating and disseminating environmental information throughout the state, conducting scientific research on environmental issues, and taking "whatever legal action is necessary to protect the Florida environment."[14]

With a new name and organization, the Florida Defenders of the Environment now felt emboldened to confront the Corps head on. That would take some time, however, as a legal case had to be developed against the canal. During the late spring and summer of 1969, the collaboration between EDF and FDE proved fruitful. Marjorie Carr's dictum that all you had to do was "get out the facts" meshed perfectly with Victor Yannacone's belief that in order to win in court, you have to have persuasive evidence on your side. For Yannacone, it was imperative that FDE convince him of the validity of their claims before he risked taking the case to court. To achieve that end, he summoned David Anthony to New York. According to Anthony, EDF "sort of put me through a cross-examination to see how I would handle a question, to see if I really knew what I was talking about." He performed marvelously. Drawing not only on the breadth of his expertise but also on the wealth of research provided by other FDE scientists, Anthony overwhelmed Yannacone and his associates. "He was an impressive guy," the attorney remembered years later; "I knew we had a case."[15]

Now convinced that it was worth the effort, Yannacone traveled south to prepare the legal brief against the canal. Forced to endure the stifling heat and humidity of a Florida summer, Yannacone toiled over a typewriter in FDE's un-air-conditioned office in downtown Gainesville, carefully crafting his argument. The suit, filed in September of that year, would come directly from the attorney's sojourn to the Sunshine State. As Carr and FDE saw the first benefits of their EDF partnership, the Army Corps of Engineers appeared oblivious to the impending confrontation. In fact, they seemed positively ebullient, at least publicly, over the progress of their project.

With the summer of 1969, boosters touted the first commercial use of the canal. More than one thousand tons of Kraft paper were transported from Palatka's Hudson Pulp and Paper Company Mill to a mill in Pine Bluff, Arkansas. Recognizing the integration of the waterway to a broader national system, the cargo was shipped from the western terminus of the canal, across the Gulf of Mexico and up the Mississippi and Arkansas rivers to its final destination. Ironically, since the canal was incomplete, the paper was trucked 95 miles from Palatka to the Inglis Barge Port, where it was loaded on a converted hopper barge for the brief trip through the canal to the open waters of the Gulf. Though a modest achievement, expectations remained high. The weekly *National Observer* commented with bemused detachment: "To a state that boasts a Presidential retreat, a moon port, spas for the super rich, a Disney World in the making, and Jackie Gleason, a partly finished barge canal might not seem like much to get excited about." However, "there was a certain amount of elation when the Cross Florida Barge Canal carried its first official cargo—even though nine-tenths was by truck." Though lauded as a significant first step, the procedure was both longer and more expensive than the standard method of rail transportation. By barge, the journey amounted to seventeen days, whereas by rail, travel was completed in four. According to the *Jacksonville Seafarer*, an obviously strong advocate for waterways transportation, Hudson "did not expect to actually save money over the all-rail route from Palatka to Pine Bluff." The experiment yielded few, if any, immediate economic benefits. Yet boosters clung to the promise of greater growth in the future.[16]

Meanwhile, canal advocates congratulated themselves on the national recognition of another important benefit of the project: outdoor recreation. In a 1969 editorial, the *Ocala Star-Banner* praised the Corps "for winning the Award of Natural Beauty in the Chief of Engineers' 1969 landscape architectural design competition for . . . Rodman Reservoir." Judged by independent landscape architects, the award—only in its third year—was designed to encourage engineers and architects to develop "functional and attractive" projects that were "more harmonious with the surrounding environment." In the case of Rodman, it involved balancing "land clearing for the unobstructed and continuous navigation channel through the dense vegetation of the Oklawaha flood plain" with "major consideration . . . given to the conservation of forestry, fish and wildlife resources, outdoor recreation, preservation of scenic values, and noxious aquatic plant control." Recognizing the project as an "excellent example of collaboration by engineers and landscape architects," the judges praised Rodman as "a creative solution of a functional inland waterway project enhanced by recreational area development."[17]

Though Corps officials excitedly extolled the award, in truth the competition was designed more for public relations than conservation. As an internal competition of Corps projects, it considered only seven entries nominated nationwide. Still, Corps officials seized the opportunity to promote its achievement. According to a Corps publicist: "the Corps should take pride in the works which have been accomplished at Rodman. It is felt that in the years to come the public will be justifiably proud of the reservoir and its recreational opportunities." Also recognizing the increasing controversy over Rodman, he felt it necessary to address the "press criticism [that] has been directed at the award." He concluded that this opposition "was triggered by the final remnants of the 'Save the Oklawaha' forces." The irony that such an eyesore as Rodman Pool could win a national beautification award was lost on canal proponents. Convinced that the Army Corps of Engineers was the true agent of conservation, boosters could easily dismiss their critics as self-serving doomsayers. As Giles Evans put it: "no single agency has done more for the overall cause of conservation. . . . Long before the general public became concerned about natural values, the Corps pioneered the development of fish ladders and related devices; it initiated the first federal beach erosion control programs, and the first such programs to assist in regulations of flood areas."[18]

Critics begged to differ. Anticanal forces considered the Corps and its

SAVE THE THINGS YOU LOVE FOR THOSE YOU LOVE

"The Oklawaha, the sweetest water-lane in the world," wrote poet Sidney Lanier.

The river, once recommended to be a National Scenic River, undulates through a mile-wide valley forest, the heart of one of Florida's few remaining unbroken ecosystems. It adjoins the Ocala National Forest on the east and pine flatwoods, cypress heads and hammocks on the west.

FLORIDA'S CROSS-STATE BOONDOGGLE

MASSES OF CRUSHED FOREST BEING REMOVED FROM THE RODMAN RESERVOIR OF THE CROSS-FLORIDA BARGE CANAL, ONCE A PART OF THE OKLAWAHA WILDERNESS.

THE CORPS OF ENGINEERS SAYS:

"It is inaccurate to think of the river as being despoiled or that a great wilderness area will be destroyed."

BUT THERE ARE THOSE

In a 1970 pamphlet, FDE used visions of both nineteenth-century wilderness and twentieth-century destruction to mount their campaign against the canal. Courtesy of Florida Defenders of the Environment (FDE), Gainesville.

canal as a "pollution horror tale." Throughout the late summer and fall of 1969, FDE launched a new offensive. Part of it was diligently working the press to publicize the plight of the Ocklawaha. Their efforts paid off in the changing tone of news stories. Soon headlines blared that Rodman was "a sickening sight to view," and that the canal was a "crime against nature." At the same time, FDE relied on their increasingly influential contacts in Tallahassee to lobby legislators and agencies to withdraw state support for the canal. Finally, drawing upon the research of Marjorie Carr's scientific community, FDE's legal ally, the Environmental Defense Fund, filed suit in federal court to halt the project.[19]

Much of this success was surprising considering the precarious standing of the new organization. Though publicly the Florida Defenders of the Environment appeared like a well-oiled machine, it actually was run "from a tiny $40 a month office in a building with a broken elevator." The downtown Gainesville office was certainly humble, but it was a step up from Marjorie Carr's kitchen table. There was little room for comfort considering the operating budget of three thousand dollars a month, the bulk of which went to photocopying, long-distance phone calls, and funding scientific research. What held all of this together was the talent and tenacity of FDE's leadership. Remarkably able to work together without the petty bickering and egotism that often plague such organizations, Marjorie Carr, David Anthony, Bill Partington, and John Couse became a formidable cadre of citizen-activists. Partington, Anthony, and Couse quickly recognized Carr's critical role. Calling her the "coordinating genius of the Florida Defenders," Partington understood the importance of her ability to connect the university community to broader environmental issues. With those connections, many made through her husband's colleagues and students, Carr could draw upon a large pool of scientific expertise. In many respects, it was a case of perfect timing. The University of Florida—no longer a provincial institution, but not yet a modern multiversity—proved the perfect breeding ground for such a movement. The campus, and Gainesville as well, was still small enough for professors, and not incidentally their wives as well, to socialize professionally and personally. That created a climate conducive for sharing ideas and expertise across disciplines. Carr flourished in that kind of environment. Never willing to accept her role as a demure faculty spouse, Carr's scientific pedigree gave her credibility in the world of her husband's colleagues. Apocryphal stories still abound about Carr's inability to take "no" for an answer, even resorting to grabbing men by their ties and ordering them to "get the facts." At the same time, she was more than willing to play her card as a "mere Micanopy housewife." As in the case of Tom Adams in the Tallahassee showdown, the public berating of citizens appeared bad enough, but when those citizens were female it created an image of heartless bullying that Marjorie Carr was more than willing to exploit.[20]

Complementing Carr's strengths, David Anthony provided "a quiet dignity [that] translate[d] frequently into sabre-like sentences about the intelligence, if not the veracity, of officials in the Corps of Engineers." As a University of Florida biochemist whose research experience reached back to the Manhattan Project, Anthony proved an unimpeachable source. Involved in the struggle against the canal since the fateful Audubon meeting in the fall of 1962, he provided continuity as well as expertise. His clear and resolute public image gave instant credibility to the movement, especially when counterpoised to the clumsy public-relations efforts of canal boosters. As a result, he became the point man in most public confrontations with the Corps. Even those who had yet to take sides in the controversy agreed that Anthony almost always carried the day. Seemingly unflappable in the face of often angry debate, Anthony's calm demeanor showed confidence in a cause he knew to be right.[21]

In private moments, Anthony revealed a more passionate, even righteous, anger in his defense of environmental issues. Writing to Colonel John McElhenny in September 1969, Anthony berated the Corps official in language he rarely used in public debate. Responding to McElhenny's complaint that FDE had not bothered to consult the Corps before making its public claims, Anthony charged: "You have not even bothered to contact us. You have not shown enough concern over your organization's activities to even check the possibility that we are right and you are wrong. . . . Accordingly we have concentrated on the ecological horror you have already perpetrated rather than on meaningless public statements or 'plans.'" That same month, Anthony's ire boiled over publicly when he suggested that the best solution to stopping the canal would be to "blow it up during dedication ceremonies." Such a flippant comment reveals a side of Anthony far different from that indicated by his conventional demeanor as a research scientist. "We were the counter-culture," Anthony recalled years later, fighting "a kind of guerilla campaign" against the canal. That sometimes required reconnaissance into enemy territory. On a quiet Sunday afternoon, Anthony visited Rodman Pool and came across the crusher, the very symbol of environmental destruction. With no one there to bother him, Anthony climbed on the machine and gave it a single-fingered salute. A photograph of the incident would become legendary and raise the status of Anthony as a cult hero in the eyes of many committed activists.[22]

Few could imagine Bill Partington or John Couse participating in such a prank. Yet their contributions to the anticanal struggle were just as important as Carr's and Anthony's. With his experience as assistant director to Florida Audubon, Partington brought stability to an infant organization struggling for an identity. Far from a titular figurehead, Partington used his position as president to spread the word throughout the state and even the nation of the significance of the anticanal movement. Using his contacts in more established environmental organizations, he tapped into resources previously unavailable to Carr and Anthony. As a result, he dramatically expanded the power and influence of FDE. As for John Couse, he operated even further under the radar, serving as FDE's "money man." With EDF demanding up-front funding to pursue their legal strategy, it was incumbent upon Couse to draw upon his contacts with the South Florida business community to support the effort. On October 14, 1969, FDE issued a press release for a "nationwide appeal for funds." One day later, the organization received a check for three hundred dollars from George Remington of Palm Beach, representing the International Order of Old Bastards, a fraternal organization for World War II veterans.[23]

While the Old Bastards may have provided some of the money, FDE had to also rely on a very different group of volunteers to achieve their goals. These individuals carried the water for the leadership. Though hardly dramatic, as one observer has noted, "making arrangements for speakers, maintaining files and records, and answering letters . . . is an essential obligation for a citizens group." Over the years, Carr had created a network of volunteers not just in Gainesville, but throughout the state to keep the pressure on the Corps. Working from their homes, they wrote letters to newspapers and in turn clipped articles and dutifully mailed them to FDE's Gainesville headquarters. The resulting flood of information enabled the group to keep tabs on their adversaries and to gauge the shifting sentiment of newspaper writers and editorialists across the state. What made FDE so effective was not only its volunteers' commitment of time and effort to the cause, but the fact that Carr, Partington, Anthony, and even Couse would always join in to lick the stamps and seal the envelopes.[24]

In the dwindling sunlight of a late September 1969 day, Victor Yanna-cone walked into the U.S. District Court in Washington, D.C., and filed suit against the Army Corps of Engineers on behalf of the Environmental Defense Fund. Calling for an immediate halt to canal construction, the suit demanded that the court should determine "the total social cost and real social benefits of the proposed Cross Florida Barge Canal." Suing on behalf of the Environmental Defense Fund rather than Florida Defenders of the Environment, Yannacone announced that "this is the first time that citizens have ever challenged the Army Corps of Engineers in an action to compel them to fully evaluate the social cost of a proposed so-called 'improvement.'" The move was a sign of frustration, as FDE saw little hope of stopping the canal by any other method. EDF's Charles Wurster put it best: "The people and independent scientists of Florida have long predicted that this barge canal would be an environmental disaster. They appealed to the Corps and to the state, and did everything in their power to bring this to the attention of the proper authorities, who looked the other way or told them that their recommendations were impractical. The project has progressed to the point where the scientists, if anything, seemed to have underestimated the damage. There is only one thing left to do at this late date, and that is to take legal action." Filed out of legal necessity in the District of Columbia (federal statute mandated any challenge to the Corps must originate in the District), the action had an additional benefit in immediately getting national attention.[25]

Declaring that the "Oklawaha Regional Ecosystem is a national natural resource treasure," the fiery New York lawyer argued that the rights of the people to use and enjoy the land would be irrevocably harmed by the "degradation resulting from the construction of the Cross Florida Barge Canal." He also asserted that canal construction would violate the fundamental rights of citizens guaranteed by the Ninth Amendment and the due process clause of the Fifth Amendment. To cover all the bases, Yannacone also filed the case as a class-action suit, "on behalf of all those entitled to the full benefit, use, and enjoyment of the particular natural resources." Quick to fend off any accusation that EDF was a bunch of hired gunslingers shopping for cases, he asserted "no trustee of the Environmental Defense Fund, Inc., including its legal counsel, receives any fees for services rendered as a trustee." Rather, "any legal action brought by EDF is founded on broad ecological grounds and asserts the fundamental constitutional right of all the people to a salubrious environment undiminished in quality by the actions of the Defendants." Building on the notion of a scientific-legal partnership, Yannacone wrote that "EDF only takes legal action where its position has broad support within the scientific community."[26]

Yannacone then laid out specific charges. Chief among them was that, in pursuing the barge canal, the Army Corps of Engineers had "consistently failed to report . . . the objections of scientists" and other experts. Moreover, the Corps had "reckless[ly] persisted" in misrepresenting cost and benefit data, often "grossly" underestimating project costs, especially with regard to maintenance. Additionally, construction methods were criminally negligent, for the Corps' use of the crusher and hydraulic dredges seriously damaged the river and the surrounding forest. Cataloging the potential harm the canal had unleashed, Yannacone hammered on the issues of "pollution, eutrophication, and siltation, along with damages to migratory fish populations." Most important were the two issues that galvanized FDE and their supporters: floating logs and weeds. A waterways project in the name of economic efficiency created snags in the river "which would adversely affect recreational and navigational values and benefits." On top of that, it increased uncontrollable aquatic vegetation, especially "known nuisance water weeds," in spite of the continuous efforts of state officials to eradicate them.[27]

Amid the legal language and list of accusations, Yannacone simultaneously drew from two centuries of naturalist imagery of an "undisturbed" Ocklawaha River valley. Starting with William Bartram, Yannacone wrote in terms more reminiscent of Sidney Lanier than Clarence Darrow. Making the case for preservation, he noted, "today, when so many of the diverse original Florida landscapes are threatened with obliteration, the Oklawaha, in its lower reaches, remains as it was, a dark, beautiful stream, clear and free-flowing, and now as in past times, noted for its fine fishing." In some respects, it was as if time had stood still as "most of the wild things Bartram saw are still to be seen along the lower Oklawaha today." Concluding that "this is the country made to glow in the writings of Marjorie Kennan Rawlings," Yannacone made the case for nature having an inherent value that must be included in any calculus of human activity. Juxtaposing this image of a natural river to the "economic, biological, flood control and navigational problems of increasingly massive proportions for which there is no permanent and satisfactory solution," he savaged the rationale for canal construction. Not content with attacking the Corps strictly on ecological arguments, Yannacone accused it of a "callous disregard for suggestions by citizens . . . that would have minimized the damaging effects" of the project. Arguing for an economic value for nature itself, Yannacone accused the Corps of failing to "determine the real value of the existing Oklawaha Regional Ecosystem in spite of the declared policies of both state and federal governments seeking to protect significant areas of unique and unspoiled native wilderness." Summing up seven years of opposition to the Corps' actions, Yannacone articulated the touchstone idea of Marjorie Carr's activism: "Such damage can cause serious, permanent, and irreparable damage to the Oklawaha Regional Ecosystem, and there is no evidence that the Defendants have even seriously considered the effects of this action, much less evaluated the dollar value of the damage."[28]

It was a bold statement. FDE now had its day in court, but winning was another matter. EDF chairman Dennis Puleston provided a cautionary note, warning "the entire might of the U.S. Department of Justice will probably be directed to preventing this case from reaching trial." Locally, canal supporters viewed the legal action with disdain. A November 1969 editorial from the avidly procanal *Ocala Star-Banner* dismissed the lawsuit as "questionable," and concluded "there is no valid reason . . . to halt construction of the project." With decades of momentum on their side, government agencies and canal boosters were not going to sit idly by and let a group of bird-watchers, butterfly catchers, and a Micanopy housewife impede progress.[29]

Now tied to EDF and its lawsuit, FDE felt itself on a roller coaster. By the middle of 1970, FDE members had to wonder whether the suit would ever go forward as shakeups within EDF rocked that organization. Charles Wurster became embroiled in scandalous charges that he was more concerned with the environment than human life itself. In the midst of that controversy, Victor Yannacone was either summarily fired by the EDF trustees or resigned in disgust at the board's lack of commitment to the cause. Deemed a "poor team player and an abrasive egomaniac" by EDF's trustees, Yannacone continued his activism after he left EDF, later initiating a class-action suit on behalf of Vietnam veterans afflicted by Agent Orange. By the end of November, EDF had replaced the colorful New York lawyer with the staid and cautious George Kaufmann. Still, Yannacone's legacy was profoundly important. Not only did he provide a blast of energy to the fledgling organization, he laid the foundation for FDE's groundbreaking environmental impact statement, which would be published in early 1970. Kaufmann's first action as lead attorney entailed amending the suit to include FDE as a party to the case. As a result, FDE reorganized by redrawing its charter and eventually incorporating to prepare for the long struggle ahead.[30]

That task became easier as the canal debate continued. In addition to the lawsuit, FDE spent months trying to convince newspaper writers of the merits of their fight. A major portion of Bill Partington's job revolved around public relations. He spent much of the fall issuing press releases and sending information "all over." As he recalled years later: "every week I'd be sending out new newsletters or releases to all the papers, sometimes two or three times a week quoting different authorities, congressmen, economists, scientists. . . . Our materials were coming to reporters quoting this authority and that authority and all of a sudden all the experts seemed to be against the barge canal, and now reporters were saying, 'Wow, this is pretty impressive stuff.'"[31]

The results of Partington's efforts became clear in early November 1969, when Associated Press reporter Ben Funk wrote an article picked up by every major newspaper in the state. It highlighted FDE's concerns and validated many of their arguments. In one article, Funk now articulated all the points that canal critics had been stressing for years. He quoted scientists like David Anthony and George Reid. He also cited Bill Partington and Victor Yannacone, whose comments sounded straight from an FDE press release. "We've got to take the bull by the horns," noted Yannacone; "anything that needs to be done morally can be done legally." The tone of the article, relying on terms like "mangled and dying forest," "log and weed cluttered reservoir," and "rape of natural resources," was a sharp departure from the procanal articles that had filled newspapers only a few years before. Underscoring both the environmental and economic arguments that critics used against the project, Funk conveyed a sense of urgency regarding canal construction. Quoting Lyman Rogers, chairman of Governor Claude Kirk's Natural Resources Committee, the article left little doubt as to who was winning the public-relations battle. "This biological idiocy," Rogers opined, "will destroy forever one

Canal construction near Palatka. Canal opponents used pictures like this to illustrate the devastation wrought by the canal. Courtesy of State of Florida Photographic Archives, Tallahassee.

of the most beautiful remaining wild rivers of this nation." To highlight the Corps' weakened public position, it could only meekly reply that its figures estimating transportation savings of $8 million annually to shippers could not be verified because they were "based on confidential business information that cannot be disclosed to the public." As to the issues raised by canal opponents, Funk concluded "the Corps admits that log jams will be a perpetual headache, that the weed problem is not solved, that there is a danger of oil spillage, and that the possibility of contamination of surface and ground waters is being reexamined." Put simply, the Corps and canal supporters were now on the defensive.[32]

Now, state officials started to publicly question the canal. The same month Funk's article appeared, the Florida House Conservation Committee authorized a study to determine if the state should continue to back the canal. Contending that "the cost of weed control alone could overshadow its value," the committee contacted state agencies for their position. As a result, the state's Game and Fresh Water Fish Commission submitted a damning report on the ecological impact of the canal. Commission staff produced a judgment significantly at odds with decades of state decision-making. Using language that sounded more like FDE and Victor Yannacone than the Canal Authority, the report pointed to the by-now ubiquitous concern over crushed trees and aquatic vegetation and illustrated the "drastic changes" the canal would make in the Ocklawaha Valley. "The eco-system which formerly supported high quality fishing, hunting and aesthetic values is in jeopardy," warned the report, "because the new system . . . functions similar to a sewage treatment polishing pond." Far from the recreational playground envisioned by Corps engineers, commission staff predicted near septic conditions that could produce the astounding quantity of 75 million tons of algae per year. Though never explicitly supporting a halt in construction, the report concluded that "a sufficient number of problems have been pointed out to indicate the complexities of the project and the need for consideration of all aspects."[33]

The ensuing backlash was palpable. Canal supporters both in and out of government railed against what they saw as a betrayal. *Jacksonville Seafarer* columnist Dave Howard denounced the "rabid conservationists and Audubon Society members" who, "despite their good intentions, are placing stumbling blocks in the path of any and every project which might change the face of the earth in the smallest degree." Former president of the Florida Waterways Association Robert Thomas could not understand how anyone could reject "an absolutely beautiful work of art in service to mankind." But no one defended the canal and attacked its detractors as vigorously as Secretary of State Tom Adams. Claiming the mantle of conservation as "the planned management of a natural resource to prevent exploitation, destruction, or neglect," Adams contended that the Ocklawaha "will be improved, not destroyed" by the canal. Overtly rejecting the preservationist vision of Carr and her associates, he asserted, "both intelligence and . . . the Bible dictate that man is to have dominion over all the resources of the earth." Adams went on to challenge the integrity of experts like David Anthony, proclaiming "professors are for hire, just like attorneys, and they will say and do what is in the best interest of the guy that is paying them." "In an effort to stop the canal," he went on, "a platoon of paid professors has descended upon the project with a smokescreen of innuendo, half-true press releases, and incorrect statements." While his tone may have been consistent, Adams's comments played out much differently in 1969 than in 1966. Increasingly, the canal's most ardent defender appeared shrill and cranky. The world was indeed changing.[34]

By 1969, Americans had become increasingly aware of the environment as a host of events intersected with deeper patterns of social and historical change throughout the decade. In many respects, Rachel Carson initiated the process in 1962 as *Silent Spring* raised troubling questions about technology and human interdependence with nature. Departing from the traditional naturalist literature that valued an idea of wilderness divorced from society, Carson's analysis of birds and pesticides placed the idea of ecology front and center, demanding that people needed to coexist in harmony with the world around them. Over time, as evidence of dying wildlife, polluted lakes, and smog-filled cities kept piling up, Americans saw an earth increasingly out of balance. As a result, many called for immediate concerted action to protect the planet from human endeavors. Willing to push beyond the agenda of such organizations as the Sierra Club and the National Audubon Society, activists took their concerns to another level, raising fundamental questions about the nature of progress and prosperity itself.

Interestingly enough, it was the very affluence of 1960s America that allowed environmentalists to address these concerns. The United States could now afford to clean up its problems. In short, a movement was developing in an already heated political atmosphere where radicals and countercultural critics routinely questioned both civil and corporate authority. Whether they were responding to something as dramatic as Vietnam or something as relatively mundane as dumping garbage, people were becoming increasingly skeptical of established power. By 1969, these rumblings provided the backdrop for a dramatic series of visual images that captured the public's imagination. The sight of sewage and soapsuds choking Lake Erie, as well as the news of an oil-slicked Cuyahoga River catching fire, persuaded many people their world was in peril. The only picture that could rival such striking degradation came from outer space as *Apollo 8* orbited the moon. It was this iconic photograph of

As the canal split Florida, the Army Corps of Engineers had little concern for the questions of opposition groups. Courtesy of *St. Petersburg Times* and Don Addis.

a solitary fragile planet floating in the vastness of space that convinced many Americans that something had to be done.

By mid-August, *Time* magazine observed, "there has not been a topic for such worried conversation since James Baldwin forecast the fire next time." The weekly anticipated that "pollution may soon replace the Viet Nam War as the nation's major issue of protest." Concern for the planet was so widespread that "suburban matrons predict the melting of the polar icecaps . . . [and] busy executives and bearded hippies discuss the presence of DDT in the flesh of Antarctic penguins." More striking was the sharp departure from conservationists of a previous generation, categorized as "nature lovers and esthetes who often seemed devoted to fencing off nature for themselves. Today's ecologists," much like Marjorie Carr and other activists fighting against the barge canal, "are scientists who know that all nature is interconnected and that any intervention has far-reaching effects. They are moved to action not only by considerations of beauty and sentiment, but also by growing knowledge of the possibly disastrous consequences of unthinking intervention." *Time* ended its appraisal of both the movement and the state of the nation with a biblical warning from the prophet Jeremiah: "I brought you into a plentiful country, to eat the fruit thereof and the goodness thereof: but when ye entered, ye defiled my land and made mine heritage an abomination."[35]

Thus it was not coincidental that FDE filed its suit in federal court in 1969, a year that marked a watershed in environmental consciousness. As Richard Nixon's chief environmental advisor, Russell Train, put it: "At the outset of 1969, the Santa Barbara oil spill fouled California beaches. The proposed Miami Jetport threatened the Everglades National Park. The proposed Alaska Pipeline raised critical environmental issues. Lake Erie was said to be 'dying.' Smog in the cities, phosphates in detergents, lead in gasoline, fish contaminated by mercury, bald eagles threatened by DDT—these were some of the environmental problems that excited public concern." Relying on traditional political strategies—letter writing and lobbying legislators—as well as more confrontational direct-action protests often borrowed from the civil rights and antiwar movements, environmentalists forced the issue into the mainstream. Membership in the nation's top-twelve environmental groups grew dramatically from 124,000 in 1960 to 819,000 in 1969 to 1,127,000 in 1972. With such numbers representing potential voters, even Richard Nixon, no friend of liberal causes, quickly recognized the political implications. As Nixon staffer John Whitaker explained, "When President Nixon and his staff walked into the White House on January 20, 1969, we were totally unprepared for the tidal wave of public opinion in favor of cleaning up the environment that was about to engulf us." Far from being swept away, Nixon instead rode this wave to establish a significant, albeit ambiguous, environmental legacy.[36]

The environmental movement had achieved some significant legislative successes during the previous Kennedy and Johnson years: the Clean Air Act of 1963, the Wilderness Act of 1964, the Endangered Species Preservation Act of 1966, and the Wild and Scenic Rivers Act of 1968. This legislative momentum continued into the new administration of Richard Nixon. Building upon these precedents, Congress passed another series of laws that addressed environmental concerns well into the next decade. While most of the legislation extended federal authority in the name of clean air and clean water, the National Environmental Policy Act (NEPA) of 1969 represented a fundamental shift in government environmental policy. Shepherded through Congress by Democrats Henry "Scoop" Jackson and John Dingle, NEPA initially received lukewarm support from the Nixon administration. Sensing strong public support for its passage, Nixon eventually enthusiastically embraced the measure, even going so far as to sign it on live television as a symbolic act to mark the 1970s as the decade of the environment.

Though Nixon's environmentalism came more from political expediency than heartfelt belief, NEPA stood as a landmark in American environmental policy. It established a Council on Environmental Quality (CEQ) in the White House, making its chairman directly responsible to the president's chief of staff, allowing for a permanently institutionalized voice in Washington. More importantly for FDE, it also established a formalized process that incorporated environmental considerations into federal public-works projects. In addition to mandating public hearings, NEPA also required the development of a new procedure to determine the viability of any federal project—the environmental impact statement (EIS). By calculating environmental considerations into the cost of any project, environmental impact statements became a key weapon in the arsenal of environmental preservation. FDE would craft one of the first environmental impact statements in their March 1970 report on the canal's effect on the Ocklawaha.

Emphasizing this dramatic shift on the environment, Richard Nixon significantly elevated its importance in his State of the Union Address of January 22, 1970. Sounding more like Marjorie Carr than Tom Adams, he proclaimed: "the environment may well become the major concern of the American people in the decade of the seventies. . . . Shall we surrender to our surroundings, or shall we make our peace with nature and begin to make reparations for the damage we have done to our air, to our land, and to our water?" To Nixon, the answer was clear: "Restoring nature to its natural state is a cause beyond party and beyond factions. It has become a common cause of all the people of this country." Using the language, and even the tone, of environmental activists, Nixon declared that "clean air, clean water, open spaces—these should once again be the birthright of every American. If we act now, they can be. We still think of air as free. But clean air is not free, and neither is clean water. The price tag on pollution control is high. Through our years of past carelessness we incurred a debt to nature, and now that debt is being called." Environmentalism was now a permanent part of the political landscape.[37]

As if to underscore the mainstream nature of the president's pronouncements, *Reader's Digest*, the very voice of Nixon's silent majority, published a stunning feature on the Ocklawaha that same month. Provocatively titled "Rape on the Oklawaha," the article transformed what was a local issue into one of national importance. By 1969, even *Reader's Digest* had gained an appreciation for environmental issues and was publishing a series of articles on the disturbing condition of America's air and water. Tied to the publication's inherent fiscal conservatism, these articles often blamed profligate federal projects for the destruction of the nation's resources. Building on those pieces, Boston Whaler president Richard Fisher suggested that an article on the Ocklawaha would fit well. The magazine agreed and sent roving editor James Nathan Miller to Florida in the summer of 1969. After spending several weeks with Partington, Anthony, and Couse, Miller crafted an exposé that articulated FDE's critique of the Corps' project. Hardly an objective reporter, Miller wrote Couse in September, complimenting him on his crusade: "Am delighted to see you people are not letting up." Months later he would continue his praise, admitting that though he had much respect for the power of the Corps, the heat FDE was "managing to keep on them makes it look as if maybe—just maybe—there's a chance" for success. Such sentiments were reaffirmed by the magazine's press-relations manager, who made the editorial position of the article crystal-clear in another letter to Couse. "When the *Digest* comes out on December 29," he crowed, "millions of Americans will read about—and, one hopes, share your concern for—the fate of the Oklawaha. . . . Obviously we'd like to help generate a substantial public-opinion campaign aimed at saving the river."[38]

Reflecting both Marjorie Carr's preservationist ethic and John Couse's obsession with governmental waste and abuse of power, the article read like another FDE press release. Starting with an obligatory paean to "the magnificent primordial river," Miller painted an idyllic picture reminiscent of Sidney Lanier: "When you are floating along the river's surface, with Spanish moss hanging above you and the tall spikes of wild rice by the bank stirring slightly, it is cool and dim and almost always still—just the slap of a fish clearing the water, the slurp of an alligator slithering off a log, the squawk of a heron. Along most stretches there are no houses or roads—nothing but trackless forest for many miles." The meat of the article, however, was a lengthy and rather tendentious indictment of the canal on primarily economic grounds. It traced the history of the project in terms of developing a positive benefit-cost ratio, a procedure that Miller deemed as "merely guesswork masquerading as a mathematical formula." Using language like "pork barrel" and "statutory worthlessness" to illustrate his points, he asserted that "the Oklawaha had finally been shot dead by a boondoggle."[39]

Miller saw hope, however, in the form of citizen activism: "We are still, after all, a democracy, and we do have the right to shout thief when we see bureaucrats making off with our landscape." To Miller, FDE embodied the struggle to set things right. Calling them a "horde of unarmed amateurs . . . working in their spare time, with their own money," he maintained their struggle would be difficult. Fighting against "professionals who are equipped with the latest weapons" as well as "well-paid businessmen and bureaucrats, tightly organized in trade association and lobby groups," FDE faced an uphill battle. Still, Miller demanded a call to action. First, the public needed to apply political pressure in both Tallahassee and Washington. Miller also saw the need for fundamental reform in the way the Corps of Engineers conducted business. Echoing the sentiments of Marjorie Carr, Miller called for a complete reevaluation of the assumptions behind federal and state public works projects. Officials needed to consider a "respect for the intangibles and un-priceables— beauty, relaxation, clean water and air—into the decision-making process." Reminding readers "there is no market price for a wild river," he concluded that "we must inject *human judgment* into a formula that now accepts only dollar signs" (emphasis in original).[40]

For FDE, the article was a public-relations coup. All of Partington's hard work had yielded an extraordinary dividend. Calling the article "superb," he suggested FDE needed to send reprints to government officials and newspaper writers throughout Florida and the nation. Florida Audubon joined in by creating a rubber stamp, affixing the statement "Don't

Miss Reading 'Rape on the Oklawaha'" to all of its outgoing mail. For Partington, "the *Digest*'s efforts are wonderful encouragement—we're not saying 'if we stop the canal,' we're now saying '*when* we stop it.'" John Couse was equally effusive. Writing directly to the magazine, Couse exclaimed: "Oh Boy! That is a Jim Dandy, completely factual with PLENTY of PUNCH. James Miller did a thorough and beautiful job" (emphasis in original).[41]

Canal supporters begged to differ with Couse's editorial judgment. Well before the article's publication, Corps officials were skeptical of Miller's objectivity. While noting he "had done extensive homework on the subject before coming to the Jacksonville District," public-relations officer Gene Brown concluded, "he had gone with arch conservationists to the Rodman Reservoir to see for himself the object of their criticism." A quick glance at the article's title was all that was necessary to confirm Brown's suspicions. Boosters soon flooded the magazine with letters. Former Canal Authority chairman Bill McCree lambasted the article for its "irresponsible journalism and the half-truths and mis-information contained" in it. Following a point-by-point rebuttal of Miller's assertions, he asked, "why would a publication like the *Reader's Digest* try to tear down a great economic and cultural development which had been dreamed of and worked for so hard by so many well-meaning citizens?" Ten days later, Mississippi Valley Association vice president Robert Shortle joined in: "It can be seen that Miller saw only what he wanted to see and wrote the story with a preconceived notion. . . . Additionally, he has maligned a respected and visionary man—Bert Dosh. . . . To so malign him is unconscionable." Not one to give up, Shortle wrote eight months later to continue his complaint. Announcing that he was canceling his subscription, he was offended that "*Reader's Digest* feels that it can print what it pleases regardless of content and by this printing it become correct or fact." One canal advocate even submitted an entire alternative article for publication in the magazine, entitled "Metamorphosis on the Oklawaha," which was summarily dismissed by the *Digest*'s editors.[42]

The article made such a strong impression that canal supporters continued to argue about its credibility years later. As late as January 1974, Canal Authority manager Giles Evans condemned Miller's article "as replete with glaring inadequacies and distortions." To Evans, the Ocklawaha was "no primordial river, but rather a partially developed, repeatedly timbered, and historically navigable swamp river . . . [that] has been canalized, deepened, widened, straightened, and impounded." Taking issue with the article's portrayal of the river as a veritable Eden, he asserted that after a "dozen or so personal trips on the Oklawaha" he had "yet to hear the 'slurp' of any alligator slithering off logs." If anything, the Corps should be commended for creating a new environment. Rodman Reservoir, what Evans called Lake Oklawaha, was now attracting "a wide variety of relatively rare bird species—the Osprey, Bald Eagle, Kite, Limpkins, and Wood Ibis." For Evans, the "whole effort was a hatchet-job aimed at the Corps of Engineers, with the Oklawaha issue as a convenient vehicle. The language is simplistic and inflammatory and full of misinformation." Evans dismissed the *Digest*'s reporter as a journalist with an agenda; someone who was "pre-committed and merely went through the paces of showing up in order to say that he had achieved in-depth coverage."[43]

"Rape on the Oklawaha" cast a pall over what should have been a celebratory moment for canal supporters. In January 1970, canal construction reached a milestone with the completion of Inglis Lock on the Withlacoochee River. On January 11, local boosters from Levy, Citrus, and Marion Counties held yet another celebration, complete with the by-now obligatory fish fry, to honor the occasion. Many of north central Florida's leading politicians, including three congressmen and former governor Farris Bryant, gathered on a cold Saturday morning to mark a proud moment. Instead, because of the article's bad publicity, speakers were forced to spend more time pummeling their critics than praising the Corps' accomplishments. Rather than celebrating the achievement, local newspaper headlines underscored the sense of hostility and defensiveness that permeated the ceremony. Left unsaid was anything positive regarding the lock completion itself. Still, Congressman Bill Chappell of Ocala confidently stated: "the decision to finish this canal has been made. We must preserve all of nature but we must finish this job primarily for the defense factor and shipping." Bryant in particular attacked *Reader's Digest* for publishing a seemingly gratuitous assault on Bert Dosh, lauded as the grandfather of the canal. The eighty-three-year-old Dosh attended the event, but was so ill that he was forced to remain in his car throughout

Constructing the U.S. 19 bridge over the canal, with Inglis Lock in the foreground, 1968. Courtesy of State of Florida Photographic Archives, Tallahassee.

the entire proceedings. Little did everyone involved realize they were attending the last major celebration of canal construction.[44]

While canal supporters were content to stand in the cold before a crowd of more than two thousand, FDE's leadership was pushing their issue toward a larger stage. For them, Miller's provocative article provided a rare opportunity to reach a national audience of more than 18 million readers. Eager to capitalize on the moment, Ken Morrison, an early canal opponent who had become a major player in Florida's growing environmental movement as a member of the Governor's Natural Resources Committee and vice president of Conservation '70s, wrote Bill Partington on December 26, 1969: "After reading . . . Miller's terrific article in the *Reader's Digest*, I have come up with an idea that I hope will prove to be a belated Christmas present to FDE." Morrison suggested an open letter to Richard Nixon, signed by leading scientists from Florida and around the nation, to apply political pressure on governmental policymakers. Buoyed with optimism in the face of a new year, he closed by writing "best wishes for a canalless decade." Partington responded promptly and positively, concluding, "Marjorie and I think that this is a great idea." Between the three of them and David Anthony, they drafted a letter and quickly solicited scientists across the country. The response was overwhelming. Out of 175 scientists contacted, 162 agreed to sign the letter. University of Florida scientists, of course, dominated the list, with thirty-eight faculty members from a variety of disciplines. Though most of the signatories represented the state of Florida, many of the nation's leading ecologists and biologists, including such eminent scientists as E. O. Wilson, Paul Ehrlich, and Barry Commoner, affixed their names to the document. Though the Florida scientists were well versed in the details of the controversy, many of the outsiders knew little of the particulars of the case and signed out of deference to their colleagues' credibility and a commitment to environmental activism. This would later be a point of contention among canal supporters, who saw FDE's assumption of the mantle of scientific objectivity as disingenuous at best, and downright fraudulent at worst.[45]

Written on January 27, 1970, the letter directly called for the president's "support in declaring a moratorium on construction of the Cross-Florida Barge Canal" to avoid "a major, national ecological disaster." Appealing first to Nixon's inherent fiscal conservatism, it asserted the canal was a waste of federal money, and "suppressing it would be consonant with the anti-inflationary policy of your administration." Moving from economic to ecological arguments, it went on to warn that canal construction "will drastically alter ecosystems associated with two major river valleys." Noting recent developments in the study of ecology had altered the way people think of their world, the letter called for a holistic approach to understanding environmental disruption. "Changes inevitably ramify widely and usually produce damage far in excess of what could have been forecast a few decades ago." Written within days of the State of the Union address, the letter sought to hold Nixon to his soaring rhetoric regarding environmental protection. It concluded on a congratulatory note, applauding him and his administration for the "wisdom in resolving another recent threat to the Florida environment—the Miami Jetport."[46]

Though located hundreds of miles from north Florida, the controversy over the Miami Jetport mirrored many of the same issues associated with the canal. Questions of growth versus preservation and the inherent value of natural ecosystems permeated both controversies. South

Florida officials proposed the jetport in late 1967 to ease congestion at the overcrowded Miami International Airport. With an eye to the future, they planned the world's largest airport to be located in the middle of the Big Cypress Swamp, astride the Tamiami Trail. Equidistant between Florida's coasts, this location would encourage unlimited growth and change the population dynamics of south Florida forever. More than an airport, developers imagined a community of 150,000 people to be carved out of the wetlands south of Lake Okeechobee. Much like canal boosters extolling the virtues of their project, jetport backers saw the landscape as something to be conquered. "The wilderness has been pushed aside," noted one developer, "with calipers and slide rules, draglines and dynamite rigs, we are literally changing the face of Florida." Initially, state and federal conservation agencies supported the project with little objection. However, opposition coalesced against the airport as scientists and environmental activists raised serious questions about the project's ecological damage. Crucial to their argument was a nascent understanding that the destruction of these "useless" swamplands would wreak serious havoc upon Everglades National Park, the crown jewel of South Florida's unique environment.[47]

By the spring of 1969, national conservation organizations established an Everglades Coalition to halt construction. Supported locally by Joe Browder, a Miami television reporter and Audubon officer, and Nathaniel Reed, the chief environmental officer of the Kirk administration, the movement also had considerable political influence on its side. In June, Walter Hickel, Nixon's secretary of the interior, authorized a scientific committee to study the environmental effects of the project—in essence inventing the environmental impact statement. The idea was inspired, but what granted it immediate visibility were the two remarkable scientists appointed to investigate the issue. Luna Leopold, son of the renowned ecologist Aldo Leopold, and Arthur Marshall, a noted expert on Florida's environment, gave the report instant credibility. In terms of substance and procedure, the report was a tour de force. The science was unimpeachable, but more important was the way Leopold and Marshall framed the issue. For them, "the south Florida problem was merely one example of an issue which sooner or later must be faced by the nation as a whole." Representing a new vision of conservation that incorporated Rachel Carson's thinking, Leopold and Marshall argued that "the effects of the Jetport and the surrounding development should not be thought of in terms of the possible elimination of some rare and endangered species such as alligator [and] wood stork. These, however, can be thought of as indicators or touchstones as to what is happening to the total ecosystem." The project threatened to destroy not only what remained of the Everglades but possibly Florida's tourism and fishing as well. Within weeks after its findings were leaked to the press in August, both state and federal officials were already working on a way to extricate themselves from the project. By January 15, 1970, the Nixon administration arranged the signing of the Jetport Pact, which provided a clear victory for environmentalists.[48]

FDE's leadership followed the jetport controversy closely. Though primarily concerned with saving the Ocklawaha, the organization strongly identified with its fellow activists to the south. Carr and her associates took notice of the strategies employed in the jetport controversy and found them remarkably similar to theirs. Appealing to both political leaders and the general public to halt the project, the movement's accomplishment rested on the strength of its science. Building upon the scientific basis of Leopold and Marshall's analysis, FDE would craft a similar document concerning the Ocklawaha. As early as the summer of 1969, Marjorie Carr had outlined the need for a scientific investigation of the canal's impact on the Ocklawaha and its surrounding environment. Released in March 1970 to coincide with a new round of public hearings, the final report, a 115-page compendium entitled *Environmental Impact of the Cross-Florida Barge Canal with Special Emphasis on the Oklawaha Regional Ecosystem*, marked a watershed in citizen activism. The scientists and other academic researchers who wrote the report produced a model environmental impact study, reaching far beyond empirical science to incorporate historical and economic arguments that made a persuasive case for preservation.

Born of Carr's persistent commitment to the idea of "getting out the facts," the report also doubled as supporting evidence for the Environmental Defense Fund's legal battle against the Army Corps of Engineers. The document's contributors represented some of the University of Florida's most talented academicians, as well as scientists from across the state and nation. As a collaborative effort, it reflected the widest and most diverse ways of thinking about ecology. Filled with detailed sci-

Marjorie Carr at her Micanopy house, working to save the Ocklawaha. Courtesy of Florida Defenders of the Environment (FDE), Gainesville.

entific data including graphs, charts, tables, and sophisticated flow diagrams that, page after page, overwhelmed even the most dutiful reader, the report read much like a traditional academic treatise. The breadth of its vision, however, set it apart from other scientific studies. Sections addressing the history and economics of the project neatly complemented the often dry scientific text and conveyed a passionate defense for preserving the river. Always with an eye on the big picture, the report grappled with the broader implications of governmental policy and environmental protection.

The report, representing the culmination of years of research and analysis, attacked the project's impact from a number of angles. Examining the region's geology, hydrology, zoology, and botany, it illustrated the canal's devastating effect on the natural environment of the Ocklawaha Valley. To FDE, problems began with the very dig itself. The disruption of the area's underlying strata of porous limestone threatened the quantity and quality of the local water supply. Over the long run, pollution from routine canal activities would find its way into Florida's groundwater and flow to natural discharge points throughout the region. Herbicides, insecticides, industrial wastes, and toxic spills from routine leakage or shipping accidents would pass through the area's permeable rock, jeopardizing not only the water supplies of local communities but the "unique recreational qualities of Silver Springs and of whatever sports fishing the canal impoundments might afford." Also buried in the report was an allusion to the problem that halted the first iteration of Florida's canal—saltwater intrusion. Though admitting the subject warranted further investigation, it claimed the project so disrupted the flow of the region's water supply that salt water would eventually encroach upon the St. Johns River.[49]

Another water-quality issue centered on the project's reservoirs. For centuries, the Ocklawaha River valley existed as an ecologically sensitive but stable aquatic system that supported a wide variety of plant and animals. The canal, however, was dramatically "converting a natural, nutrient-rich flow-through river ecosystem into a series of interconnected impoundments." Far from the Corps' vision of beautiful artificial recreational lakes, the impoundments created by the Rodman and Eureka dams would lead to expansive stagnant pools that would within fifty years become lifeless swamps. The problem stemmed from destroying the river's forest canopy to create artificial lakes. With increased sunlight and decreased flow, the impoundments became nutrient traps, which resulted in aquatic weeds that in turn depleted the water's oxygen and accelerated the buildup of organic mud and debris. This argument restated a long-held objection of canal opponents that was already seen in the luxuriant growth of hydrilla and water hyacinths in Rodman Pool. Control of such weeds could only be addressed by continual chemical treatments that would ultimately destroy the very ecosystem they were designed to protect. Building on the recent research of state game and fish officials, the report warned the excessive growth of water weeds was "emphatically NOT a transient phenomenon that will disappear in a year or two as has been suggested. The fundamental environmental changes . . . will persist. . . . The only solution to this dilemma is to drain the Rodman pool and never construct the Eureka pool" (emphasis in original).[50]

With respect to fish and wildlife habitat, the report asserted canal construction would destroy the Ocklawaha's distinct environment. The natural river hosted more than a hundred different species of fish, including bass, sunfish, and pickerel, which represented "the stability of the ecosystem and serves as a measure of the quality of the environment." With canal construction, this rich population would rapidly diminish and be replaced by a "population of trash fish—gars, bowfin, shad, and bullheads," which undercut the Corps' claims for improved sportfishing within artificial lakes. The hardwood forest that surrounded the river supported a similarly large and diverse wildlife population. For larger species like turkey, bear, and panther, as well as smaller ones like frogs, shrews, and songbirds, the canal would eliminate habitat and significantly reduce their numbers. While concluding that some animals would remain, the area's wildlife "would no more be representative of the original community than a city park with its tame squirrels is of the original forest." More than the loss of individual species was the violent disruption of something that was more than the sum of its parts. "The ultimate tragedy will be the destruction of an entire ecosystem."[51]

In addition to the science, the report's reliance on history bolstered its case against the canal. Quoting Sidney Lanier, Harriet Beecher Stowe, and Marjorie Kinnan Rawlings, it asserted that "if the whole history of the area has a moral it is that the tradition of the region is more than anything else a wilderness tradition." Since the steamboat days of Hubbard Hart, when tourists captured a glimpse of the disappearing "exotic beauty" of a "paradisal jungle," the river and its environs "remained quite literally a frontier area . . . until well into the present century." Over time, the region's hard-bitten inhabitants had learned to live and work with nature. All of this would be irrevocably altered, in the poetic words of Gordon Bigelow, the University of Florida English professor writing this section, by the "all-conquering bulldozer."[52]

The economics portion of the report directly assailed the logic and credibility of the canal itself. In clear and unambiguous language, UF economist Paul Roberts charged that the Cross Florida Barge Canal "is a classic example of a long standing national disgrace—pork-barrel legislation," and should thus be stopped in the name of good government and fiscal responsibility. Drawing from Elizabeth Drew's recent article in the *Atlantic Monthly*, the report cataloged the Corps' close relationship with various lobbying groups behind public waterways legislation. It also charted the course by which seemingly "local" interests successfully secured federal funding and congressional authorization to begin constructing projects of dubious national merit.[53]

Taking aim at the canal's benefit-cost ratio, the report claimed that the Army Corps of Engineers' primary method for evaluating projects was hardly an objective measure of public-works projects. "Highly arbitrary choices as to what constitutes primary benefits and/or collateral benefits" may create firm numbers, Roberts wrote, but they are all the result of "subjective value judgments." In the case of the canal, the economist charged, more than twenty years of arcane statistics issued by the Corps rested on dubious assumptions that failed to consider its actual cost. Newer estimates, Roberts continued, also failed to account for approximately $20 million in interest charges for capital and land costs during construction. Neither did the Corps' budgets include expenses for such ancillary projects as the Interstate 75 bridge and designated recreational sites. Moreover, nowhere did the Corps mention the cost of managing the waterway's weed problem. Nor for that matter did it adjust the ledger to take into consideration the consequences of aquatic weed infestation. Over time, the environmental degradation that accompanied the growth of these weeds would dramatically undermine the project's supposed recreational benefits. Finally, tying the economic and environmental arguments together, the report concluded that at no time did the Corps's benefit-cost analyses consider the price of destroying the Ocklawaha River valley. Taking the full measure of the instrinsic value of nature, Roberts boldly asserted that the "Oklawaha ecosystem is an irreplaceable economic resource. It derives its value by its very existence. The argument that non-commercial use of such natural resources represents no economic value is simply an erroneous assumption. Much re-thinking concerning how such resources are valued seems in order." Roberts ended with a recommendation for a restudy of the economic justifications for the canal.[54]

The report's publication helped make the beginning of the new decade quite exciting for anticanal forces. All of their legal, political, and public relations efforts appeared to be making headway. More importantly, the struggle seemed less isolated than before. In March, a segment broadcast on the national *CBS Evening News* described the battle in stark terms: "this is one area where government, free enterprise, and those white knights of the seventies, the ecologists, have laid down the battle line." One of those government officials was Florida House of Representatives member Gerald Lewis. Writing in *Florida and Tropic Sportsman*, Lewis, an influential south Florida Democrat, lambasted the canal in an article entitled "Death on the Oklawaha." Obviously a play on the *Reader's Digest* piece, it was a clear sign that FDE's message was reaching a larger audience. To Lewis, Rodman Reservoir was nothing more than an "eyesore," clogged with logs and filled with weeds. Calling for a meaningful state conservation policy, he suggested policymakers needed a greater sense of balance: "We must recognize the value of Florida's natural resources—our number one asset—and take affirmative action to create as much interest in protecting, preserving, and developing our natural environment as we do in promoting new industry."[55]

Lewis's stand represented a broader shift in the political dynamics of the canal issue. Following Lewis's lead, state senator Warren Henderson of Venice called attention to FDE's struggle by forwarding Congress a copy of its letter to Richard Nixon. Henderson agreed that a moratorium was necessary, and that in light of "outspoken opposition" to the project, Congress should step up and intervene. For decades, the canal was associated with the Democratic Party and the strength of the north Florida Pork Chop Gang. A new generation of Democrats, which included not only Lewis and Henderson but such future Florida leaders as Lawton Chiles and Reubin Askew, became increasingly devoted to environmental issues. More inclined to take FDE's concerns seriously, they seemed willing to publicly challenge the legacy of their party's old guard.[56]

Republicans, especially members of Nixon's administration, also recognized the importance of environmentalism as a hot political issue. Chief among them were Russell Train, chairman of the newly formed Council on Environmental Quality (CEQ); Walter Hickel, the secretary of interior; and John Whitaker, Nixon's domestic advisor for environmental affairs. Train, an urbane Ivy League–educated easterner, reflected the party's long-standing commitment to conservation that began with Theodore Roosevelt. As the former head of the Conservation Foundation, Train's appointment to CEQ garnered nothing but praise from the nation's leading environmentalists. The less polished Hickel, on the other hand, met significant opposition in his confirmation hearings as most conservationists considered the former governor of Alaska a tool of oil and gas interests. During his brief tenure, however, Hickel would both confound and delight his critics with his increasingly strident support of environmental issues. Whitaker helped make the environment a major interest for Nixon and pushed him to spotlight it in the 1970 State of the

Union address. All of these men played a significant role in halting the Miami Jetport and thus received high marks from environmentalists for their willingness to take on such controversial issues.

Train in particular recognized the potential political capital to be gained by pursuing environmental issues. Writing to presidential counsel John Ehrlichman in November 1969, Train, then undersecretary of the interior, laid out both the deep commitment and craven calculation of Nixon's environmental policy. He asserted, "when a community is unable to fish or swim in a river, it is paying a very real cost in terms of foregone benefits." He also warned that "much of the supposed economic basis for decisions in the past has been false economics. . . . [T]he costs are merely passed on to someone else or are hidden or perhaps shifted to another generation." Signaling a position that would be fully articulated in Nixon's State of the Union address two months later, he asserted, "the president should speak of the need to reexamine, and where necessary, to change our values." Conversely, Train was not hesitant to view environmental issues in pragmatic political terms: "Environmental quality is rapidly emerging as a major political issue, which I believe the president should and can make his own." Asserting that "the Republican environmental image (and prospects) is . . . complicated by the Party's identification with business interests . . . there is widespread uncertainty and doubt among the general public as to the Republican Party's commitment in the field of conservation." To overcome these obstacles, Train proposed Nixon should "identify the Republican Party with concern with environmental quality [and] take the initiative away from the Democrats and make it our issue." Train's suggestions found a relatively sympathetic audience in the White House. Not to be outdone by its Democratic adversaries, the Nixon administration would attempt to take the lead on environmental issues throughout the next several years.[57]

As a recent convert to the issue, Interior Secretary Walter Hickel zealously pursued environmental causes. Less than a year into Nixon's tenure, Hickel publicly pushed the issue: "The administration must face up to its natural resources and environmental responsibilities. We must plan and act in terms of altered priorities." His enthusiasm pleasantly surprised grassroots activists, but raised alarm bells with the president, who viewed the issue through a more political lens. As early as November 1969, Nixon confessed to Ehrlichman: "this worries me. He may be caving in too much to his critics." Still, the environmental focus of Nixon's January 1970 State of the Union address reflected the views of both Hickel and Train.[58]

By spring 1970, the White House was listening to FDE. In a March 16 internal memo to Russell Train, Whitaker succinctly summarized the path the Nixon administration would take on the canal. Citing the critical importance of this issue both politically and environmentally, he boldly predicted that "no question about it—the Cross Florida Barge Canal will be this year's Jetport, at least in Florida, although it will lag behind [the Alaskan oil pipeline] as a national environmental issue." Though Whitaker laid out a series of scenarios that included continued construction, stopping the project was ranked first on his list of possibilities. Indeed, he asserted that "the Canal should be stopped cold." The question remained how to do it. Whitaker thought that Train and Hickel should play critical roles in shaping any administration policy. He suggested Train chair an interagency task force much like the one that had resolved the jetport controversy a year earlier. Sensitive to the political ramifications of halting the project, Whitaker understood that such a course of action would result in many Florida politicians going "out of their mind[s]." With so many interests at play, timing for such a decision was everything.[59]

Considering FDE had only been established in July 1969, the organization had gained significant influence in Washington. Much of that came from the efforts of Joe Browder, who had left his Miami television job for a permanent position as conservation director of Friends of the Earth. Though technically not affiliated with FDE, Browder became the organization's liaison in Washington, working especially with John Whitaker. Those connections paid off by February 1970, when a presidential assistant informed David Anthony that the organization's open letter to the president would be forwarded to the Council on Environmental Quality. With Russell Train as its chairman, the canal issue would get more than a cursory overview. A month later, Train asserted that the project was within the council's purview and worthy of reappraisal. Simultaneously, Hickel charged his agency with considering the case as well. By the end of March, Hickel had received yet another review of the canal issue, this time from the Atlanta Division of the Bureau of Sports Fisheries and Wildlife. Bureaucratically, this report played it safe, cautioning that it was not "a full assessment of the environmental impact of the Cross Florida Barge Canal." Drawing heavily from FDE's research and the November 1969 study from Florida's Game and Fresh Water Fish Commission, it seconded many of their findings and indeed concluded that their "biological information appears to be sound." As a result, enough questions had been raised to convince the agency that "a thorough review seems appropriate." On the basis of this review, Hickel had enough evidence to write Secretary of the Army Stanley Resor on June 5, 1970, to suggest that canal construction be halted for fifteen months while further studies were undertaken to determine the project's full environmental impact. Adding insult to injury, Hickel requested that funding for these studies come directly from the Corps' budget and not the cash-strapped Department of the Interior. The Corps seemed hardly enthusiastic in its response to Hickel's request. Not only was he trying to use the Corps' money to undercut its own project, but he was overstepping bureaucratic boundaries by interceding in an issue in which his department had little authority.[60]

Hickel took the lead on the canal controversy at a significant political cost. Already under siege at the White House for his public criticism of the American incursion into Cambodia and his strident defense of Earth Day, the canal issue further marginalized Hickel's influence. In calling for the construction moratorium, Hickel blindsided the administration by failing to consult Train or even to notify the president. Hickel's office issued a press release on the decision only three days after the secretary sent his letter to Army Secretary Resor, even before the army secretary crafted a reply. His action also alienated Florida politicians, who viewed his decision as capricious and irrational. On June 11, Congressman Don Fuqua wrote the president to complain that he was "not consulted in these determinations" and that Florida's delegation "should have had the opportunity to discuss such a drastic decision." Four days later, Senator Spessard Holland mailed a blistering five-page letter to Hickel, asserting he was "unalterably opposed to any moratorium." Citing the long history of government support for the project, Holland repeated the by now-usual litany of rationales for canal completion. Taking issue with Hickel's reliance on FDE as a primary source of information regarding the project, Holland advised the interior secretary to turn to former Canal Authority chairman William McCree as the leading authority on the issue. For Holland and other canal boosters, Hickel's reliance on FDE's scientific claims was but the latest example of their dissipating influence in Washington.[61]

Using public-relations blitzes, letter-writing campaigns, and the editorial pages of sympathetic newspapers, advocates for canal completion now worked diligently to get the message out that the canal would benefit Florida and the nation without damaging the environment. But by early 1970, they recognized they were fighting an uphill battle. FDE's attack on both scientific and aesthetic grounds had proved remarkably successful and put advocates in the unenviable position of always reacting to criticism. As Farris Bryant remarked in 1971: "When environmentalists get after you, you have been got after [sic]. They put on us the burden of proving that the canal would not provide a hole in the aquifer through which all the water in Florida would run out. You can't prove a negative. That's the job we had." Their responses proved cranky and unsatisfactory, at times making their position even more untenable.[62]

Directly after the publication of the FDE impact statement, the Florida Waterways Association, led by Orlando power broker and former Canal Authority chairman Bill McCree, developed a point-by-point refutation for public dissemination. Though some contentions were scientifically valid (such as debunking FDE's rather dubious assertion that the canal's locks and dams would increase the risk of earthquakes in Florida), much of the document relied on castigating FDE for telling "only part of the truth." Its attack on the state Game and Fresh Water Fish Commission's report criticizing the canal proved even weaker. Concluding "we are at a loss to know why such an important commission of State Government should change its mind in such a short space of time," the report made little effort to contradict claims. If the object was to debate the scientific validity of FDE's claims, the document failed miserably. Garnering only a small portion of the publicity of the impact statement, the refutation made little impact in the ongoing debate.[63]

Canal Authority manager Giles Evans also endeavored to defend his bailiwick from the FDE interlopers. Both publicly and privately, Evans railed at his critics, castigating them as "self-styled ecological experts." As in the case of the Waterways Association, the responses to FDE's attacks may have done more harm than good. Certainly verifying David Anthony's later recollection that "Colonel Evans was really our best ally,"

Evans attacked scientific concerns with a dismissive wave of the hand. Testifying at a Florida State Senate hearing on the canal in March 1970, Evans authoritatively asserted that FDE's claim of the "purported 'ruin' of the Oklawaha River is a distorted fantasy." Concluding that "NO single agency has done more for the overall cause of conservation" than the Corps of Engineers, he called for "all possible speed in completing this project without delay" (emphasis in original). The Tallahassee audience seemed unimpressed with Evans's performance, suggesting that many state officials now favored stopping canal construction.[64]

If Evans's public testimony proved less than stellar, his letter-writing campaign seeking allies beyond the Florida's borders completely backfired. In April 1970, Evans seemed so desperate to ward off FDE's assault that he wrote to, of all places, the Dyersburg Tennessee Chamber of Commerce, asking for information on recreational fishing in shallow lakes similar to the Rodman Pool. Denouncing canal critics for their bad science and intemperate accusations, Evans assumed the Chamber would confirm his belief that the project's impoundments would one day be idyllic fishing holes. Seeking confirmation, he must have been stunned with the Chamber's response. Instead of extolling the virtues of Reelfoot Lake, a managed reservoir in northwest Tennessee, they cataloged decades of destruction wrought by efforts to "improve" what was once an ecologically balanced environment: "Reelfoot Lake appears to be a perfect example of the fact that man is still not able to outsmart nature; and our efforts to regulate and improve on nature by building dams, canals, and artificial impoundments merely add to our ecological woes, rather than alleviating them." Sounding not at all like local economic boosters, the Chamber continued in a vein that would have been unimaginable only five years earlier: "Like so many other areas of nature, man's efforts to improve his economic lot appears to have brought ruin to a great natural phenomenon, and the economic benefits, if any, have helped only a few people to go to Memphis and buy a new Cadillac every year." "It is our firm opinion," they concluded, "that the construction of the Cross-Florida Canal is simply another effort of vested interests to realize economic benefits while the remainder of the population, upon whom they call upon to pay for it, suffer from results of ecocide." The letter closed with, "I hope this gives you the information you need." Even worse for Evans, the Chamber forwarded his letter and its response to the National Audubon Society. They warned that "the [Canal] Authority is apparently trying to use groups outside of Florida to demonstrate that there are no ecological consequences to a canal such as they propose." Within two weeks, headlines around the state exposed Evans's awkward effort to shape the canal debate. Amid the insular world of military bases and corporate boardrooms, Evans, an astute bureaucratic player, had few rivals. However, his gruff military bearing no longer played well in a rapidly changing world. The future now belonged to people like Marjorie Carr and Bill Partington, not Giles Evans and Tom Adams.[65]

Blithely unaware of the shifting tides, or perhaps unwilling to concede, canal boosters remained positive about the future of their project. Always looking for opportunities to publicize even the most modest of gains, they eagerly promoted the passage of one of the canal's first commercial shipments through the Inglis Lock. In late June, the Dixie Lime and Stone Company of Ocala loaded 6,000 tons of dolomite, a mineral used for fertilizer, on a 350-foot barge designed for overseas trade. Confident that this event marked the beginning of viable commercial traffic on the canal, the Corps hired a public relations firm to issue news releases and photograph the auspicious event. On July 1, the William Cook Advertising Agency issued a laudatory statement declaring the shipment "couldn't have gone smoother." But far from a public-relations coup, the shipment turned into what one newspaper deemed "a comedy of errors." Too large for the canal, the *Aiple* could not sufficiently navigate either the waterway itself or Inglis Lock. It remained stuck crossways on the canal's bottom just west of the lock through three successive tidal changes. A deep draft, high winds, and snapped lines exacerbated the problem. To make matters worse, the barge was loaded only to half capacity, leaving critics to ask, "What happens when we're carrying something big?"[66]

The incident alone was embarrassing enough. However, through sheer coincidence, the *Orlando Sentinel* was simultaneously holding a meeting to reevaluate its editorial support for the canal. Representatives from both the Corps and FDE gathered in the paper's downtown office to make their case. An emergency phone call for David Anthony interrupted the proceedings. Bill Partington was on the other end of the line, full of excitement with news that the *Aiple* had run aground. What made this news all the more devastating was that no one in the boardroom knew about it. As Corps officials gloated over their apparent success, Partington informed Anthony that he was standing next to a reporter who had witnessed the entire debacle. Anthony confidently returned to the room and dropped a bombshell on the unsuspecting audience. "I can tell you where your barge is," he recounted years later, "it is not on the high seas with a mythical load headed to a mythical port somewhere at sea. It is stuck hard aground and has been there through three changes of tide at the mouth of the Withlacoochee River." Corps officials and boosters answered with dead silence. Momentarily stunned by Anthony's disclosure, they left the room in a huff. Out in the hallway, one turned to another and angrily whispered, "find out about that fuckin' barge."[67]

Floridians soon found out about the barge from headlines across the state. Far from the positive tone of earlier canal articles, press coverage of the incident revealed a broader skepticism about the project. The lead from the *Gainesville Sun* gleefully mentioned that the barge had run aground on the same day the Corps' new public-relations firm, hired for a fee of $1,600 a month, heralded the event. When Giles Evans asserted that the barge "didn't get stuck in [the] canal," the *Tampa Tribune* mockingly suggested that it must have been "resting between tides." The editors of the *St. Petersburg Independent* went several steps further, deeming the incident a "comic hangup" and a "real-life fiasco." According to the *Independent*, "the episode was pregnant with possibilities for a Keystone Kops–like film, but it also drove further wedges into the ever-widening credibility gap that belies the feasibility of the $220 million canal." Though the *Aiple* seemed too large for the canal, it was still a small barge by modern standards; "in fact, she was obsolete in World War II." Perhaps more damaging, to successfully negotiate the canal such craft had to navigate half-loaded to avoid running aground on the shallow bottom. This undercut any claim for efficient transportation. Citing FDE's charge that the locks were already "obsolete and inadequate," the editors, seeking balance, concluded with an observation from Florida Canal Authority chairman L. C. Ringhaver, who called the shipment "the forerunner of things to come." Unwilling to give Ringhaver the last word, the *Independent* added sardonically, "that's what worries us."[68]

The incident resonated throughout the state for over a month. Seemingly oblivious to the controversy, Evans simply noted in his manager's report that "Dixie Lime and Stone successfully initiated two shipments of dolomite from Citrus County to South America." The last word on the affair, however, came in a letter to the *Tampa Tribune* on August 26, 1970. The letter writer condemned Evans for using "canal propaganda financed by taxpayers money," and concluded that he could not "understand a man of his background stooping to such tactics. It casts a pall of doubt over the entire canal operation."[69]

Despite all the setbacks, canal proponents steadfastly refused to believe they were losing the public relations battle. They took it upon themselves to go on the offensive and attack their opponents with heated rhetoric and ad hominem arguments. Not coincidentally, three of the most vociferous onslaughts came in April 1970 as the nation, recognizing the growing importance of the ecological imperative, stood poised to celebrate the first Earth Day. Canal Authority chairman L. C. Ringhaver stridently dismissed FDE as "a clamorous gaggle of self-styled experts [who have] seized upon this . . . opportunity to assault the project on every conceivable grounds—whether factual or imaginary." While castigating canal critics, "Ring" lauded the Corps' record on the environment. Capitalizing on the growing interest in Earth Day, he declared that "no single agency has done more for the overall cause of conservation." Authority manager Evans continued the hot talk before the Melbourne Kiwanis Club, an audience of like-minded businessmen. Dismissing FDE members as "emotionally disturbed people," he denounced them as uninterested in the welfare of the people of Florida. "Look at the number of people coming to Florida," he charged. "We must accommodate all these people, not just a few people looking at wild animals." That same month, the trade paper for the nation's barge industry, the *Waterways Journal*, joined the fray. In an April 4 editorial entitled "Ecological Hysteria," the publication accused FDE members of "using the current hysteria for their own selfish ends." In contrast, it praised the Corps for having "taken the lead for a number of years in urging the control of water pollution." Again invoking the image of Earth Day, the editorial claimed that "instead of destroying an ecological balance, the [canal] will provide valuable ecological benefits." Two weeks later, a second editorial continued the attack. Vilifying ecologists as radicals across a broad political spectrum, "from the John Birchers to the communists, and from 'Ramparts' magazine to *Reader's Digest*," the piece asked its readers to consider the Corps as the true environmentalists. Asserting that the Corps had "done more to . . . protect the ecology than any other single agency in the country," it plaintively questioned why "none of these positive steps to preserve the ecology attracted the notice of those who attack the government."[70]

Four years earlier, opponents had requested the rerouting of the canal directly east from Silver Springs, connecting with the St. Johns River at Lake George, effectively saving the Ocklawaha. State officials rejected the suggestion out of hand at the infamous January 1966 Tallahassee showdown. In the spring of 1970, as if to illustrate how dramatically things had changed, Corps official Robert Jordan proposed another alternate route designed to protect thirteen thousand acres of the river and its floodplain. It would cost an extra $3.5 million dollars to route the canal parallel to the river from the still-to-be-built Dosh Lock at Silver Springs to the Eureka Dam. The river and its floodplain would be divided from the canal by a levee and "would permit flexible operation of the both Oklawaha River and the barge canal as separate systems." The Corps' new plan would allow for the canal's completion while saving most of the river. The devastation of Rodman Pool, however, would remain. By the time the compromise measure had reached the White House for serious deliberation, any consideration of alternative routes was beyond question. The canal would stand or fall on the original design to cut through the Ocklawaha Valley.[71]

Room for compromise increasingly diminished as FDE and its allies continued the drumbeat of criticism. Building on their successes throughout the rest of 1970, they kept the pressure up on the press, politicians, and the courts to stop the canal. Particularly effective were dozens of editorial cartoons throughout the state that lampooned the efforts of the Corps and the Canal Authority. Drawn by such gifted cartoonists as Don Addis of the *St. Petersburg Independent* and Sam Rawls (Scrawls) of the *Palm Beach Post*, these cartoons reduced state and federal authorities to mere caricatures for supporting a project that was useless, wasteful, and destructive. While these may have proved influential in shaping public opinion, shifting editorial policies among the state's important newspapers reflected a sea change in perceptions of the canal. In July, for example, the *Gainesville Sun* lambasted canal construction in a piece entitled "The Big Ditch." Assailing the project on the usual economic and environmental grounds, the editorial writers went further this time to accuse both the Corps and the state Canal Authority of secrecy and duplicity in their canal dealings. Taking aim at recent announcements that the Corps and canal boosters were increasing their public-relations expenditures, they pointedly asked, "if the canal is so good, why does it need all that promoting?" Less than a week later, a headline in the formerly procanal *Orlando Sentinel* announced, "Let's Decide If Canal Needed Before Debating a New Route." Echoing John Couse, the editorial staff focused on the economic viability of the canal. "Economic questions must be answered first," they asserted, "for if the canal won't pay its own way by the time it opens in 1977, the project must be abandoned and therefore a debate over ecology is pointless."[72]

Florida's changing political climate proved even more crucial than newspaper cartoons and editorials, for 1970 marked a watershed year in state politics. With reapportionment, the retirement of longtime senator—and longtime canal supporter—Spessard Holland, and a governor's race that featured maverick incumbent Republican Claude Kirk, the potential for a massive realignment in the state's political climate seemed very likely. The canal became an important, though not pivotal, issue during the campaign season. Yet, political observers considered the issue portentous as, for the first time, some statewide Democratic candidates opposed the project. In August, the *St. Petersburg Times* reported that "the coming election could be the most important for the canal since Andrew Jackson first proposed it 130 years ago."[73]

In the governor's race, Kirk rarely mentioned environmental issues generally or the canal in particular. Lyman Rogers, the chair of Kirk's Natural Resources Committee, expressed the feelings of his boss in a July letter to Richard Nixon: "I have held mixed emotions regarding the building of the Florida Cross State Barge Canal." Yet Rogers, and certainly Kirk, understood the political implications of completing construction. "The citizens of Florida," he continued, "are more aroused than our congressional delegation realize. . . . I plead with you to use the office of the presidency to support the position of your Secretary of the Interior [who, of course, called for a moratorium on construction]." In August, Kirk issued a press release favoring a rerouting of the project away from the Ocklawaha, much like the idea of Marjorie Carr in 1966. When Kirk revealed his opposition to the canal's route through the Ocklawaha Valley, his major opponents in the Republican primary, Skip Bafalis and Jack Eckerd, went a step further to call for abandoning the project entirely. Upon his primary victory, Kirk then faced Democratic opponent

Reubin Askew in the general election. Though originally from Pensacola, Askew's policies seemed more in line with liberal Democrats from Miami and south Florida than the Pork Choppers traditionally associated with his region. In a strange but politically astute move, the party chose Tom Adams to run as Askew's lieutenant governor. Apparently, Askew saw no problem in running alongside the canal's most ardent proponent, especially considering that he himself favored Hickel's proposal for a moratorium. Responding to an FDE questionnaire on the issue, he promised that, if elected, he would "assert every available means at my disposal to affect a rerouting of those portions of the canal necessary to preserve the Oklawaha Valley." Moreover, Askew added that in the future he would "expect to rely on leaders from recognized conservation organizations such as the Florida Defenders of the Environment." For Evans, Ringhaver, and other ardent canal supporters, this proved ominous news indeed.[74]

The Senate race to fill the seat of retiring canal advocate Spessard Holland featured Democrat "Walkin'" Lawton Chiles and Bill Cramer, Florida's first Republican congressman since Reconstruction. Both candidates overcame crowded primary fields to achieve their party's nomination. Chiles defeated canal proponent and former governor Farris Bryant, emphasizing, among other issues, his antagonism to Bryant's unwavering support for the project. Though most political observers saw Chiles as a vigorous opponent of the canal, his position only seemed that way in comparison to Bryant's. Instead, he played it safe by simultaneously encouraging "continuing construction . . . pending completion of Hickel studies," recommending an "alternative route," and recognizing the importance of "non-Corps ecological studies when construction continues." Showing the complicated convergence of multiple issues, Chiles gladly accepted a strong endorsement from Holland, in spite of his halfhearted support for the canal. At the same time, he prominently highlighted his backing of Hickel's moratorium in ads that associated his campaign with law and order, fiscal responsibility, and opposition to forced busing. Of the three major Republican candidates for the seat, only Cramer failed to take a stand on the canal issue. Both of his opponents favored either abandoning the project or halting it pending further ecological review. Thus, while pundits began the political season with claims that environmental issues would be crucial in the 1970 elections, Cramer's primary victory suggested that concerns over school desegregation and busing, as well as the Vietnam conflict, held more importance to voters.[75]

Still, the Florida Defenders of the Environment considered the 1970 campaign season as a momentous opportunity in their struggle to stop the canal. In a September 4 press release marking the beginning of the fall campaign, the organization trumpeted its newfound political muscle. Proclaiming that "the Cross-Florida Barge Canal will never be finished if Florida politicians keep their promises to the public," FDE revealed the results of a lengthy survey it submitted to politicians from across the state. Of the 123 candidates for statewide and federal office who returned their questionnaires, fully 81 percent favored either Hickel's moratorium or completely stopping the canal. As if to emphasize how much FDE had altered the terms of debate, even Congressman Charles Bennett, a stalwart supporter of canal construction, seemed to back away from his previous positions. While holding that "the present plan, in my opinion, will enhance, not destroy, the environment" he also conceded that "if another plan be found, I'd be for it 100%." Using the tabulated results, Bill Partington ratcheted up the political pressure on Richard Nixon. "In view of this unprecedented expression of opposition to continuing the Cross-Florida Barge Canal," he wrote the president, "we respectfully call on you to implement a moratorium on construction." Knowing his audience, he concluded by tying his request to issues dear to the president's heart: "we cannot equate the continuation of this project with your administration's avowed concern for fiscal responsibility and for conservation."[76]

The fall campaign continued to buoy FDE's optimism. As the election neared, Kirk increasingly distanced himself from any support for the project. Campaigning with Nixon in mid-October, Kirk made it clear where he stood on the issue. Facing pickets protesting canal completion at a rally in Clearwater, Nixon asked Kirk pointedly: "They want to stop a canal. Are you building it?" Kirk answered, "No, you are." Reported in papers throughout the state, the incident again vividly revealed FDE's growing influence. Kirk's checkered gubernatorial record and increasingly strange personal antics steered many away from supporting his reelection. But his strong environmental commitment, particularly in appointing Nat Reed and Lyman Rogers to positions of importance in his administration, convinced many canal opponents to hold their nose and support Kirk anyway. Askew seemed to oppose the project, but the mere fact that Tom Adams also ran on the Democratic ticket gave pause to those who would oppose Kirk. Both Askew and Chiles eventually won impressive victories over a GOP in disarray. And while FDE and other canal opponents may have preferred Kirk as governor, neither Askew nor Chiles seemed hell-bent on canal completion. In the end, concerns over the barge canal did not seem to be a deciding factor in either race. Yet FDE and other activists saw hope in that they did not lose ground. More importantly, as long as the issue remained in the press, their position seemed to gain more legitimacy among policymakers and the general public. As a headline in the *Tampa Tribune* announced, "Canal Foes Believe Election Strengthened Their Hand."[77]

As FDE garnered rather small political successes, it continued its legal assault on the canal. The organization's leadership considered the courtroom crucial because their political struggle—regardless of the increasing public support for stopping the project—could always be stonewalled by the entrenched power of politicians like Tom Adams and Charles Bennett. Yet, this avenue of attack proved to be problematic. Victor Yannacone's ugly departure from EDF in the final months of 1969 left the September lawsuit in a tangle. Always more showman than scholar, Yannacone had crafted his suit on rather shaky legal grounds. While both EDF and FDE seemed genuinely pleased at his departure, Yannacone's advocacy for the persistent and strident use of the media established an important legacy in FDE's battle against the canal. In early 1970, George Kaufmann, EDF's replacement counsel, who was as diffident as Yannacone was boisterous, began working on an amended suit that promised to "be much better than the one that was originally filed." Brought to court in Washington, D.C., in February, this first amended complaint added both FDE and directly affected individuals, including John Couse, as plaintiffs to the action. Within a month of the new filing however, Kaufmann too had been replaced; this time by Jon R. "Rick" Brown. The third time proved to be the charm, as Brown's legal expertise and winning personality endeared him to the FDE leadership. He immediately filed a second amended complaint, significantly strengthening Kaufmann's arguments and continuing FDE's and Couse's participation in the suit. Key to Brown's strategy was the use of the recently passed National Environmental Policy Act (NEPA) legislation to claim relief from the Corps' actions in building the canal. In the suit, Brown argued that the Corps had "refused even to consider seriously the suggestion that the canal construction be held in abeyance to consider . . . mandates of the NEPA." Using this avenue, as well as the by-now usual anticanal complaints, he asked the court to enjoin the Corps from continuing canal construction until it complied with NEPA and to direct it to immediately drain Rodman Pool.[78]

Judicial enforcement of NEPA held considerable promise for environmental protection. Yet its implementation depended on the whims and willingness of government officials and political appointees. For much of 1970, Walter Hickel, John Whitaker, and Russell Train struggled with the canal issue within the Nixon administration. Hickel, of course, took the lead with his request to halt construction pending further environmental review. Succeeding only in exacerbating an already tense situation, Hickel's decision resolved nothing. As Claude Kirk complained in a letter to Whitaker, "the current controversy over the Cross State Barge Canal is like a cancer gnawing at the citizens of my state." By the summer, as Hickel's ham-fisted politics relegated him to the margins of the administration, it became apparent that Whitaker and Train would play a decisive role in determining a solution. At that point, Carr and FDE had already made their case. And all of their efforts were about to pay off. In confronting the big budgets, political clout, and institutional support of the Army Corps of Engineers and local boosters, canal opponents had to simultaneously be disciplined and nimble. Still, for all their tenacity and abiding faith in the persuasive power of scientific evidence, the success of anticanal movement depended as much on politics as science.[79]

For the last half of the year, John Whitaker wrestled for a solution that would split the difference between politics and science. As the administration's chief environmental aide, he was well aware of the rare and exceedingly narrow window of opportunity to press for environmental issues. As an astute political insider, he also understood the importance of political calculation in any White House decision. For a brief moment, self-interest and principle conveniently overlapped. Much like the Miami Jetport a year earlier, this moment pushed the canal to the forefront of environmental issues. And it gave activists like Carr and Partington a real voice in government decision-making. Increasingly aware of the political implications of the canal in Florida, as well as the scientific evidence warning of its devastating impact on the Ocklawaha, Whitaker slowly and cautiously moved the administration toward abandoning the project.

In a series of memorandums to White House staff from July to December, Whitaker weighed the consequences of the possible alternatives to resolving the canal issue. As if to emphasize FDE's growing influence, continuing the project as originally planned was never seriously considered. In July, Whitaker wrote a memorandum to John Ehrlichman that laid out three simple options: "Kill it. Study it and stop construction while we study—what Hickel proposed. Engineer and construct a bypass around the environmentally fragile and beautiful area." At that point, he favored the idea of a bypass because it would mollify both sides. Under the plan suggested by the Corps and Army Undersecretary Robert Jordan, the canal could be completed with much of the Ocklawaha intact. By doing so, the administration could score a public-relations coup and "be visible on a conservation issue." With a vigilant eye on the fiscal constraints taking shape amid a faltering economy, Whitaker added, "if the budget problem is really rough I could shift to option one, 'kill it,' very easily." In short, he remained relatively detached from the issue. As with the president, politics drove much of the decision-making process. Indeed, Whitaker even suggested a month later that, regardless of the outcome, the Corps should continue subcontracting construction on the western region of the canal, far from the Ocklawaha, through the upcoming electoral season. This would eliminate concern that the administration was "deliberately stalling these contracts and thereby sharpening the whole issue of the canal in Florida as the primary election comes to a head." Mindful that any decision should not jeopardize partisan gains in the Sunshine State, Whitaker recommended delaying action on the canal until after the November elections.[80]

On Wednesday, November 4, the day after both Republican candidates Cramer and Kirk went down in defeat, Whitaker resumed the drumbeat on the canal issue. Still holding to the bypass as a serious solution, he now asserted that Nixon should make maximum use of his choice for an alternative route and "go look at the Oklawaha while he is in Florida next week and announce the decision." For Whitaker, Nixon had to keep a promise made before Claude Kirk and the people of St. Petersburg in a campaign stop less than a week beforehand. Responding to Kirk's accusation that the barge canal was Nixon's problem, the president had vigorously asserted that "any program by the federal government . . . whether it is a barge canal or anything else, will not go into a place unless we are given assurance that it is not going to affect the environment." This short remark, no doubt made off the cuff in the middle of a campaign rally, proved critical in determining the fate of the canal. Afterward, Nixon staffers reached a consensus that the president was now publicly tied to the issue. By mid-November, Donald Walden, a Water Resources staffer, reported that administration officials had decided to "draft a presidential message for the president's consideration that would announce that an alternative route would be used to save the river." Whitaker and other staffers seemed wedded to the compromise course. Yet they did not count on the intervention of Russell Train and his Council of Environmental Quality, which encouraged Whitaker and Nixon to reject a compromise and abandon the canal completely.[81]

On December 1, 1970, Train drafted a memorandum explicitly stating, "I recommend termination of the Cross-Florida Barge Canal." Such a dramatic gesture offered Nixon the best of all possible worlds. According to Train, the "termination of the project would bring maximum political benefits, would prevent potentially significant environmental problems and would save a great deal of federal money for a marginal project." Outright cancellation would yield far more than safely splitting the difference with an alternate route. As Train put it, "the benefits nationally of dropping the project would greatly outweigh the benefits of continuing it. I further believe that a bypass over only twenty miles of the project would be considered 'tokenism' by conservationists and many others." Train supported his decision with a twenty-four-page report prepared by the Council on Environmental Quality. The report virtually mirrored FDE's environmental impact statement. Indeed, in many respects it was a summation of nearly a decade of scientific labor to save the Ocklawaha. It was all there. Stopping the canal would protect sportfishing and preserve wild-turkey habitat. It would retain a canopied free-flowing river. Otherwise, with the river's flow obstructed by dams and locks, "the entire ecosystem is drastically disrupted, producing a warm water, highly enriched, unshaded, shallow water course, with little or no flow." The result, of course, would be the ubiquitous presence of algae, water hyacinths, and hydrilla. And regardless of the scientific debate, Train concluded, "the project is marginal from an economic point of view and hence very undesirable."[82]

Such a report, coming with authority from the likes of Train, forced Whitaker to rethink his position. In a new memorandum dated December 23, Whitaker advised Nixon to "stop the canal completely." Addressing the issue in explicitly political terms, he reminded the president that "we have a situation where you have committed yourself to environmental concern over the Cross Florida Barge Canal. Your former Secretary of the Interior has publicly urged that the canal be stopped and yet there has been no public action to actually alter plans for canal construction." Warning that Nixon "will have a serious credibility problem" if canal construction continues, Whitaker insisted that either realignment or halting the project were the only viable options. National political considerations, however, now made the former impossible. "It is time for the conservationists," Whitaker continued, "to win a clean cut symbolic victory like stopping the barge canal." This was especially important in light of "the ongoing SST battle and an announcement next month that we are proceeding on the Alaska Pipeline (both symbolic anti-environmental moves by the Administration)." Whitaker closed with a recommendation that the administration take full advantage of the decision and turn it into an elaborate media event. Nixon could "make the announcement at the beautiful Ocklawaha River (when returning from Key Biscayne) with Morton, Train, and Ruckelshaus and if the time is right, swear Morton in as Secretary of Interior on the site."[83]

To seal the deal, Whitaker attached three potential draft announcements for the president. The first draft emphasized the political and scientific dimensions of the issue. Writing for the president, it began with an anecdote: "When I was in Florida recently, I was greeted by a number of persons carrying placards opposing the Cross Florida Barge Canal. I promised that, upon my return to Washington, I would get the facts concerning this project. . . . I now have the facts. And I can see why I was greeted with placards." Here was the ultimate vindication of Marjorie Carr's "get the facts" strategy. The second possible draft went a step further and validated Carr's broader environmental vision. It had Nixon calling for the end of canal construction as part of a broader reevaluation of government water-resource projects. "A similar project," Whitaker wrote, "could not today be developed with such indifference to environmental values. But it is not enough to deplore the shortsighted-ness of our government and our society in the past. Something must be done—and can be done—to save the Oklawaha, despite the fact that nearly $50 million has already been committed for construction of the canal." The final possible announcement, and the one that most closely reflected Nixon's brief public statement, approached the issue as directly as possible. To protect and preserve the Ocklawaha River, canal construc-tion would end immediately.[84]

Though it would have made for a nice Christmas present for the Flor-ida Defenders of the Environment, Nixon's decision had to wait until the end of the holiday season. On January 9, 1971, John Ehrlichman in-formed Whitaker that the president had approved his recommendation, but would not "make the announcement from the beautiful Oklawaha River." Instead, ten days later, following a preliminary injunction issued by the federal courts, Nixon issued his statement with little fanfare. While the *New York Times* gave it considerable coverage, the *Washington Post* buried the news in an article on the formation of a new foreign pol-icy council. It was easy to do so for the press release was short and to the point: "I am today ordering a halt to further construction of the Cross Florida Barge Canal to prevent potentially serious environmental dam-age. . . . A natural treasure is involved in the case of the Barge Canal—the Oklawaha River—a uniquely beautiful, semitropical stream, one of a very few of its kind in the United States, which would be destroyed by construction of the canal." After acknowledging the CEQ's role in deter-mining his decision, Nixon declared, "the step I have taken today will prevent a past mistake from causing permanent damage." In addition to rectifying a gross injustice, Nixon implicitly tied his order to future gov-ernmental action: "We must take timely account of the environmental impact of such projects—so that instead of merely halting the damage, we prevent it."[85]

Little did anyone outside the corridors of power know how close the president later came to changing his mind. After releasing the state-ment, Nixon traveled to south Florida to visit his friend and confidant Charles "Bebe" Rebozo. While there, Rebozo, having caught hell from his Jacksonville friends and business associates embittered by the deci-sion, pressured the president to reconsider. On the return flight, Nixon told Ehrlichman to reverse course. "When Ehrlichman told me of the

Presidential decision, I did what all good Presidential aides do," Whitaker recalled years later. "I sat on it and did nothing, hoping the storm would blow over." About a week later, on board a helicopter flying to Andrews Air Force Base, Nixon asked for an update on the canal situation. When Ehrlichman and Whitaker informed him that they had yet to act, he agreed to stand by his initial order. "Too bad," Nixon added with an ironic smile, "Jacksonville *was* a great town" (emphasis in original).[86]

For anticanal activists—blissfully unaware that the fate of the canal depended on such a close call—celebration was in order. Years of hard work had seemingly come to a positive conclusion. As the news spread, most of Gainesville's FDE membership gathered at Carr's country house in Micanopy. Highlighting the boisterous party that lasted long into the night was the sight of UF biologist George Cornwell, a bear of a man weighing over three hundred pounds, dancing on Carr's kitchen table—the same table that had long stood next to the overworked Xerox machine so crucial to the victory. In a far corner of the globe, a more surreal recognition of the moment took place. Stephen Carr, Marjorie's second-oldest son, was serving as a forward artillery observer for the United States Army in Cambodia. In the midst of the American incursion into that war-torn land, Carr heard the news of Nixon's decision over Armed Forces Radio. "I was excited for my Mom," he remembered years later, "but nobody else in my unit had any idea why I was so happy. I had to explain to them about the Cross Florida Barge Canal."[87]

Death of a Dream

Issued on one of the coldest days on record for north central Florida, Richard Nixon's decision to cancel the canal elicited a firestorm of protest from both supporters and those working on the canal. Upon receipt of Nixon's edict, the Corps ordered all construction halted immediately. Having received the command, dragline operator Nick Head, working on the western edge near Yankeetown, simply "shut down his rig. He left a bucket of rock hanging high on the boom and went off to get a case of beer and some fresh beef liver to bait his trot lines." By the next day, he had a bucket of fish, "a hangover, and no job." Head's anger was palpable. "If they ain't going to finish the canal," he fumed, "they ought to just cover the damn thing up. Maybe with Nixon in it." Others who were more indirectly affected by canal construction also felt the economic pain of Nixon's decision. "Boots," the proprietor of a Yankeetown bar, conceded that "there are plenty of people around here who are opposed to the canal. But," she railed, "they're the ones who don't have businesses here." For well-established entrepreneurs like William McCree and L. C. Ringhaver, the canal's demise marked a serious setback, but one that could be overcome by other business opportunities. For workers like Nick Head, Nixon's decision was both more personal and devastating: "I'm going home this evening and put a new set of tires on that old car of mine. Tomorrow I'm going to load up the old lady and the two kids and ride for a while. I might go up to New York, where I heard operators draw $9 an hour. . . . But I wouldn't go back to work on that canal even if they started up again. . . . To hell with the government."[1]

Other Floridians similarly blamed the federal government for their disappointment. A front-page headline from the *Palatka Daily News* said it all: "Nixon Nixes Canal Dream." Not content to editorialize on its op-ed pages, the paper instead pronounced its position boldly on the front page. Arguing that "we all stand to lose by Mr. Nixon's rash decision," the paper rallied "every citizen, every club, every official body; to write or wire President Nixon direct, protesting this one-sided official decree and urging that reconsideration be given immediately." Former Canal Authority chairman McCree took the request to heart and wrote the president immediately, arguing that "the order raises some very serious moral and legal questions." One Bradenton attorney was so outraged that he forwarded his correspondence with Nixon to twenty-one government representatives in both Tallahassee and Washington. "I had feelings of disbelief, dismay, and disgust," he complained. "Thousands of us here in Florida who have worked diligently for many years to make this long-dreamed-of project a reality are sick at heart over this latest tragedy." Since Nixon issued his decree without consultation from local or state officials, Canal Authority manager Giles Evans publicly announced that he was "as surprised as everybody else about the order. I'm bouncing around like a yo-yo." In light of the setback, the usually forceful official remarked, "We are lying low and waiting for further developments." In a more formal report to the authority's board on February 8, 1971, Evans admitted that Nixon's decision represented "by far the most serious obstruction imposed" on the canal. As a result, most of Evans's efforts in light of the decision "centered on picking up the pieces and considering how to proceed from here."[2]

In early February, John Eden, a farmer from Inglis, crystallized the sentiments of canal supporters in a full-page open letter to the *Citrus County Chronicle*: "It seems utterly inconceivable that the President of the United States would categorically halt all work on a duly authorized state and federal public works project solely on the emotionally charged advice of misled opponents." An ad placed in the *Florida Times-Union* and sponsored by a Jacksonville barge towing company executive echoed Eden's rage and proffered a political remedy. "Boosters, keep your cool," the ad cautioned. "By 1973 we can have a new president and a new Council on Environmental Quality." Others laid the blame solely on environmental activists. In March, one Silver Springs resident wrote a blistering letter to the Corps, complaining that "these damned ecologists and environmentalists get me so hot." Pointing the finger firmly at an increasingly vocal academic community, he went on to say that "I sincerely believe that 75% of these University of Florida ecologists don't know their ass from a water hole."[3]

Representative Charles Bennett quickly became the point man for the opposition to Nixon's decree. Bennett distilled his disappointment into one simple principle: Nixon "violated the Constitution and ended the canal." To him, Nixon's order was a serious abrogation of the separation of powers at the expense of Congress and the citizens of the state of Florida. Testifying before the Senate Judiciary Committee in March 1971, Bennett asserted that Nixon's decree was "an edict not unlike a Catherine de Medici decision of the sixteenth century." He elaborated with the charge that Nixon had "destroyed the delicate balance between the federal government and the state government by cavalierly breaking a contract between the United States government and the state of Florida, and he also dictatorially repealed an authorized law of Congress by permanently halting the Cross-Florida Barge Canal."[4]

It seemed only fitting that the most vitriolic assault on Nixon's decision would come from the always-combative Lieutenant Governor Tom Adams. In preparing a May 1971 speech for the dedication of Palatka's St. Johns River Barge Port, a facility now rendered essentially obsolete by Nixon's order, he expressed anger at what he considered an outrageous betrayal. Unable to deliver his remarks in person because of a sudden illness, Adams had his staff assistant Jack Brandon read his vituperative attack. "Let me recall for you," he exclaimed, "that five presidents of the United States have supported, have had faith, have pressed ahead with the Cross-Florida Barge Canal. President Nixon broke that chain of faith, that continuity of purpose stretching across five administrations and he did it because it was politically opportune and for no other reason. . . . President Nixon threw the Cross-Florida Barge Canal to the phony ecologists as he would a bone to a pack of hungry dogs. They took it and are now howling for more." For Adams, Nixon's willingness to yield to so-called environmental extremists established a dangerous precedent. "If they take over," he warned, "they will halt [other projects] one after another in a kind of domestic domino continuity, where if one falls they all fall."[5]

Environmentalists were indeed howling—with the joy of success. All of their hard work appeared to have paid off, and they were more than happy to show their appreciation for Richard Nixon. The Tropical Audubon Society, speaking on behalf of the south Florida environmental community, wired Claude Pepper to announce they were "gratified that the barge canal controversy has been decided by the President." Tom Pankowski, on behalf of the Izaak Walton League, congratulated Nixon directly for "his decision to stop further construction." The league encouraged the president to "hold firm in your resolve and to resist ill-advised pressures to reconsider your decision." Even the forty-thousand-member Oklahoma Wildlife Federation (OWF) chimed in. Enthusiastically endorsing "your halt in construction," they cautioned Nixon that "real estate and speculative interests will apply all possible pressures to reverse this decision. The OWF will stand behind you and help in any way possible." And the popular radio and television personality Arthur Godfrey, an early and vocal supporter of FDE, took time from his busy schedule to inform the White House "how ecstatically happy we all were when the President so courageously stepped in and issued the executive order permanently halting the Cross Florida Barge Canal."[6]

By February 24, 1971, Nixon had received 5,010 letters, cards, and telegrams on the canal. By one staffer's count, it was the third-largest response to any issue since the president's inauguration. To the chagrin of those like Tom Adams, convinced that Nixon was out of step with public opinion, fully 81 percent of the correspondence favored the decision. One White House aide examined the reactions of Nixon's critics and concluded that of the 967 opposition letters, an overwhelming majority "seemed to have been inspired by Jacksonville groups, and the others include messages from citizens of towns on or near construction sites who object to the loss of the construction payroll." Recognizing the broader policy implications of the issue, the staffer argued that these "letters suggest that the environment . . . has clearly emerged as a major national issue. While this one issue does not make the case, it does imply that there is a possibility for 'trade-off' in environmental issues and that favorable Presidential decisions on conservation questions could help offset foreseeably unfavorable reaction to difficult but necessary decisions in other areas." The analysis led John Whitaker, Nixon's chief environmental aide, to commend the president: "I think you made the right decision . . . . To continue the canal would have given you real problems."[7]

Newspapers and magazines from around the state and even the nation almost universally supported Nixon's order. The *Miami Herald* called the decision a "timely and proper move." The *Stuart News* went a step further and exclaimed, "Nixon hit the jackpot, engineers should hear those cheers." The decree seemed so popular that the *Orlando Sentinel* declared, "Nixon could capture Florida in '72 by saving the Oklawaha." Echoing the sentiments of most Florida media outlets, the *New York Times* hailed the "Death of a Boondoggle." The *Washington Post*, no friend of the president even before the Watergate revelations, commended Nixon for his "courage . . . in calling off the bulldozers." *Time* magazine joyously announced the end of the barge canal. Its competitor *Newsweek*, recalling the voice of Arthur Vandenberg, deemed the decision a "victory for economic sanity."[8]

Such high praise no doubt pleased Nixon, who considered his relationship with the press at best lukewarm. Almost universally, the presidential order to halt the canal seemed like a solid political calculation. Thus in many respects Tom Adams was on target with his accusation that the move was more about politics than environmentalism. This observation was maybe the only thing that two people like Tom Adams and Nathaniel Reed could agree upon. When asked years later why Nixon issued the January order, Reed replied with only one word, "Democrats." He continued: "it had nothing to do with ecology. It was uproarious. . . . He [Nixon] said, 'Damnable Democrats, this Jack Kennedy gave to Florida, this is Claude Pepper's canal, this is George Smathers' canal—we got them guys, we got them.'" When pressed whether politics had really been the only consideration, Reed responded, "Yes, I am sorry to tell you." To Nixon, the order represented retribution for years of New Deal Democratic dominance. For decades, Republicans who had long been out of power resented the Democratic Party for relying on the largesse of the state, through such ambitious projects as the Tennessee Valley Authority and the barge canal itself, to garner public support and maintain political dominance. From that perspective, killing a public-works project so associated not only with Franklin D. Roosevelt, but also with the man who defeated Nixon in the 1960 presidential election, must have been the sweetest form of revenge.[9]

For staffers like John Whitaker and Russell Train, stopping the canal signified something more than mere political spite as it had the potential to establish an emerging environmental vision within the Nixon White House. Almost immediately, Nixon would test that assumption with his response to another southern waterways proposal—the remarkably similar Tennessee-Tombigbee Waterway. More than 230 miles long and involving five dams and ten locks, this massive canal—the largest single project ever undertaken by the Corps of Engineers—joined the Tennessee and Tombigbee rivers and allowed Mobile, Alabama, to become a major rival of the port of New Orleans. Though authorized in 1946, Congress had not allocated one cent for construction until 1971. Many in Alabama saw the "Tenn-Tom," as the project quickly became known, as an avenue to rapid economic growth in a region beset by poverty and economic underdevelopment. At the same time, environmentalists considered the project a potential ecological disaster. In May, activists wired Richard Nixon, asserting that "we do not believe sufficient evidence has been presented to assure the public this project can be undertaken without the risk of doing serious damage to the environment. The exemplary position you took in regard to the Cross-Florida Barge Canal is firm evidence you have zealous regard for our natural resources." Russell Train also associated the two projects. He concurred that "the President has received broad support across the country for his decision halting further construction of the Cross-Florida Barge Canal," and warned that most of that "support he could quickly lose by any public display on behalf of the Tennessee-Tombigbee."[10]

Alabama was not Florida, however, and the political dynamics behind the Tenn-Tom were dramatically different. While a pro-environmental stance could garner votes in Florida, the opposite applied to Alabama. Moreover, Nixon also had to contend with the state's overwhelmingly popular former governor, George C. Wallace, who was considering another run for the White House in 1972. What better way to convince Alabamans to vote for the current occupant than to throw money at them? Therefore, on May 25, 1971, Richard Nixon traveled to Mobile and gladly presided over the dedication of a project that was at least as destructive as the Cross Florida Barge Canal, if not more. Critics railed against the blatant inconsistency between halting one project and then, only months later, wholeheartedly endorsing another. Administration officials countered, however, that the two cases were completely different. The groundbreaking for the Sunshine State's project began in a world before NEPA and its obligatory environmental impact statements (EIS). Moreover, Florida's canal threatened a uniquely wild river, one of the few remaining in the United States. The Tenn-Tom was beyond comparison, especially as the Army Corps of Engineers submitted an EIS that fully weighed the cost of construction against modern environmental values. That the president grabbed the chance to personally deliver the keynote speech vividly underscored how the window of opportunity for an expansive environmental policy was quickly closing. For Nixon, as one observer bitterly noted in 1972, "the politics of the environment had been considered, found less significant than the politics of the pork barrel, and rejected." Within the White House, John Whitaker clearly recognized the environmental moment had passed. Whitaker viewed Nixon's appearance at the groundbreaking as a "black day for the environment," a stunning rejection of principled policymaking that embraced a greener world. On May 18, a despondent Whitaker, who clearly felt the influence of people like himself and Russell Train had begun to wane, wrote John Ehrlichman. He advised that the administration should best avoid, at least for the time being, discussions of environmental issues as "the President is digging his ditch in Mobile and fouling up the environment."[11]

While the Nixon administration wrestled with the political implications of halting the canal, a larger struggle loomed within the judicial system. On January 15, 1971, four days before Nixon's announcement, the Federal District Court in Washington, D.C., issued a preliminary injunction halting canal construction. Ruling on the suit brought by EDF against the Corps of Engineers, Judge Barrington Parker, a "cantankerous" jurist appointed by Nixon in 1969, concluded that EDF's attorneys had made "a strong showing" in their contention that the project would irrevocably damage Florida's environment. Denying the Corps' motion that the court summarily dismiss EDF's claims on the grounds of sovereign immunity, Parker announced that "the Plaintiffs [EDF and FDE] would suffer irreparable injury for which there is no adequate remedy at law by the continued construction of the Cross-Florida Barge Canal project." Additionally, he determined that the Corps had "not met all the requirements of certain of the statutes relied upon by the Plaintiffs, particularly the National Environmental Policy Act of 1969 [NEPA]." As a result of these findings, Parker ordered the Corps, "their agents, officers, servants, employees, and attorneys," not to engage in "clearing, cutting, damaging, dredging" along the Ocklawaha River, as well as "excavating any portion of the Cross-Florida Barge Canal." On a final note, Parker raised the question of Rodman Reservoir, which would become the most contentious and enduring of all the issues concerning the canal. Announcing that the court would retain jurisdiction over Rodman's future, Parker informed both EDF and the Corps that they needed to report back to the court "with respect to whether they have been able to agree to plans and procedures for consultation regarding the future of Rodman Reservoir." Issued in its entirety on January 27, Parker's edict, in tandem with Nixon's decision, brought canal construction to a standstill.[12]

Canal opponents were obviously ecstatic over Parker's ruling. EDF chairman Dennis Puleston wrote Marjorie Carr to offer his "hearty congratulations for a well-deserved, hard-earned victory." He especially acknowledged the importance of Carr and her allies. "It was FDE," he conceded, "with its many dedicated scientists and conservationists, who are primarily responsible for the outcome. Without the sound science that was assembled by you to rebut the baseless claims of the Corps," the case could not have succeeded. For EDF, Parker's decision marked the end of the struggle. "It has been an unusual privilege to have been associated with FDE," Puleston wrote as EDF prepared for another fight, this time with a "case against TVA." William Partington also closed his active participation in FDE's anticanal struggle. Though he would remain a major player in Florida environmental politics, his official hands-on involvement in the canal issue would be over. A long chapter had come to an end. Now Marjorie Carr, more than anyone, would move even further to the forefront of a struggle that increasingly centered on the remains of canal construction, notably the future of Rodman Reservoir.[13]

Despite the twin blows from the courts and the White House, canal boosters refused to admit defeat. Some followed the well-worn path of Giles Evans and Tom Adams and remained intransigent. Chief among them was the Water Resources Congress (WRC), the most recent heir of a long line of regional trade associations for an integrated national waterways system. As in the case of the early twentieth century, the WRC had ample Florida representation; its hierarchy included hard-liners like Randolph Hodges and Tom Adams. Thus WRC president Dale Miller gladly raised the standard for another round of battle. At an organizational meeting only days after Nixon's order, the Texas businessman denounced the president for his arbitrary decision. This was not only about Florida and the canal as "this unprecedented action jeopardizes all public works projects in the nation." Taking a stab at the environmentalists who bedeviled the association's interests, he continued, "we are beset by hostile, often vindictive forces, by blind and unreasoning forces, thrusting toward the obstruction of our nation's progress." Nixon's announcement represented a failure to stand up to those "deliberately malicious" individuals who believed in "the mysticisms of ecology." Their corrupting influence, through Washington insiders like John Whitaker and Russell Train, reached its "most spectacular" high point "when the President of the United States preemptorily [sic] halted all work on the Cross-Florida Barge Canal." Having heard Miller's blistering words, the editors of the WRC's newsletter maintained that "it will be Miller's job to whomp up political opinion and sheer political power hoping for a graceful retreat by the President." It looked like he was off to a good start.[14]

Compared to Miller's bluster, other canal advocates preferred some sort of accommodation. Though still relying on words like "capricious," "unwise," "unfair," and "unconstitutional" to characterize Nixon's decision, these proponents recognized their relatively weakened position and thus were more willing to find a compromise. On the same day that Miller assailed Nixon, Charles Bennett met with representatives of the Council on Environmental Quality (CEQ) in Washington and presented an offer that would soon gain support from other influential canal advocates. The Jacksonville congressman recommended continuing the project but along an "alternate route suggested by the Corps of Engineers." The route would save twenty miles of pristine river by creating a parallel bypass from the Silver River to Eureka Lock. Located immediately west of the Ocklawaha, the route relied on nothing more than a dike, decorated with "extensive landscaping to maintain a natural appearance," to separate the straight cut from the winding natural river. Illustrating just how much ground the Florida Defenders of the Environment had gained since its inception, Bennett also sought the establishment of "an Oklawaha National Park to include not only the river valley but also the present Ocala National Forest in its entirety." Understanding the importance of this recommendation, the *Jacksonville Seafarer* dedicated its entire April 1971 issue to promoting the new plan.[15]

During the infamous Tallahassee Showdown only five years earlier, Marjorie Carr and her allies had called for a different alternate route, located southward toward Lake George, similarly designed to save the Ocklawaha. At that time, Tom Adams and other state officials, confident in their power, imperiously dismissed the idea as unnecessary and expensive. Bennett's change of heart revealed how much the political dynamics behind the canal controversy had changed. Environmentalist critics would now accept nothing less than a complete halt in construction. This was a far cry from their relatively weak position in 1966, when they were willing to live with any canal so long as there was an alternate route that could save the river. Then their offer stemmed from the cold fact that, in the words of David Anthony, "it was politically impossible to stop it [the canal] in 1966." So much had shifted by 1971 that Anthony—now standing on five years of FDE's scientific research—flat out rejected Bennett's compromise proposal. "With what we now know," Anthony explained, "no construction route through Florida can be justified. . . . It's all academic now anyway. The President has ordered the canal stopped. An injunction halts construction anywhere. It would be in contempt of court to touch that ground with a shovel." As Paul Roberts ruefully observed, "the chain of events from the 1966 Tallahassee hearings ends where it began but with positions reversed on the alternate route question."[16]

In early 1971, debate over the Florida canal increasingly centered on the murky intersection between politics and the environment. For members of the president's Council on Environmental Quality, however, the conclusion was beyond question: environmental issues trumped everything. On February 2, Dr. Lee Talbot, CEQ's senior scientist, "listened attentively" to Bennett's proposal and summarily dismissed it. Announcing that "the President made his decision, and we are working within the framework of that," he asserted that any future considerations regarding the Ocklawaha River valley would "absolutely not include any form of a canal." A week later, an official twenty-four-page CEQ report formally reinforced Talbot's judgment, and thus slammed the door on an alternate route. Though the bypass would not directly destroy the river, it would "sap the underground water table and damage vegetation . . . in the surrounding area." By the end of February, Talbot made the first of many trips to Florida to both examine the project and meet with disaffected citizens and canal officials. Addressing a hostile crowd in Dunnellon, Talbot insisted that "no discussion would be allowed on continuing the canal project." Such pronouncements invited trouble in a place that identified itself as "The Only City on the Cross Florida Barge Canal."

Olive Wulfing, a "sweet but peppery gray-haired lady," confronted Talbot and "launched into a rapid-fire account of the innumerable reasons why citizens in this area are generally in favor of the canal." Reporting on the level of rancor in the room at the Inglis Community Hall, the editor of the *Dunnellon Press* sardonically observed, "I'm almost sure I heard Mrs. Wulfing add that she would be tempted to do something rather drastic if she was sure that Dr. Talbot was the man who had persuaded President Nixon to kill the canal." Considering the CEQ's position, Washington seemed to offer little recourse to ordinary citizens like Olive Wulfing. Thus all eyes increasingly turned toward Tallahassee.[17]

Faced with the apparent finality of Nixon's decision, boosters presumed that if they rallied a statewide united front, they could convince the president to change his mind. Yet all they would find in Tallahassee was, at best, vacillation and indecision. Most striking was the position, or lack thereof, of newly elected Governor Reubin Askew. Fresh off a convincing victory over incumbent Republican Claude Kirk, the Panhandle Democrat had tepidly opposed the canal during his campaign. In so doing, Askew embraced a modern vision of Florida far different from that exemplified by the Pork Chop politics so identified with his area of North Florida. Being the ultimate political player, however, Askew could not abandon the traditionalists entirely. Indeed, he called on none other than the incumbent secretary of state, Tom Adams—the archetypical Pork Chopper—to join his ticket as lieutenant governor. This created a palpable tension in the Capitol, especially considering Adams's ardent support for the project. Adams's presence allowed Askew to play both sides of the controversy. In some respects, Nixon's decision should have forced the issue, but Askew—ever the pragmatist—adeptly avoided taking a stand. His letter to the president a week after the pronouncement was a remarkable example of bureaucratic fence-straddling. Never stating explicitly whether he agreed with the decision or not, Askew harkened back to the classic fallback position of requesting further information. He asked Nixon to "share with me the studies and other materials or reports upon which you based your decision to halt work. . . . Before I articulate a position or attempt to answer inquiries, I would very much appreciate the opportunity to share any information that has been available to you." Publicly, the governor was even less clear. As north Floridians clamored for political leadership, all Askew could say was "the cross-Florida barge

With construction halted in 1971, there were no easy answers for Florida's future. Courtesy of *St. Petersburg Times* and Don Addis.

canal is going to remain a difficult question whether we continue it or not continue it."[18]

Askew's waffling may have derived from an astute political calculus, but it did little to resolve the canal issue. Hard-line opponents, including Attorney General Robert Shevin and State Senator Gerald Lewis, both south Florida Democrats, lobbied furiously for the project's demise. Shevin pleaded for Askew and his fellow Cabinet officers to strongly oppose any effort to continue the project in any form. Lewis even called for the dismantling of the state's Canal Authority after he discovered the agency had initiated a lawsuit against the Army Corps of Engineers. This bizarre twist came in mid-February as the Canal Authority raised questions about the legality of Nixon's order and went even further to claim that any halt in construction would itself harm Florida's environment. In short, the authority filed suit, in the words of member Louis Smith, "to prevent the status quo of the project from being changed without notice." Lewis and Shevin would have none of that, however, and charged that "the state of Florida looks ridiculous to be pursuing this lawsuit." Calling it a "waste of $50,000 in taxpayers' money," Shevin claimed the legal action would amount to "sending a fox into the hen house to see what the chickens are doing." Regardless of Shevin and Lewis's opposition, the authority's suit proceeded and become part of the one the most tangled legal proceedings in the state's history.[19]

While this comedy of errors was playing out in the Capitol, canal proponents arranged a high-level meeting of the governor, the state cabinet, and the Canal Authority to resolve the controversy. Randolph Hodges, the ardently procanal state natural resources director, invited advocates from both sides to attend the meeting scheduled for March 1, 1971. Though the atmosphere was much less rancorous than the Waterways meeting just five years earlier, both sides left little room for compromise. Jack Lucas, from the Jacksonville Area Chamber of Commerce, testified "in support of continued construction" because the canal was a "vitally needed, important national defense project, economically justified." He went on to berate Nixon for his decision, which struck "at the democratic process of separation of powers and basic fair play." To rectify the situation, he "strongly urge[d] the Cabinet to continue to support the efforts of the Canal Authority." Moreover, he appealed for congressional funding for the alternate route if necessary. From the other side, David Anthony, speaking as the newly elected president of FDE, announced that he came to Tallahassee "to speak to the future of the Cross-Florida Barge Canal lands and waters—as something other than a canal." Overwhelming everyone with evidence gathered by FDE's scientists, he recommended the state pull the plug on the project and "return the flow of the lower Withlacoochee River (below the Inglis Pool) to approximately pre-project normal rates." As for the Ocklawaha, he made an immediate request for "national scenic river status."[20]

Following the testimony, the recommendations of the Florida Game and Fresh Water Fish Commission strikingly revealed just how much FDE had shaped the debate. Though commission director Earle Frye announced that "the Commission . . . does not now take the position that it is against the canal," it would have been hard to reach that conclusion listening to the commission's recommendations. Noting that the agency's new guidelines were "based upon the assumption that the barge canal project has been terminated," Frye's presentation appeared as if it had been pulled straight out of FDE's fact book. He began with the flat-out assertion that the creation of Rodman Reservoir "created ecological problems almost beyond comprehension." Assailing the creation of "3700 acres of water hyacinths" and the need for the "removal of logs that were crushed into the mud," Frye called for "a gap . . . in the earthen Rodman Dam so the river can follow its original course." Any breach of the dam "should be wide enough so that the natural sheet flow through the flood plain could be restored. . . . While it is tempting to retain a reservoir that has produced good fishing in its three years of existence, we feel obligated to evaluate it in terms of its ecological impact over a period of more than its short lifespan." Therefore, Frye concluded, "the Rodman Pool should be drained to restore the Oklawaha River flood plain to its original scenic qualities." On that afternoon, a state agency established the agenda that would consume Marjorie Harris Carr until her death in 1997.[21]

As with so many meetings concerning the canal, little was resolved on that March Monday. Governor Askew remained noncommittal, the cabinet refused to take a stand, and both sides left the Capitol Building feeling that the state government supported their position. The lack of political leadership encouraged years of wrangling and political gridlock on what was becoming an essentially dead project. In spite of the rhetoric of those like Tom Adams, most informed people understood the canal was never going to be finished. But the lack of concerted action on the part of the government offered boosters just enough hope that somehow, someway they could convince someone to continue construction. In many respects, newly elected Secretary of State Dick Stone embodied this sense of ambivalence. One of a new generation of south Florida politicians, Stone was a close friend and ally of Attorney General Shevin, a staunch enemy of the canal. Yet Stone, a determined canal opponent himself, could not bring himself to directly recommend the project's termination. Instead, he tried to resolve the conflict with the suggestion of a $70 million compromise that he felt "could satisfy all sides" and "benefit everyone." Announced before his fellow cabinet members at the end of the March 1 meeting, Stone's proposal called for a fifty-five-mile rail line connecting the St. Johns River at Lake George with the completed canal at Dunnellon. Designed to carry barges across the state in "gigantic railborne tubs," Stone's fanciful idea was dubbed the Florida Overland Railway Barge System (FORBS). Based on a Belgian design, FORBS would use huge incline planes to lift the barges from the water, place them in large portable dry docks, and carry them "across the Ocala National Forest" by rail. The technical details of engineering and construction mattered less for Stone than the political benefits of the project. He felt he had found a solution that "could end all the shouting matches." With FORBS bypassing the Ocklawaha and using the very same route suggested by Carr at the 1966 Tallahassee meeting, Stone assumed his idea would mollify both sides of the debate. On the day he unveiled his plan, Stone drafted a letter encouraging Nixon's support for it as well. "Mr. President," he wrote, "conservationists are right in opposing the canal, the Chambers of Commerce are right in supporting the canal, and most of all you are right to stop work on the canal. I know that implementation of this new system will preserve the ecology as the conservationists desire; will provide economic benefits as the Chambers of Commerce desire; and preserve the Oklawaha River as you desire." Though the *Jacksonville Seafarer* responded favorably to Stone's compromise solution, FDE dismissed it out of hand. Government officials, including CEQ officials and Nixon himself, simply ignored it. With little resolution in Tallahassee, all eyes now turned, again, to the courts.[22]

The legal wrangling to settle the fate of the barge canal and the Ocklawaha took more twists and turns than the river itself. The state Canal Authority quickly intervened in the ongoing EDF suit against the Corps, alleging that "as a result of the President's order the federal defendants were no longer able to defend the canal project." On February 12, 1971, the state agency took their legal offensive a step further by suing the Corps itself for wrongfully terminating the project. Authority lawyers cleverly filed the suit against Secretary of the Army Stanley Resor, rather than the agency itself. This technicality allowed them to bring the case before a potentially more sympathetic judge in the Federal District Court for the Middle District of Florida (in Jacksonville) instead of Judge Barrington Parker in Washington, D.C. The Authority based their claim against Resor on the assertion that the Corps, acting under the secretary's direction, was "under a duty to complete construction of the Cross-Florida Barge Canal project in accordance with plans approved by Congress." The order to stop construction and "abandon the Canal project is an illegal attempt to exercise lawmaking." Though not addressing Richard Nixon directly as a defendant, the suit alleged that his order was unconstitutional since the president does not have "any authority to override the will of Congress expressed in the project authorization and subsequent appropriation bills." The suit finally sought to compel Resor and the Corps "to fulfill their legal duties" to complete the project.[23]

Adding to this increasingly confusing legal situation, private interests like the Jacksonville Port Authority and the newly established Cross Florida Canal Association filed similar suits against Resor, seeking an immediate resumption of canal construction. On February 22, the waters became even muddier as FDE, along with EDF, announced it would intervene in the Canal Authority suit on behalf of the Corps. Environmentalists based this apparent reversal of direction on the contention that the Corps' halt in construction was predicated upon the legitimate authority of the president himself. Thus FDE felt the Canal Authority's lawsuit had little legal standing and should be summarily dismissed. At the same time, FDE continued their own legal battle *against* the Corps in the case before Judge Parker in Washington. That suit still remained important in spite of Nixon's edict since, in the words of one political observer, "a future President might well reconsider and give the canal another lease on life."[24]

As suit followed suit, every case involving the Cross Florida Barge Canal became a tangled legal morass. Within weeks of Nixon's decision, the situation became so confusing that FDE, the Corps of Engineers, and the state Canal Authority could simultaneously be both defendants and appellants, allies and adversaries, across a variety of cases. The situation perplexed even the most seasoned legal minds. Thus, by July 1971, the federal Judicial Panel on Multidistrict Litigation, a bureaucratic agency established in 1968 to handle such complicated legal issues, ordered all suits consolidated into a single expansive case under the jurisdiction of the Federal District Court for the Middle District of Florida in Jacksonville. Supreme Court Chief Justice Warren Burger designated longtime federal appeals judge Harvey M. Johnsen, a hard-bitten, poker-faced Nebraskan appointed to the Eighth Circuit Court of Appeals by Franklin D. Roosevelt in 1940, to oversee the case. With his long years of service, and hailing from a region far removed from the concerns of either party, Johnsen embodied a sense of impartiality and judicial propriety that could perhaps settle the controversy once and for all. That would not, however, be the case. As Johnsen explained at the trial's conclusion: "I had hoped, as I am sure had everyone else, that when the President intruded into the Canal situation, with its long underlying history and its strongly divided public sentiment, with the matter then moving into its technical and torturous judicial course, some judicial solution would be capable of being found that could put the controversy at rest. I can, however, find no immediate solution"[25]

Though the contentious legal battles of the next three years ostensibly centered on whether the canal should be built, with additional questions concerning Nixon's authority to halt construction, the issue of Rodman Dam and its attendant reservoir soon became the primary focus of the case. By 1972, with the exception of truly diehard canal supporters, most observers agreed that the project was not going to be completed. Thus the future of the canal's existing structures became paramount. Most important among those structures was Rodman Dam on the Ocklawaha. Judge Johnsen put it best in his 1974 decision: "The Rodman Pool appears to be the most symbolic portion or element of the Canal Project as thus far constructed." For Carr and FDE, stopping the canal had only been part of their goal. They would not rest until Rodman was removed, thereby letting the Ocklawaha flow freely as a wild and untamed river to the St. Johns.[26]

Canal proponents embraced the dam for two important reasons. First, its very presence allowed them the slimmest hope that the project might someday be completed. Second, if they could not have the canal, they could at least have the reservoir, the impoundment they called "Lake Ocklawaha." Referring to the body of water as Lake Ocklawaha was more than semantics for people like Giles Evans: it granted greater credence to the Corps' project. The term "lake" implied a naturally functioning ecosystem that essentially worked as well as the free-flowing river it replaced. Underscoring the importance of the nomenclature, in December 1970 Congress passed a bill, introduced by Charles Bennett, that officially renamed the reservoir "Lake Ocklawaha." For FDE and other critics, however, a simple name change could not alter the fact that the body of water would always remain artificial, and thus they continued to call it Rodman Pool, Rodman Reservoir, or simply "the impoundment." For canal advocates, the water represented a vision of nature, but one much different than that espoused by Marjorie Carr and her allies. Their view of the environment accepted, indeed glorified, human improvements that created a lake where the fishing was good and, more importantly, accessible to all. If giant barges could not be a vital part of the region's future, then bass boats would suffice.

Even with the double victories of January 1971, FDE lacked the political capital to remove Rodman Dam outright. As the court case dragged on, they settled on a legal strategy to draw down the water level of Rodman Pool and save whatever they could of the pristine forest. Lowering the impoundment's water level by five feet, as suggested by FDE scientists, would not only recover more than one thousand acres of trees, it would also profoundly weaken the case of anyone seeking a reversal of Nixon's decision. As presidential advisor John Whitaker observed in the spring of 1971, "conservationists see that the pool has not been drawn down and see this as a possibility that the President's decision isn't final." By the time the canal issue reached Judge Johnsen, FDE and their allies in both Tallahassee and Washington viewed the potential drawdown of Rodman Reservoir as the fundamental issue of the controversy. Arthur Godfrey, by this time the honorary chairman of FDE, expressed it best in an April 1971 letter to Whitaker. Appealing for Nixon to stand firm in his decision to halt the canal, Godfrey asserted, "the longer the reservoir stays full of water, the more complete and irreversible will be the desecration of the entire area."[27]

FDE once again turned to the scientific community to make its case. In May, David Anthony wrote Governor Askew requesting state support for "drawing down the Rodman impoundment to save the remaining trees there and speed the restoration of the forest and its wildlife." In a manner similar to the 1970 FDE letter to President Nixon, Anthony bolstered his petition with the signatures of 126 scientists. In addition to the imprimatur of scientific respectability, FDE had powerful allies within the federal and state governments. Later that month, the Bureau of Sports Fisheries and Wildlife called a drawdown a major step in restoring the river to its natural state. Another federal agency, the U.S. Forest Service, proposing to designate the Ocklawaha as a National Scenic River, came to a similar conclusion in a draft environmental impact statement. By late May 1971, Florida Attorney General Robert Shevin, who proposed eliminating state support for the Canal Authority, introduced a motion to the Florida cabinet "to begin immediately the draw-down of the Rodman Pool." Basing his decision on the "recommendations of 145 Florida scientists, the U.S. Forest Service, the Florida Audubon Society, Florida Wildlife Federation, the Florida Game and Fresh Water Fish Commission, Izaak Walton League and others knowledgeable in the field,"

Shevin's request vividly demonstrated the growing influence of FDE and its scientific research.[28]

Shevin's proposal initially went nowhere at a June 1 cabinet meeting. The drawdown did not even come up for a vote. Chagrined by the lack of action, Marjorie Carr wrote Governor Askew and expressed her disappointment, "particularly when it was evident that a majority of members were prepared to cast an affirmative vote." To buttress her case, Carr attached more scientific data and waited optimistically for the next meeting. On June 15, the cabinet finally took up Shevin's proposal and discussed the Rodman drawdown. This time, canal opponents held the high ground. Jack Kaufman, University of Florida zoology professor and FDE scientific advisor, presented testimony regarding the scientific rationale for reducing the pool's water level. Claiming that "a five foot drawdown [from the lake's present level at 18 feet above sea level to 13 feet] will save the remaining live trees, . . . permit the regeneration of about 3000 acres of floodplain forest, . . . [and] restore about eight linear miles of terrestrial wildlife habitat," he asserted that "it is absolutely essential that the drawdown be made immediately." A quiet man who shunned the spotlight, Kaufman was shocked at the response from those who passionately considered the drawdown proposal as another attempt to kill the canal. "I was almost attacked on the floor of the Cabinet Room," he recalled, "by a pro-canal advocate because of something I had said in my testimony. He rushed up to the front of the room at the end of my testimony and I really thought he was going to start a fight right there in front of the Governor and the Cabinet." Fisticuffs were averted, however, and Kaufman's testimony helped convince the cabinet to support the drawdown by the narrow margin of 3 to 2. FDE's substantial scientific evidence encouraged Governor Askew's deciding vote. Writing to David Anthony the next day, Askew expressed "thanks to the many scientists who have used the Florida Defenders of the Environment as a conduit to those of us holding high-elected office." The cabinet decision, Askew continued, was heavily influenced by the "compelling argument" made by the "environmental scientists, representing the cream of Florida's academic community."[29]

Though FDE's scientific evidence held considerable weight, the opponents of the drawdown proposal marshaled their own set of facts to maintain Lake Ocklawaha's current water level. In late March 1971, Colonel Avery Fullerton, the Corps' district engineer in charge of the canal, wrote Governor Askew and addressed several issues raised by Nixon's order, including concerns about Rodman. Fullerton assailed the position taken by both FDE and the state's Game and Fresh Water Fish Commission. "The lake is there now and people are enjoying it in large numbers," he wrote. "Considering the extensive recreational use of this lake and the fact that the entire Eureka reach will now be preserved for trees, the possible saving of a few added trees would not seem to be worth the considerable loss of other values." Two months later, a resident of the Ocklawaha Valley "for forty years or more" also wrote the governor with a request to reconsider the drawdown. With a reduction in water levels, he claimed, the lake would become "a series of low, marshy pools, breeding grounds for mosquitoes, and a secondary wood cover of cat-briars and dog-fennels." Furthermore, the lake was now being used by "more and more people coming to boat and camp in an area where before they had very limited access." Within Tallahassee's halls of power, Secretary of Education Floyd Christian similarly opposed the drawdown. As a member of the Florida cabinet, he attended the June 15 meeting but had to leave before casting his vote. Nine days later, he explained his position in a letter to Art Marshall, the ecologist who had worked on the Everglades Jetport controversy. After listening to all the testimony, Christian concluded that "most of the trees that are going to die will die even if the water level is lowered. If this is true, then it seems only logical that the next best service to the public would be to leave the water level where it is."[30]

Though the state cabinet favored the drawdown, if only by the slimmest of margins, its decision could only be advisory. The ultimate fate of Rodman's water level, as with all the other contentious questions regarding the disposition of the canal, at this point lay in the courts. On September 29, 1971, after hearing testimony all summer, Judge Johnsen granted the Canal Authority's request for a preliminary injunction to prevent the proposed drawdown. FDE felt betrayed; all their scientific evidence seemed to serve no purpose as Judge Johnsen had paid it little heed. The door was not closed completely, however. Though his ruling favored the Canal Authority, Johnsen "specifically invited subsequent applications for 'vacative or modificatory' relief from the preliminary injunction if facts could be developed to justify such relief." Once again, if Marjorie Carr and FDE could just "get the facts," they could get their way.[31]

For the next ten months, Marjorie Carr and anyone else even remotely tied to the canal would have more "facts" than they could imagine. By the spring of 1972, a variety of federal agencies—including the Forest Service, the CEQ, and the Corps of Engineers itself—conducted a series of studies centering on the environmental effects of Rodman Dam and the ultimate disposition of the land surrounding canal construction. These investigations raised a host of questions concerning the future of the Ocklawaha Valley. Should the river or portions of it be designated a wild and scenic river? Should Lake Ocklawaha be managed as a fishery, or should it be drained and the reservoir area be restored to its original forested condition? Should Rodman Dam and Buckman Lock be removed? If not, would these structures be acceptable under the National Wild and Scenic Rivers Act? The final reports from these inquiries provided overwhelming evidence that tended to support FDE's position. On May 18, 1972, representatives of both the CEQ and the Corps—the most unlikely of bedfellows—issued a joint interagency proposal to establish the Ocklawaha as a wild and scenic river. This meant, of course, that the canal had no future in the area. They also recommended lowering the Rodman Reservoir to the level requested by FDE. To show how far the debate had shifted, Corps General A. P. Rollins announced that "it is possible that Rodman Dam, which forms the reservoir, will be removed." A month later, CEQ solidified its position by announcing that it had "agreed to drawdown the level of Rodman Reservoir. . . . This measure has been recommended by the U.S. Department of Interior, the Forest Service, the Florida Governor and Cabinet, the Florida Game and Fresh Water Fish Commission, and numerous scientists, other citizens, and citizen organizations."[32]

Canal stalwarts scoffed at the alleged scientific objectivity of the reports. Lynwood Roberts, president of both the Jacksonville City Council and the Cross Florida Canal Counties Association, railed against the newfound alliance between CEQ and the Corps. "They're at it again," he charged; "here come the carpetbaggers again from Washington, D.C. . . . They are determined to stop the barge canal and all its beneficial assets for whatever reason. . . . Both their goals and their methods indicate they do not care for the real facts." Jack Lucas, representing the

Jacksonville Chamber of Commerce, put his faith in the legal system. "We trust that Judge Johnsen will yet again deny this obvious effort to kill the barge canal under the guise of concern for already dead trees."[33]

Lucas's trust in the court was misplaced, however, for on July 21, 1972, after a week of hearings, Johnsen reversed course and ordered the reservoir's level reduced to thirteen feet. By doing so, the 10,000-acre impoundment would wither to a mere 4,300 acres. Johnsen's decision underscored the importance of the growing scientific evidence gathered by various federal agencies. "The array of the task force personnel and its expert consultants," Johnsen wrote, "is so formidable and the strength of their opinion is such that the court does not feel entitled to allow such layman's doubt" to challenge their conclusions. Johnsen's ruling led the procanal *Jacksonville Journal* to proclaim a "Major Barge Canal Defeat." Indeed, it seemed like a convincing victory for Marjorie Carr and the Florida Defenders of the Environment. There was, however, a problem in the fine print. Johnsen's drawdown was only temporary, with a deadline of December 1, 1972.[34]

Almost immediately, the Army Corps of Engineers began lowering the reservoir's water level at a rate of about three inches a day. By the end of August 1972, the drawdown was complete, and both sides took the time to assess the impact on the river valley. As expected, canal opponents applauded the beginnings of forest growth. Understanding the trade-off that accompanied the drawdown, Bill Partington argued: "I think it is worth the sacrifice of a few hundred fish to save the trees." Considering that the reservoir, after only three years in existence, contained what environmentalists considered undesirable "trash fish" like shad, few of them lamented the loss. With the drawdown, the river, "if left alone, would return to its natural beauty and would support much more desirable species of fish." Canal proponents saw the drawdown area as a visual blight, with hundreds of dead trees and little opportunity for regrowth in the near future. They also tried to claim the environmentalists' mantle by arguing that the loss of any fish (even those labeled as "trash" species) was problematic. Both sides would take those arguments back to court to determine whether the temporary drawdown be reversed, extended, or even made permanent.[35]

In late November, federal officials, under the auspices of Nathaniel Reed, now working as assistant secretary of the interior, requested an

The drawdown of Rodman Reservoir revealed thousands of dead trees killed by construction. Courtesy of Florida Defenders of the Environment (FDE), Gainesville.

extension for the drawdown. They bombarded the judge with evidence of the drawdown's success and the environmental devastation that would accompany an expanded reservoir. Concluding that almost nine thousand additional trees would die if the water level rose, Reed pleaded with Johnsen to make the extension permanent. Johnsen granted two temporary extensions, first to December 15, and then into January. On January 12, 1973, however, he confounded both sides once again with a reversal of the drawdown. Johnsen ordered the Corps to raise Rodman's water level back to eighteen feet. Though recognizing the validity of the environmentalists' scientific claims, he based his decision "on a balancing of other factors, such as the adverse effect of a drawdown on recreation, wildlife habitat, fisheries, and aesthetics." More importantly, Johnsen's decision stemmed from a sense of frustration with the lack of progress on the larger ramifications of the case. Fully two years after Judge Parker's decision, no firm resolution of the canal controversy seemed in sight. Instead, all energies had been focused on a seemingly obscure issue that increasingly taxed the patience of the seventy-seven-year-old jurist. By settling one controversy, however imperfectly, Johnsen thought his

January decision would offer the chance "to bring the trial situation to a head." With Rodman behind him, Johnsen could now proceed with the final hearing on the canal itself.[36]

Johnsen's decision elicited predictable responses from both sides of the controversy. Editors from the *Orlando Sentinel* best articulated FDE's sense of disappointment. "With signs of vegetation revival evident along river and lake shore, it is difficult to understand Judge Johnsen's recent order that the water be returned to eighteen feet by February 1st." Raising the water level "would sign a death warrant on a vanishing segment of Floridiana." The *Ocala Star-Banner*, of course, had a different perspective. Lauding the judge's opinion, the paper stated that "there really was only one logical conclusion and fortunately Judge Johnsen recognized this. . . . It makes no sense at all to play yo-yo with the reservoir and it is incumbent upon the courts . . . to leave the water level in the lake alone." Happiest of all were local fishermen, who saw the drawdown as a direct threat to their way of life. Salt Springs fishing guide Carson Johntry put it best when he proclaimed, "With the Miami Dolphins winning the Super Bowl and this [Johnsen's decision] . . . it's been the best weekend I've had in years."[37]

With the apparent resolution of Rodman's water level, Judge Johnsen could now pursue the questions that brought the case to court in the first place. At the same time, FDE appealed his decision to the Fifth Circuit Court of Appeals in New Orleans. That case would remain a sideshow, however, as the center of legal action returned to Jacksonville in July 1973. There Judge Johnsen presided over a five-and-a-half week trial he hoped would finally settle the issue. Scientists and economists from FDE as well as the Canal Authority appeared as expert witnesses in a face-off over the canal's continuation. As if mirroring the tedious, torturous path of the controversy, the case immediately fell behind schedule after the first day of testimony. After several hundred exhibits and twenty-six volumes of trial transcripts, Johnsen closed the case and gratefully returned to Omaha to ponder his decision. Five months later, he returned to Jacksonville and issued his judgment on February 4, 1974.

Johnsen wrote a forty-page decision that, in the end, resolved little. With regard to the constitutionality of Nixon's decision to halt construction, Johnsen came down squarely on the side of the Canal Authority. "I am unable to regard the President as having any general executive power," he wrote, "to terminate whatever legislatively authorized public-works projects he chooses.... The President's order of intended termination of the barge canal project must be held to be invalid." While it appeared the Canal Authority had secured success, Johnsen also handed the Environmental Defense Fund some good news. According to Johnsen, Nixon's order, though illegal, should remain in place until an appropriate environmental impact statement (EIS) was conducted. "The general claim of EDF to a right to have construction halted until an EIS under NEPA [National Environmental Policy Act] had been provided will . . . be permitted to stand." To meet EDF's demand, Johnsen also ordered Nixon to release $150,000 appropriated by Congress for the Corps to conduct an environmental impact statement. Finally, Johnsen added that his decision on raising Rodman's water level would remain indefinitely, "as the few months' drawdown permitted in 1972 . . . left the area with aesthetic conditions repellant to general visitors." Since Johnsen ruled the president lacked the authority to stop the project, it became abundantly clear that the canal's future remained in the hands of Congress. "The situation is going to have to be allowed to grind its way along," he bemoaned, "until Congress sees fit to provide the final answer."[38]

Canal proponents joyfully responded to Johnsen's ruling. The relentlessly procanal *Ocala Star-Banner* proclaimed in huge headlines above the fold: "Judge Revives 'Dead' Canal." George Linville, chairman of the Cross Florida Canal Association, one of the private organizations involved in the suit, hailed the decision as a great victory and claimed it "was not for any individual but . . . a service to mankind." No one could have been more pleased than longtime Canal Authority manager Giles Evans. "The Canal Authority is gratified," he wrote in a February 8 memo, "that the court has upheld the Authority's position to such a large extent, and on so many of the key legal issues." Looking ahead to Johnsen's demand for a restudy, he asserted that "we believe that the evidence already on record will continue to refute most of the blatant and ill-founded antiproject histrionics."[39]

Anticanal activists had at best mixed feelings about the decision. For Marjorie Carr, Johnsen's ruling was but another round in the battle against the canal. "I am sure we'll just go ahead and continue to oppose this," she noted with a slight bit of resignation. The judge's opinion that Nixon had acted illegally in halting canal construction troubled Carr and other environmentalists. They also took a jaundiced view of who Johnsen placed in charge of the new environmental impact study—the Army Corps of Engineers. Writing to the judge in a tongue-in-cheek editorial, the *St. Petersburg Independent* asserted, "it must be [that] nobody told you the corps is the very agency that has been itching to build that big, $200 million ditch for the past 40 years or more. And you want THOSE people to decide the canal's environmental acceptability. . . . We never thought foxes should guard hen houses, or goats the cabbage patches" (emphasis in original). Similarly, Marjorie Carr had great cause for skepticism about the Corps' objectivity in drafting an EIS. Two years before, she had warned that "a request for another environmental impact study is without justification and therefore a waste of taxpayers' money." For Carr, FDE's 1970 report provided more than enough evidence to justify the canal's demise. Another study would "only serve to delay the restoration . . . of the Oklawaha river area."[40]

Even worse, a Corps study could provide enough evidence to resurrect the canal itself. Proof of this possibility came in a press release from Canal Authority chairman L. C. Ringhaver in June 1972, who took delight in the recent legislation appropriating the $150,000 for a new EIS. Initiated in Congress by canal stalwart Bob Sikes, the new study represented "a goal we have been seeking ever since the President issued his press release/stop order." The possibility of a Corps-conducted EIS filled Carr with fears of a bureaucratic whitewash with the Corps co-opting environmental rhetoric to validate canal construction. Despite their misapprehensions with the EIS, canal opponents were not completely dejected with Johnsen's ruling, especially the contention that Congress was the ultimate arbiter of the canal's fate. Florida Audubon's Hal Scott took solace in the fact that Johnsen saw fit to kick the issue back to Congress. "We ought to be able easily to get through the Congress a bill that will deauthorize it once and for all," he said. With the apparent conclusion of the protracted legal struggle, politicians would now ultimately determine the future of the canal. And by 1974, the political winds were blowing in FDE's favor.[41]

Politically, the balance of power that had long favored canal construction had shifted by the early 1970s. In many respects, Richard Nixon's order created a new political climate. With each day's passing after construction had ground to a halt, more and more state lawmakers became willing to challenge the efficacy of the project. With the legislative reapportionment breaking up the Pork Chop Gang, north Florida canal backers found fewer allies in the state house. As early as 1970, south Florida politicians, particularly Gerald Lewis and Robert Shevin, led the charge to completely dismantle the project. Moreover, under the leadership of Governor Reubin Askew, state cabinet officials no longer routinely deferred to the demands of the likes of Giles Evans and Randolph Hodges. This marked a significant break from the past, especially considering Florida's unique governmental structure, in which cabinet officers were elected rather than appointed by the governor.

By mid-summer 1972, while the fate of the canal languished interminably in the sweltering courtroom of Jacksonville's Federal Building, the cabinet took its first tentative steps toward pulling the state's endorsement for the project. On August 1, the cabinet voted on Richard Stone's resolution to "rescind its previous policy supporting completion of the Cross Florida Barge Canal . . . and hereby suspends support for construction." Though somewhat tempered by its demand for yet another round of studies, the cabinet's message was clear—if the canal was not dead in Tallahassee, it was definitely on life support. Even Tom Adams, the proverbial pit bull of procanal advocates, recognized the signs of change. In the months following Nixon's decision, Adams, much to the chagrin of Governor Askew, traveled the state berating the president for stopping the canal. A year later, however, he had dramatically tempered his remarks. At the May 1972 dedication of the new State Road 40 bridge over the Ocklawaha, just east of Ocala, Adams's tone was much different from that of his fiery remarks at the opening of the St. Johns River Barge Port only a year earlier. "The promise here for the future could be even greater than the now, well-faded promise of the barge canal," he announced. Facing "the ecologic [sic] and environmental awareness which contributed inevitably to the untimely halt of the Cross-State Barge Canal," Adams admitted defeat at the hands of Marjorie Carr and her allies. He wearily concluded, "there is an unmistakable futility in simply posturing, on and on, with impotent gestures of mere defiance." Within two years, when Reubin Askew ran for reelection, Lieutenant Governor Adams would be removed from the ticket and relegated to the political wilderness. His fate was a metaphor for the cause he so diligently supported.[42]

By 1973, many politicians had moved past simply seeking a compromise solution. Fueled by constituent concerns over the project's environmental degradation and exorbitant costs, they actively called for complete congressional deauthorization. A November 1971 survey of seventy-five thousand residents of Paul Rogers's Ninth Congressional District, located along Florida's southeast coast, attested to the public mood on the canal issue. Fully 62 percent of those polled recommended ending the project and turning canal lands into a recreation area. Less than a quarter favored continuing construction, and that was with an alternate route that saved the Ocklawaha. With such figures, it was no longer politically risky to support termination, particularly for newly elected Republican congressmen like Bill Young and Skip Bafalis. Young, the former minority leader of the Florida State Senate, was elected to Congress in 1970 to represent the St. Petersburg area. In June 1972, he engaged in a vigorous debate on the House floor with Bob Sikes and Charles Bennett over the canal. Concluding that "too much of the taxpayers' money already has been wasted on this discredited project," Young exhorted Congress to provide no further funding for anything to do with the canal. A year later, Bafalis, first elected in 1972 to a seat representing southwest Florida, went even further when he authored a bill to "deauthorize permanently the recently halted" canal. Though Congress voted down Bafalis's proposal, his concerns more than balanced the advocacy of supporters such as Sikes and Bennett.[43]

In Tallahassee, members of both parties in both chambers of the legislature moved closer to the once-radical position of Shevin and Lewis. By 1973, State Senator Henry Sayler of St. Petersburg, a formidable force in an increasingly influential Republican Party, and Representative Lewis Earle, another Republican opposed to profligate government spending, introduced bills to cut off funding for the Canal Authority and transfer authority lands to the state Department of Natural Resources. Though the measures did not pass, they affirmed an idea that had been developing among environmental activists since the late 1960s. As Lee Ogden, one of FDE's founding members, declared in 1972, "we encourage those with jurisdiction to consider a recreational trail (hiking, bicycling, bridle) within this area . . . and provide an opportunity to view and study the broad spectrum of natural vegetation in the area." With the defeat of the canal, environmentalists saw a unique opportunity in the 107-mile swath of land across the state that had been marked for canal construction.

Where else could individuals traverse a natural area that stretched from the Gulf of Mexico to the St. Johns River? Indeed, it was quite a dream. However, there was still work to do. The canal controversy continued to linger in the courts, and until there was a clear sign of resolution, canal advocates would relentlessly lobby for their cause.[44]

By 1974, few politicians seemed willing to keep the Cross Florida Barge Canal alive, especially as its future rested in the hands of the federal courts. With Judge Harvey Johnsen's decision, canal proponents now seemed on the defensive. From Miami to Pensacola, in a series of public hearings held throughout the state between 1974 and 1976, Giles Evans, George Linville, and other canal proponents found themselves facing increasingly hostile audiences. The once tried-and-true tactics of accusing opponents of being out-of-state interlopers, tools of the railroad industry, or hopelessly naïve bird-watchers and butterfly catchers no longer held much credence. Unable to control the debate, they often found themselves bombarded with questions and accusations for which they could provide few satisfactory responses.

The increasingly marginalized position of canal supporters grew even weaker in the summer of 1976 as organizations completed their studies as part of Judge Johnsen's 1974 court order. In July, the University of Georgia's Institute of Natural Resources submitted its "restudy report" to the Corps. The seventy-nine-page document concluded with the worst possible scenario for canal proponents: "the highest and best uses for the Oklawaha Valley and Lake Rousseau are to restore the CFBC [sic] area to its original condition." A month later, on August 20, the Corps issued its own draft environmental impact statement and economic restudy. Though the report made no definitive recommendation on whether the canal should be completed, its impenetrable statistics, graphs, tables, and charts could not obscure the fact that the Corps now predicted the canal would cost $350 million to complete. And that figure did not include the approximately $70 million already spent on the project. FDE, while acknowledging that the Corps' statement confirmed their belief "that the canal is economically and environmentally undesirable," took particular issue with the agency's economic projections. Basing their estimates on an interest rate of 2 $^{7}/_{8}$ percent, the Corps announced the canal would return $1.24 in benefits for each dollar spent. With a current interest rate of 6 $^{1}/_{8}$ percent, however, only an unacceptable sixty-six cents of benefits would be realized. Such a wide discrepancy was untenable and raised more questions than it answered.[45]

The preliminary report provided the final straw for Representative Claude Pepper, whose support for the waterway stretched back to the days of Franklin Roosevelt's Ship Canal. Only six days after its publication, Pepper announced, "I see no prospect of this project being approved by Congress in the foreseeable future and, as the situation now stands, I would not feel justified in supporting it." Within weeks, other former supporters started to distance themselves. In September, WJXT, Jacksonville's premier television station, issued an editorial announcing: "there's no economic justification for resuming construction of the Cross Florida Barge Canal. . . . In other words, [the project] is a bad investment for the taxpayer." That same month, officials from Silver Springs responded to the Corps' report as well. Apart from the dollars and cents, they accounted for the ecological costs of the project: "The Army Corps of Engineers has shown a complete disregard for the hydrological influences and pollution threats this project would have on the one most spectacular natural feature in the southeastern United States: Silver Springs." Finally, the once resolutely procanal Ocala/Marion County Chamber of Commerce announced on September 30 that it also could no longer support the project. Board member John McKeever concluded: "the economic benefits are debatable, while there are definite ecological problems. I am not willing to risk my future and my children's future in Marion County." Jacksonville district engineer Colonel Donald Wisdom understood perfectly the implications of these changes of heart. For Wisdom, it all came down to Florida politics. "If we had no support from the state of Florida," he stated right after the publication of the draft Corps report, "we'd have no local sponsor. Unless the state of Florida approves the canal, we aren't going to build it." With this understanding, the seven-member Florida cabinet took up the canal issue once again in December 1976.[46]

As they had ten years earlier, hundreds of pro- and anticanal advocates made their way to Tallahassee to convince the cabinet to finally take a definitive stand on the project. Held in the Capitol's Hayden Burns Auditorium, the meeting featured two days of testimony from dozens of witnesses, including, as the *St. Petersburg Times* reported, "a congressman, legislators, scientists, economists, businessmen, environmentalists,

Nov. 16, 1976

BARGE CANAL

"Don't be alarmed, Sir, I'm a treasure hunter. I heard the tax-payers have lost millions of dollars in this thing!"

By 1976, most Floridians recognized the canal was the epitome of a pork-barrel project with little return on the government investment. Courtesy of *St. Petersburg Times* and Don Addis.

engineers, state administrators, retirees, housewives and even school-children." The cacophony of discordant voices, however, merely rehashed old arguments. In a rare public appearance, the reclusive former governor Farris Bryant attacked FDE's selective reading of scientific evidence when he asserted, "we now know that the environmental damage claims were 90% illusory." Moreover, he warned that "the hysteria of other years should not guide us as we consider our course for the future. We have paid a high price to give a full hearing to the opponents of the canal; let us now get about the business of completing this great project. . . . No canal, no barges, no coal, no industrial development, no growth of our economy." Marjorie Carr's testimony was just as provocative. "Despite ridiculous claims to the contrary, the canal would not generate any significant economic growth in the canal region. . . . In view of the many irrefutable arguments against completing the canal, it is time that we get on with the important work of repairing past damage and ensuring that the canal lands are used for the greatest public good." Unlike her earlier experience in Tallahassee, there was no Tom Adams representing the state to either browbeat or ignore Marjorie Carr. With the conclusion of her testimony, hundreds of supporters stood and cheered. The applause was as much a tribute to her years of tireless effort as for her hour of testimony that December afternoon.[47]

Things had changed dramatically for the Corps as well. In a bombshell announcement that opened the two days of hearings, district chief Colonel Wisdom disclosed that the Corps had shelved plans to build a transloading operation on the west side of the canal. All sides in the controversy viewed the facility, designed to unload cargo from ships too large to traverse the waterway, as crucial to the rationale for the entire project. Without it, the canal's already marginal economic benefits would be reduced even further. Once again, the "dismal science" of economics lived up to its name as Wisdom and Attorney General Shevin sparred over tedious accounts of benefit-cost ratios. "You don't even try to explain it," Shevin chided the officer, "you just put it up on the chart." Moreover, the attorney general expressed frustration with Wisdom's continued support for the project, even though the colonel himself had pulled the plug on a key part of it. Trying to understand the awkwardly contradictory testimony, Shevin remarked in exasperation, was "like trying to nail jello to the wall." Reubin Askew himself took umbrage with Wisdom's convoluted statements. When the Corps official minimized his organization's role in the project and placed years of canal advocacy squarely on Floridians, the governor asked testily, "Are you saying the Corps has never been a strong supporter of completion of the canal; it has only been the state?" Laughter filled the auditorium.[48]

After more than twenty hours of often emotional yet tediously repetitious testimony, Attorney General Robert Shevin called for a vote late in the afternoon of Friday, December 17, 1976. With a near-unanimous verdict, the cabinet reached a historic decision when it officially withdrew its support for the canal. The lone holdout was Agricultural Commissioner Doyle Conner, one of the last remaining members of the north Florida Pork Chop Gang. After the failure of his initial compromise to indefinitely suspend (rather than end) the project, Conner voted against termination. He justified his decision on the rather specious grounds that the cabinet would need even more information before coming to such a momentous decision. A frustrated Robert Shevin icily suggested his opponent was being disingenuous. With participants having worked their way through ten years of documents and two days of testimony, the attorney general said, "I can't believe that anybody up here can honestly believe that we have had limited information." After pulling its support, the cabinet further recommended congressional deauthorization of the project. Continuing on a path that could have been hewn by Carr herself, the officials also called for the Ocklawaha's restoration, as well as its designation as a wild and scenic river. At the same time, they agreed to turn all canal lands into an expansive recreational area. Finally, in a separate 4-to-2 vote, the cabinet urged the state legislature to abolish the Canal Authority in its upcoming legislative session.[49]

Predictably, Carr and her allies viewed the cabinet's decision as "exhilarating and historic." Recognizing that this was only part of a larger struggle, most activists restrained themselves and remained relatively low-key. The lone exception was Florida Audubon's Hal Scott, who gleefully proclaimed: "we are planning an Irish wake. . . . A wake for an old devil whose demise we have been looking for a long time, and thank heavens it is finally here."

Canal boosters considered the political setback "disappointing" and "hard to understand," but they were hardly making funeral arrangements. George Linville called the cabinet meeting a "travesty. It was the most stacked deck I ever appeared before. They won one today but we are definitely not dead." Canal Authority chairman Louis Smith concurred, announcing that canal supporters needed to "do what we can to see this project through to a successful conclusion. Not only for the people of today, but I think we're talking about 20, 30, or 40 years down the road. All these jobs may not be for us but will be for our children and their children." Marjory Stoneman Douglas, the other grande dame of Florida environmentalism, was all too familiar with such intransigence. In a congratulatory holiday message to Marjorie Carr, she cautioned, "I know you will not rest easy until the final word is said . . . but surely all the preliminaries for success are in your hand." Understanding the difficulties of the continuing conflict, she concluded her letter by noting, though "your fine work is almost complete, we will keep up what pressure we can bring until it is over."[50]

Signs that Carr's long struggle may have been nearing completion appeared almost immediately. On January 6, 1977, General Ernest Graves, director of Civil Works of the Corps, submitted a disposition form that laid out his agency's position in light of the upcoming filing of the final EIS in February. Through all of Graves's bureaucratic verbiage, one thing stood out. He concluded with finality that the Corps "should not recommend resumption of construction under the current authorization." Five days later, a federal interagency task force composed of representatives from the Department of Interior (including Florida's Nathaniel Reed), the Forest Service, EPA, CEQ, and the Corps of Engineers concurred. Underscoring the importance of the December cabinet decision, it released a document that stated that the position of Florida officials "will make it difficult, if not impossible, to fulfill the elements of local cooperation required by the current Congressional authorization for the project." Sounding like Marjorie Carr, the group recommended the "Oklawaha River should be restored to a free-flowing river condition. This would involve elimination of Rodman Pool." Following Judge Johnsen's demand for a restudy, this was the clearest indication yet that FDE's vision would soon be realized.[51]

On February 24, 1977, the Army Corps submitted its long-awaited environmental impact statement to Judge Louis Bechtle, who had replaced Johnsen following his death in late 1975. A tersely worded press release categorically recommended termination of the project. Though the chief of engineers conceded construction was still feasible, he asserted that the project's "economic justification is presently marginal and, when combined with the potential adverse environmental impacts, [the Corps] does not favor completion." Twenty-four volumes of dense and at times impenetrable scientific text accompanied the announcement that in many respects reinforced FDE's claims made seven years earlier. David Anthony gleefully announced that he was "certainly gratified after over a dozen years of work that finally even the Corps of Engineers has come to the same conclusion we did, that [the canal is] no good economically, and it's environmentally damaging. We have known for some time that it's a turkey, and now the Corps admits it." While it appeared that the Corps had done a complete turnaround, other longtime activists took issue with Anthony's optimistic assessment. They felt, with good reason, that army engineers could not be trusted to halt the project. Such a radical shift in the Corps' position, especially considering its long history of environmentally destructive projects, justifiably engendered a great deal of skepticism. Unable to believe its engineers could actually terminate the project, Nathaniel Reed fulminated that the "spectacle of the Corps groping for new benefits as the EIS is 'completed' reinforces my opinion that the Corps is incapable of an honest evaluation of a project which it has promoted and supported for so many years."[52]

Reed's sense of frustration and mistrust was not unwarranted. By the mid-1970s, the Corps remained profoundly divided over its mission and how that related to the emerging environmental movement. Its response to the canal controversy reflected the schism. While many officials viewed environmental activists like FDE as nothing more than a bunch of troublemakers, some, particularly junior officers and civilian employees, actually sympathized with the concerns of Anthony, Partington, and Carr. Little of this bureaucratic infighting left the Corps' offices, but the institution was ever so slowly moving away from the "dredge and fill" mentality of the early 1960s. Indeed, by 1975 the army had created a new level of Corps' management—a civilian assistant secretary for civil works, ostensibly established to keep a watch over expensive public-works projects. It tabbed Victor Veysey, a stalwart Republican and a former congressman from California, for the position. As General Graves remembered, Veysey "was determined to turn the image of the Corps around to an agency that was among the most, if not the most, responsive to environmental concerns." Almost immediately, he clashed with Major General Frank Koisch, the Corps' district engineer for the Lower Mississippi Valley. Koisch represented the traditional vision of the Corps—let's finish this project, and the environmentalists be damned. "At breakfast one morning Veysey and Koisch got into the most incredible argument about the Cross-Florida Barge Canal," Graves continued. "As far as Koisch was concerned, get the shovels and start digging, which, of course, was the very image of the Corps that Veysey deplored." Though Veysey would eventually win the argument, Reed certainly was correct in remaining skeptical over the Corps' ability to objectively evaluate one of its own projects.[53]

The Corps' strained relationship with environmentalism reflected broader trends throughout the country. Though the environmental movement was most certainly taking root in the early 1970s, it also had to compete with innumerable seemingly intractable national concerns. Richard Nixon seemed to embrace environmental issues early in his presidency, yet his inherent conservatism and the overarching fixation on Watergate and impeachment mitigated those gains during his second term. Following Nixon's resignation in August 1974, his replacement, Gerald Ford, concentrated on healing the gaping wounds that resulted from the scandals swirling around the former president. Amid all this turmoil, the nation faced problems associated with the end of the Vietnam War, an OPEC oil embargo and subsequent energy crisis, and the stagnant wages and rising inflation of a profoundly disrupted economy. All of these difficulties contributed to a sense of national unease that seemingly relegated environmental concerns to the back burner. However, the 1976 election of Jimmy Carter, who supported FDE's position on the canal during the Florida primary, offered the chance to push environmental issues to the forefront once again.

Jimmy Carter marked a significant departure in presidential leadership with his willingness to address "unpleasant truths" concerning the environment, consumption, and the realization that people must live within a world with limits. Campaigning as an outsider far removed from the corrupting influence of traditional power, the Georgia peanut farmer appeared supportive of environmental causes like the one championed by Marjorie Carr. Sensing a potential ally in the White House, FDE seized the initiative to lobby the new president, who was already working on an ambitious environmental policy within the first few months of his administration. On May 20, 1977, drawing on a time-worn strategy, more than 180 scientists signed a letter addressed to Carter seeking the president's assistance in the removal of Rodman Dam, which would allow the Ocklawaha to once again flow freely. Safely presuming the canal would never be built, the message instead focused on the restoration of what had been destroyed. Three days later, the president went before Congress and issued a sweeping thirty-six-page message that laid out an agenda to strengthen environmental policy on several fronts. Within a statement calling for clean air, clean water, and the protection of endangered species, Carter strongly encouraged Congress to finally deauthorize Florida's canal. Following FDE's recommendations, he also called for the designation of the Ocklawaha as a wild and scenic river. Though he did not address the controversy directly, Carter's statement implicitly supported eliminating Rodman Dam. Marjorie Carr announced that she was "just delighted" with the announcement, especially with Carter's emphasis on "the restoration of the Oklawaha River. The river," she recalled, "is where it all began . . . where the public outrage began."[54]

A sense of outrage remained, yet this time it was expressed among canal proponents. Some were completely dejected. The president of the Putnam County Chamber of Commerce mournfully declared his county had "been dealt a mortal blow." In Ocala, Bill Rodgers, vice chairman of the Canal Authority, confessed with more than a hint of understatement that Carter's announcement "certainly is not very encouraging." However, he blamed the boosters themselves for their predicament since "we have not really fought hard enough for the canal—[we] did not exert enough pressure." The implication seemed clear: with more lobbying, there would be greater support for the canal, and thus the project could continue. A gloomy Putnam County commissioner saw promise in the growing distinction between canal construction and the removal of Rodman Dam. "If the canal project is scrapped," he plaintively suggested, "we hope to save Lake Ocklawaha." As usual, the most truculent response came from Giles Evans. Emphasizing the limits of executive authority, Evans minimized the threat of Carter's announcement. "The canal is a project begun and controlled by the Congress," he snarled. "Nixon tried the same thing before and the canal is still here, so far." Warning that he and other proponents were going to fight until the bitter end, he boldly predicted that the canal is "going to be built one of these days. It will just cost more money." Evans, however, was simply whistling past the graveyard. Even the procanal *Ocala Star-Banner* grasped the significance of Carter's speech. "President Carter," it announced, "prescribed the death sentence Monday for the Cross- Florida Barge Canal."[55]

As any observer of the American legal system can attest, death sentences often take tortuous paths and innumerable years to reach their conclusion. Such was the case for the canal. Although Florida's senators Lawton Chiles and Richard Stone immediately sponsored a Senate bill to deauthorize the canal, prompting the *Ocala Star-Banner* to label them as Carter's "volunteer executioners," the legislation failed to pass. For the next decade, despite an overwhelming consensus for deauthorization in both Tallahassee and Washington, recalcitrant House members—diehards like Charles Bennett, Bob Sikes, and particularly Ocala's Bill Chappell—blocked any measure to kill the project in Congress. The struggle

By 1979, the canal struggle had reached a protracted stalemate. The canal was not going to be built, but it also was not going to be deauthorized anytime soon. Courtesy of *St. Petersburg Times* and Don Addis.

became so protracted and so arcane that newspaper editors had to resort to metaphors—often awkwardly mixed—to convey the growing sense of frustration with a controversy seemingly without end. The *Orlando Sentinel* chimed in early, presciently suggesting in February 1977 that "the reason the Cross-Florida Barge Canal seems to have more lives than a cat is, it can't be declared dead by anyone but Congress and Congress has a way of keeping boondoggles attached to life machines." Most newspapers clung to this imagery of death: from the cute, "This wicked ditch is dead"; to the maudlin, "Barge Canal May Face Watery Grave"; to the utterly morbid, "Environmentalist Wants Cemetery to Mean Death for Canal." The latter stemmed from Marjorie Carr's 1983 proposal to locate a burial ground for 250,000 veterans along the path of the canal, which would in turn pressure Congress to deauthorize the project and thus once and for all kill any dream of a waterway. No matter how one looked at it, the controversy was becoming absurdly intractable. As Florida Attorney General Jim Smith noted after the problem had dragged on into early 1983, any "delay in resolving the fate of the canal will further complicate an already complex situation."[56]

In April 1979, Florida's legislature waded into the controversy by passing a bill that disbanded the Canal Authority and developed a repayment plan for the six counties that contributed land for canal construction. The law seemed rational enough, but it offered no firm solutions, since it could only take effect when Congress deauthorized the project. The impatient House sponsor, Frank Mann of Fort Myers, confidently proclaimed the law would finally "kill the cotton-picken [sic] canal . . . by sending one more message to Congress that we want the barge canal gone." Others mistakenly thought the end was near. Jacksonville Representative Tommy Hazouri conceded defeat, announcing despondently, "we recognize the fact that when the canal is dead, it's dead." The legislation may have struck a bold blow against the project, but that was only in Tallahassee. Canal opponents still had to reckon with Congress. Thus even Mann had to admit some doubt. The canal, he said, was "a snake with a lot of heads. We keep chopping them off and it keeps coming back."[57]

While the death of the canal may have been a foregone conclusion for most Floridians, a few members of Florida's congressional delegation begged to differ. For the rest of the 1970s and 1980s, Charles Bennett and Bill Chappell stubbornly kept the canal alive, if barely. With their years of seniority, which led to influential positions on such important congressional committees as Appropriations, Armed Services, and Water Resources, they commanded considerable influence. As fellow Florida congressman Clay Shaw explained with a touch of awe: "they are tough adversaries. They know how to twist arms and get votes." Few could rival Bill Chappell for his knowledge of legislative protocol and willingness to use it to support the canal. Anyone seeking legislation eventually had to horse-trade with him. And Chappell almost always got his way. His very presence seemed to stop the anticanal movement in its tracks. "During one committee vote on the canal last year," reported the *Miami Herald* in 1985, "he stood in the back of the room, arms folded, trademark cigar in his mouth. The canal survived the vote."[58]

Chappell's command of the legislative process allowed for nothing more than a series of obstructive rear-guard actions. Savvy enough to understand that Congress would never fully fund canal construction, he instead focused his energies on forestalling deauthorization, for he believed that at some point in the future, "if it is not deauthorized, it *will* be built" (emphasis added). Twice the Senate, prodded by Lawton Chiles, voted favorably on such legislation, only to be rebuffed in the House by Chappell's willingness to play hardball and manipulate parliamentary procedures. For Chappell, deauthorization represented not only the death of a dream, but the admission of defeat to those whom he considered unworthy adversaries. Moreover, he realized the legislation would hardly signal the final act of an ongoing drama. Deauthorization would simply create a whole new set of problems concerning the disposition of canal lands and the ultimate fate of Rodman Dam, all of which centered around the "restoration of the land to its original state." It would mean the destruction of Lake Ocklawaha, which Chappell saw as having developed "a unique ecology . . . abound[ing] with waterfowl, egrets, eagles, fish and the changes brought about [that] have resulted in a national recreational treasure enjoyed by thousands of families annually." The ensuing economic costs—ranging from compensation to the threat of never-ending lawsuits—would also result "in an unconscionable taxpayer burden: one beyond calculable projection." Such talk was ironic for someone who, as late as 1982, fervently demanded nearly a half million dollars for yet another Corps feasibility study. And canal opponents saw the idea for what it was—"an attempt on the part of the proponents to keep the canal alive."[59]

Chappell's political maneuvering appeared even more contradictory as support for his cause increasingly dwindled. Even James Watt, Ronald Reagan's notoriously prodevelopment secretary of the interior, declared the canal should not be completed. By 1983, the overwhelming majority of Florida's political establishment, including Governor Bob Graham and former governor LeRoy Collins, firmly supported congressional deauthorization. At the same time, public approval for the project had dramatically waned in areas that were once the very center of procanal boosterism. Even Chappell's once personal bailiwick of Marion County deserted him. On April 7, 1983, the Ocala Board of Realtors—never noted for their radical environmental views—issued a resolution urging Congress to finally pull the plug. Claiming the project would "be of no economic benefit to Marion County . . . and provide an economic threat to the Florida Aquifer," the realtors suggested the dream of Bert Dosh must come to an end. Adding insult to injury, a recent congressional redistricting plan pushed Chappell out of the central Florida county and into a new district along Florida's east coast. Now representing Marion County, freshman congressman Kenneth "Buddy" MacKay vividly demonstrated just how much the situation had changed by immediately joining with Senator Lawton Chiles to sponsor deauthorization legislation.[60]

Jacksonville's business community provided the nucleus for what little support remained for the canal. Strongly backing Chappell's fight in Congress, they found an ally in Andy Johnson, a young, strident, and vociferous state representative who enthusiastically assumed the mantle of canal boosterism once held by the likes of Duncan Fletcher and Gilbert Youngberg. In 1981, he helped organize the Coalition for Rational Energy and Economic Development (CREED) and pushed, against imposing political odds, for canal completion with an entirely new rationale. Johnson focused almost exclusively on the project's assumed importance in delivering cheap energy to Florida, a major consideration in the early 1980s as the continuing oil crisis and problems associated with nuclear power bedeviled the nation. Johnson and other members of CREED claimed the canal presented an opportunity to move coal inexpensively to Jacksonville's new electric power plants. While extolling the economic benefits of the project, Johnson sounded like Tom Adams as he blasted Marjorie Carr and her allies as "crazy phony environmentalists." He combatively offered "to go to any city in Florida to debate her and show her for the liar she is." Appealing for a broader constituency, he enlisted "Dr. X.," a Jacksonville radio talk-show host, to spread the message and distribute procanal bumper stickers. All of Johnson's frenetic activity energized even the aging Claude Pepper, who had wavered in his support for the canal nearly a decade earlier. By 1985, the octogenarian seemed once again as devoted to the canal as he was to Social Security. "Why should we throw it all away," Pepper asked rhetorically, "with a resolution to deauthorize it? The dream of centuries will be done away with by the precipitous action of the people of this generation." Despite all the posturing, Johnson's boosterism provided few tangible results. Even CREED's membership realized they were significantly outgunned. One charter member, according to a political observer, had "never attended a meeting, thinking it is a waste of time. He has told Johnson a number of times that the canal project is dead." More than anything, the basic mathematics that lay behind political calculation suggested Johnson's last-ditch efforts had little chance of success. "A pro-canal vote is worth about 20,000 votes," the observer asserted, "whereas a vote for the deauthorization is worth an easy 250,000 state-wide." With those numbers, the question surrounding deauthorization became not "if" but "how."[61]

By the mid-1980s, the stars finally began to align for congressional deauthorization. From 1983 to 1986, state legislators held a series of hearings around the state to determine the ultimate disposition of the canal and Rodman Dam, as well as the future use of lands previously allocated for canal construction. The usual cast of characters appeared and made their case. On one side, Marjorie Carr, David Anthony, and Jack Kaufman called for a measure that would end the canal controversy once and for all. For them, deauthorization had only one meaning; the removal of Rodman so the river could finally begin the long process toward restoration. On the other side, the few remaining canal supporters tried to convince Floridians that the waterway would yield untold economic benefits for the state. However, if the canal was no longer possible—which was becoming increasingly obvious—they argued Rodman Dam should at least remain to preserve the vibrant recreational playground of Lake Ocklawaha. While a forest had indeed been lost under the tracks of a wantonly destructive crusher nearly twenty years beforehand, a new environment had taken its place. They now extolled the virtues of the reservoir as a thriving ecosystem where wildlife was abundant. In many respects, arguments from both sides were simply old news; a rehashing of time-worn and seemingly intractable positions. Yet the hearings provided frustrated state officials the chance to publicly push the canal controversy toward resolution. At the Palatka meeting of June 1985, Governor Bob Graham made clear the state's unequivocal position. "We do not want this canal, period," he asserted. "We have many, many needs—we need new schools, more teachers, roads, bridges, mass transit, water and sewer lines—but there is one thing we don't need, and that's the Cross Florida Barge Canal." Furthermore, he complained that years of congressional inaction had exacerbated the problem. Understanding that the lack of resolution allowed for a continuing victory, albeit one with little reward, for procanal forces, he called on Washington for an ultimate solution. "Congressmen, how many times do we have to say no? How many ways are there to say no? Please—take no for an answer."[62]

In the Senate, Lawton Chiles remained a staunch advocate of deauthorization. And Paula Hawkins, who spent most of her term vacillating on the issue, finally came down firmly for the project's termination. Her stance came at a price, however, which profoundly troubled FDE. She could support eliminating the canal "on the condition that the locks, dams, and other canal structures now in place will be permanently maintained for the benefit of boaters, fishermen, and other sportsmen who utilize the structures." This left open the very real possibility that Rodman Dam would remain standing. On the other side of Capitol Hill, an overwhelmingly majority of Florida's representatives signaled support for deauthorization. Democrat Buddy MacKay and Republican Clay Shaw of Fort Lauderdale fought hardest for the issue, authoring bills to finally end the project. Of course, Bennett, Chappell, and Pepper stood firmly as the lone exceptions to the growing House unanimity. Here, too, though, things were changing. Claude Pepper continued his opposition, but his primary legislative interests related to health care and Social Security. Moreover, his increasing age (by the summer of 1986 he was eighty-five) and his weakening physical condition made him significantly less effective on the House floor. Bill Chappell could also no longer give the fight his undivided attention as rumors of kickbacks from defense contractors swirled around the congressman. That left Charles Bennett as the solitary stalwart, and even he began wavering in his defense of the canal. Never simply a shill for local business interests, Bennett slowly recognized the futility of continually pushing for the project in the face of significant opposition from fellow Floridians. By 1986, he began to see deauthorization as an acceptable compromise, under the condition that the property allocated for canal construction remain in public hands. Bennett's position was gradually moving toward Marjorie Carr's—if the waterway was no longer possible, canal lands could best serve as the basis for a unique linear greenspace across Florida. Under the auspices of either the state or federal government, this swath of land would both protect a unique part of natural Florida from rampant development and provide a suitable return for the millions of tax dollars expended on the project. Even as Bennett came on board, contentious questions remained, especially concerning compensation for lands taken for the canal. Would deauthorization mean that hundreds of parcels of land automatically reverted to their previous owners? Could the state use the land for other purposes, such as the park proposed by Carr and Bennett? Were former owners due reimbursement for lands no longer dedicated for canal construction? Would counties that had spent millions in tax-payer dollars purchasing property for the canal right-of-way similarly expect reimbursement? And if they were due some form of repayment, who would assume responsibility for the compensation, Washington or Tallahassee?[63]

By the fall of 1986, Congress finally addressed these issues head-on when it passed a deauthorization measure as part of a huge $16.3 billion omnibus water resources act. Signed by President Reagan on November 17, Section 1114 of Public Law 99-662 established the "Cross Florida National Conservation Area" and declared "that portion of the barge canal project located between the Eureka Lock and Dam and the Inglis Lock and Dam (exclusive of such structures) is not authorized." As the *Miami Herald* simply put it, the act "drove a stake through the heart of the monster known as the Cross Florida Barge Canal." The law answered many of the questions that had dogged the project for decades. Besides the obvious death of the canal, it allocated the return of $32 million to the six counties (Duval, Clay, Putnam, Marion, Levy, and Citrus) associated with land purchases. It mandated an interagency management plan for former canal lands, under the auspices of the Corps of Engineers, stressing the "enhancement of the environment" and the "conservation and development of natural resources." It demanded a federal presence in the new conservation area and maintained that Washington, not Tallahassee, should "operate, maintain, and manage the lands and facilities." All this sounded wonderful to Marjorie Carr and FDE. And yet, victory was not complete. The law granted the Corps of Engineers, the bane of environmental activists, a significant voice in the disposition of the canal. Worse yet, "the Secretary [of the Army] shall operate the Rodman Dam, authorized by the Act of July 23, 1942 . . . in a manner which will assure the continuation of the reservoir known as Lake Ocklawaha." The canal was finally, irrevocably dead, but the continued operation of Rodman assured that the Ocklawaha River would not yet flow freely.[64]

Major questions persisted over the relationship of the state and federal governments and their role in the new conservation area. State officials adamantly asserted that Florida, not Washington, should manage and operate the conservation area. As late as 1989, the state legislature had even issued a memorial calling for an amendment to the bill that would ensure greater state latitude in disposing of canal lands. Charles

Rodman Dam (renamed Kirkpatrick Dam in 1998) became the centerpiece of the continuing struggle to remove the last vestiges of the canal from the Ocklawaha River. Courtesy of Florida Defenders of the Environment (FDE), Gainesville.

Bennett was just as convinced that federal control, through some sort of national park designation, would best provide protection for the land. At the same time, private citizens prepared for legal battles over the taking of their lands for a project that had now been terminated without their consent or input. Not surprisingly, the Corps had yet to finalize its management plan within the stipulated one-year period. And, most important, the contentious issue of Rodman Dam remained unsettled.

Finally, in 1990, everything came into place as Congress took up the canal question once again. In the four years since deauthorization, both Claude Pepper and Bill Chappell had died, stilling the most vigorous and long-lasting voices favoring canal completion. In an ironic twist, Charles Bennett himself introduced a new deauthorization proposal in January. Under pressure from Tallahassee officials who sought state control of canal lands, the Jacksonville congressman now maintained that "the most acceptable solution is to create a state park or state conservation area [that] would be like a Phoenix bird arising from the ashes of yesterday's idea." A month later, Senator Bob Graham and Republican congressman Cliff Stearns of Ocala, who had filled MacKay's seat, offered an alternative bill that closely resembled Bennett's. "I am very hopeful something will come out of this Congress," Bennett announced, "to put this behind us and create something valuable for future generations." FDE gladly supported these measures more than the 1986 legislation because Congress had now placed the future of Rodman Dam in the hands of the State of Florida, where environmental activists thought they could exert more influence. An FDE press release applauded the bills as "a constructive and surprisingly happy solution to a problem that has been plaguing Floridians for 20 years." Carr herself was positively ebullient. "The people of Florida," she announced, "will have a beautiful greenbelt now, instead of this albatross hanging around their necks." Florida state officials joined the love fest. Republican Governor Bob Martinez remarked that the "passage of this legislation would make a fitting end of the misguided era of 'ditch and drain.'" Looking forward to a future without the same tireless annual debates about the fate of the Cross Florida Barge Canal, Fred Ayer, assistant director of the Canal Authority, announced: "now it's up to the state to go ahead and finish it off. . . . The idea of a linear park from coast to coast is pretty exciting."[65]

With Bennett's agreement to the compromise measure, the entire Florida congressional delegation threw its support behind the bill. In May 1990, the Florida legislature passed its own measure that agreed to the terms of the federal legislation, assuming that it passed through Congress and was signed by the president. Submitted that same month, the compromise proposal was folded into another large omnibus water resources act. Enduring a summer of obligatory hearings, Congress finally passed the measure on October 27 and sent the bill to President George H. W. Bush. One month later, Bush signed the comprehensive act, ending once and for all the Cross Florida Barge Canal. The legislation immediately deauthorized the canal and transferred all project lands to the State of Florida. The state, in accordance with its own statute, received the lands to "create a State park or conservation/recreation area" and preserve and manage them "for the benefit and enjoyment of present and future generations of people and the development of outdoor recreation." Like the 1986 act, the 1990 statute stipulated a reimbursement of $32 million to the six counties involved in the project. This time, however, Florida assumed responsibility for the payment, with funds originating from the assets of the state's Canal Authority and the Navigation District. Finally, the law proscribed a two-year turnover for the

Corps to transfer land and management responsibilities to the state. As the *Miami Herald* explained, the law "is akin to a team of doctors deciding to cut off life support to a comatose patient." Conspicuously absent from the legislation was any mention of the disposition of Rodman Dam and the pool/lake/reservoir/impoundment that lay behind it.[66]

Epilogue

From Canal to Greenway

With the January 22, 1991, passage of a resolution signed by newly inaugurated Governor Lawton Chiles and the cabinet agreeing to the terms
of the federal deauthorization bill, the era of the Cross Florida Canal was
finally, mercifully, over. Though Marjorie Carr and FDE could justifiably
feel proud of their efforts in that victory, they also understood that the
fight was far from over. Profound questions remained over the shape of
the seventy-seven-thousand-acre park that was to take the place of the
canal. Legal issues concerning the control of former canal lands threatened to keep Florida in a state of continuous litigation for years to come.
The rancorous debate over the fate of the Ocklawaha River loomed largest of all. If that was not solved to the satisfaction of Marjorie Carr, if
the dam still remained blocking the river, would the years of hard work
be in vain? After all, this all had started as a campaign to save the river
itself. Yet, divergent groups offered differing visions of recreation in the
space that was to be the canal. The decisions made by Florida politicians
on these concerns would determine the very nature of the entity that
would replace the Cross Florida Barge Canal.

In the summer of 1991, the Florida legislature began the process of
preparing to decide how to use and develop the swath of land now bureaucratically designated as the Cross Florida Greenbelt State Recreation
and Conservation Area. To do so, it established the twenty-one-mem-
ber Canal Lands Advisory Committee (CLAC), an advisory board composed of politicians and interested citizens from the surrounding area.
In recognition of her interest and influence, the legislature appointed
Marjorie Carr as the committee's representative of "the public at large."
CLAC's primary responsibility lay in creating a master plan for the best
use of the land. That meant balancing a variety of competing interests,

articulated during more than a year of local public meetings. For Carr and many in FDE, there was not much to debate. They felt such passive recreational pursuits as hiking, biking, and canoeing should stand alone at the center of the greenway experience. As Manley Fuller of the Florida Wildlife Federation explained at a CLAC hearing, "we get sort of nervous about building and paving within the greenbelt." Yet as early as September, the *St. Petersburg Times* pointed to radically different visions of recreational use: "Environmentalists envision picnic tables and horse trails. Developers dream of a marina and motel complex. And home owners hope to restore the original flow of the Withlacoochee River." There were even disputes within those large constituent groups. Many people from Marion County's horse country yearned for a world-class equestrian and agricultural center that seemed more like a tourist trap than a escape from the hustle and bustle of modern living. Jim Eyster, a Crystal River developer, formulated plans for a mammoth 367-slip marina near Inglis Lock on the western boundary of the proposed park, while others were more content with primitive fish camps. And to the east, many of Putnam County's residents remained steadfast in their demand for the retention of Rodman Reservoir as a bass-fishing paradise. Spending the weekend trolling on a bass boat, they saw "something magic about the shout of the adult female when she realizes she has caught her first fish. Take them to Rodman reservoir and enjoy life." All of this was rather alien to Marjorie Carr and her allies. For them, fishing was something better experienced on the free-flowing, densely canopied Ocklawaha with a "canoe or johnboat, . . . not a noisy two-cycle smoke-belching gasoline guzzling outboard engine" powering an expensive rig on the flat and unappealing waters of the stagnant Rodman Reservoir.[1]

On September 17, 1992, following more than a year of deliberation, CLAC met in Ocala to issue its final report on the future of the Greenbelt, now called the Cross Florida Greenway. As an advisory board, its recommendations held considerable weight, but the ultimate fate of the land rested in the hands of state officials. During a two-day meeting, the committee settled a host of difficult issues pertaining to park boundaries, governance, funding, and local land use. With regard to the heavy imprint of canal construction, especially the 1930s excavations and bridge stanchions, CLAC recommended leaving most of the Corps' work intact. In many respects, CLAC validated much of Carr's environmental vision.

The remains of the 1930s bridge abutments still exist in a forested setting near Santos as part of the Marjorie Harris Carr Cross Florida Greenway. Courtesy of John Moran and Florida Department of Environmental Protection, Office of Greenways and Trails, Tallahassee.

Expressing a belief in passive recreation, it rejected the Inglis marina outright and looked cautiously at other relatively invasive forms of outdoor activities. Yet, it abdicated its most important responsibility by refusing to address the controversial issue concerning the ultimate disposition of Rodman Dam and the Ocklawaha River. Instead, it voted 14 to 7 for yet another study, this time a three-year review under the auspices of the St. Johns River Water Management District. This new demand that once again examined the usual technical, environmental, and economic cost-benefits of the reservoir left many members of FDE howling in protest at what they saw as just another round of delays. With Marjorie Carr now weak with emphysema at the age of seventy-seven, FDE officials plaintively conceded their leader would not live to see her dream fulfilled. "The river will not be restored in her lifetime," announced David Godfrey, director of FDE's Ocklawaha Restoration Project. "This decision today means that action may not even begin in her lifetime."[2]

The September committee meeting represented an important transitional moment. Besides wrestling with the issues associated with deauthorization, it also introduced a new player to the debate; State Senator

George Kirkpatrick of Gainesville. A member of the state legislature since 1980, the fifty-three-year-old Democrat quickly became the face of the movement to retain Rodman Reservoir. Contentious and prickly, he reveled in his well-earned reputation as a political street fighter. "I'm someone who comes on the scene asking the questions that these frustrated rednecks have always wanted to ask," he remarked in a 1995 interview. "I keep refusing to take no for an answer. I pound and I pound. . . . I'm perceived as arrogant. But if someone manages to turn me on their side, and I know they're right, then they've got their own personal Rottweiler." As dogged as Henry Buckman, as vitriolic as Tom Adams, and as politically astute as Bill Chappell, Kirkpatrick was more than just another loud-mouthed politician. With the authority of senatorial seniority, he would become the chairman of the powerful Senate Rules Committee in 1993 and remain a bitter adversary of Marjorie Carr and other environmentalists who wanted to see the Ocklawaha flowing freely.[3]

Even before the final CLAC meeting, Kirkpatrick was instrumental in organizing a coalition of interests bent on preserving Rodman Reservoir, which had become a haven for recreational and sportfishing, even considered by some experts as one of the best bass lakes in America. In July, the senator encouraged Dan Canfield, a professor at the University of Florida's Department of Fisheries and Aquatic Sciences, to conduct yet another study—this time designed to refute FDE's claim that the reservoir was nothing more than a weed-congested ecological disaster. Funded in part by the Putnam County Chamber of Commerce, Canfield's forty-six-page report added to the furor over the disposition of Rodman. Pro-Rodman forces now went beyond traditional assaults upon FDE's research as they used Canfield's research to buttress their position to protect the lake. Canfield conceded as much when he wrote that "proponents of restoration have written extensively and eloquently about their concerns," but his study was designed "to determine if a case could be made for Rodman Reservoir." Asserting that Lake Ocklawaha was "not a 'dying' water body that is destined for 'biological senility' in our lifetime," he added the lake "would continue to serve as a refuge for not only fish and wildlife, but also anglers." With consideration of the reservoir's economic benefits for the local Putnam County economy, Canfield reached a simple conclusion: "we recommend that Rodman Reservoir be retained for now. . . . There is no compelling biological/ecological reason to rush restoration at this time." The scientific rationale behind the Canfield report soon became the basis of support for keeping the reservoir intact.[4]

Canfield's research was remarkably effective, especially as he delivered the report on the first day of CLAC's September meeting. Kirkpatrick praised the study as a significant improvement over FDE's examination of the lake, which he claimed had "numbers . . . quoted from a study done in 1988 whose numbers were collected from a report done in 1978 which had been taken straight from biased studies done . . . in the early 1970s." Not surprisingly, FDE dismissed Canfield's conclusions as "garbage." Faced with evidence that had only appeared in the final hours of more than a year of difficult meetings, and with a whirlwind of competing claims circling the room as a result of the study, the CLAC played it safe and, almost by default, concluded that further scientific investigation was necessary. Another round of delays led many FDE members to see another, more sinister reason for the decision in the very person of Senator Kirkpatrick himself. Marjorie Carr blasted him for his strong-arm bullying tactics. "Senator Kirkpatrick has clobbered them [CLAC members]," she fumed. "He has carried out the most intensive campaign of intimidation that I have ever seen. God knows he has clout, but I'd call that a misuse of power." FDE, recognizing Kirkpatrick's power as the incoming chairman of both the Rules Committee and next session's Appropriations Committee, accused the senator of threatening various state agencies with budget cuts if they blocked any effort to study the lake and dam once again. Kirkpatrick downplayed his influence. "My effort," he averred, "has been to make sure that the recommendation is based on accurate information." When asked about his alleged threats, Kirkpatrick played coy. "I didn't do any of that," he said. "I talked to DNR [Department of Natural Resources] and asked how we could come up with a compromise. There's been no threats by me." With the cockiness that would become part of his political persona, he loudly proclaimed that FDE's complaints were "just sour grapes."[5]

In December 1992, the governor and cabinet met in Tallahassee to review CLAC's recommendations on turning the former canal into a linear park. Though the public meeting dealt with many of the broader concerns related to the transitional process, contentious debate centered on the fate of Rodman. Once again, adversaries descended on the Capitol and staked out their positions, hoping to sway government officials their way. This time, however, Marjorie Carr's illness made her too weak to appear in person. Instead, her supporters brought along an emotional videotaped appeal from their leader. In it, Carr called the Ocklawaha "a natural work of art" and asked the cabinet to "restore it and care for it as if it was a Pieta by Michelangelo." She summarily dismissed the economic and recreational concerns of those who pleaded for retaining Rodman Reservoir. "I realize bass fisherman will be inconvenienced," she said. "I trust they will find good fishing in nearby lakes." Heeding Carr's words, Commissioner of Education Betty Castor offered an amendment to the CLAC proposals that overrode their call for another study of the Rodman area. Directing the Department of Natural Resources to "immediately take steps" to "complete the restoration of the free flowing Ocklawaha River," she called for the drawdown of Rodman Reservoir. Backed by Governor Chiles, who expressed frustration with the glacial pace of resolving the controversy, the amendment passed unanimously. This policy statement placed the executive branch and its agencies firmly on the side of Marjorie Carr and river restoration. FDE and fellow environmentalists were elated. Calling the amendment a "wise decision," Timothy Keyser of the Florida Wildlife Federation agreed that "restoration of the wildlife habitat is more important than maintaining a degrading [sic] system." From Gainesville, Carr concurred: "It is a giant step forward for Floridians."[6]

Not all Floridians were as sanguine as Carr. In Putnam County, local fishermen expressed disbelief as the cabinet pulled the plug on Lake Ocklawaha. "I can't imagine how anybody can go to Rodman," announced fishing guide Billy Peoples, "and see what's there and make that kind of decision." Wes Larson of the Putnam County Chamber of Commerce bemoaned the loss of 110 jobs and $7.2 million in annual fishing revenue if the dam was removed. Putnam County administrator Gary Adams concluded: "I think it is a terrible economic blow to Putnam County. I think it is the wrong thing to do for a multitude of reasons." He also assailed Carr's growing influence in Tallahassee. "It appears to me that the Florida Defenders of the Environment had enough clout that they could get their position through." In Gainesville, George Kirkpatrick seconded Adams's assessment. Embittered by the cabinet meeting's result, he took on the very nature by which the decision was reached. Claiming the cabi-

net's vote was "based on strong emotions that had very little relation-
ships to the facts," the senator concluded that the Cabinet "bypass[ed]
an appointed task force and completely rejected all their recommendations. . . . I realize that this is a very well orchestrated political decision,"
he said with no hint of irony. "I realize the people I represent will probably lose." In a moment of self-deprecating sarcasm, he took a personal
swipe at Marjorie Carr herself. "I'm not a scientist, I'm not an eloquent
speaker," he intoned, "and I don't have a T.V. video to show you."[7]

At first glance, the cabinet decision seemed to finally resolve the issue in FDE's favor. However, buried in the language of Castor's amendment was the phrase "upon favorable legislative action," which took the
controversy out of the governor's hands and placed it in the statehouse.
Even FDE recognized the tentative nature of their victory. We are "fully
aware that only half the task is done," David Godfrey admitted. "The
unanimous vote gives us momentum going to the legislature, and that's
a whole other ball game. But it sends a strong message." That message
would be countered by George Kirkpatrick, who warned, "the Cabinet decision Tuesday is far from the final say on the future of the Rodman Dam
and the lower Ocklawaha River." On the other side of the Capitol, Ocala
representative George Albright concurred: "By no means is this cast in
stone." For the next few months, Kirkpatrick and his allies prepared for
battle over the fate of Rodman.[8]

By the next legislative session, George Kirkpatrick dominated the
debate surrounding Rodman Dam. Beating back numerous efforts to
comply with the cabinet's decision, Kirkpatrick instead offered a plan
to fulfill CLAC's demand for further study. By the summer of 1993, the
legislature passed a measure allocating $900,000 for an eighteen-month
examination of Rodman Reservoir. The law called for four possible scenarios for future action—full or partial retention of the reservoir, or full
or partial restoration of the river. In many respects, the study—managed by the newly established Department of Environmental Protection
(DEP), which then subcontracted most of the research to the St. Johns
River Water Management District—was the summation of a generation
of scientific research. And given the divisive nature of much of that work,
the resulting twenty-volume report, submitted in January 1995, offered
no final resolution of the issue. Though it concluded that "no further
studies are necessary to answer the question" concerning Rodman, the

report was often so ambiguous and technically arcane that both sides saw it as confirming their position. George Kirkpatrick most certainly did. After combing the report for the slightest bit of evidence that would favor his cause, he announced that he was "elated by the findings included in the DEP report," which "gave us even greater evidence of the positive environmental impact of the [Rodman] ecosystem." Though small parts of the study may have supported his position, the thrust of the report clearly warmed Marjorie Carr's heart. Hidden in volumes of dense prose was the simple statement—"efforts should be directed instead at restoration of the Ocklawaha River."[9]

Following the report's recommendation, Governor Lawton Chiles ordered the Department of Environmental Protection to begin an immediate drawdown of the reservoir in anticipation of restoration. Kirkpatrick lashed back, informing DEP secretary Virginia Wetherell that he, representing the legislature, and not the governor, was in charge. "Any movement towards restoration on the part of the Department," he asserted, "would be highly presumptive. . . . Any movement towards restoration would presume that the Department has already determined that the legislature will eventually decide against keeping the structure [Rodman Dam]." He added presciently, "This would be highly premature." Thus began what became an annual ritual of Florida politics. With the emergence of spring, the governor and executive agencies, in addition to a majority of the state legislature, would call for the removal of Rodman Dam. And George Kirkpartrick, much like his congressional predecessors who had blocked deauthorization, stood in the way.[10]

When first examined, George Kirkpatrick's commitment to Rodman Dam appeared rather unusual. Representing a university town that stood at the center of the anticanal movement, he seemed out of sync with its environmentally conscious constituency. However, his district stretched far beyond the city limits and embraced rural areas of north central Florida, particularly Putnam County. An avid angler, Kirkpatrick had an affinity for the lake and the good ol' boys who spent whatever free time they had fishing in it. As he once noted, I "represent the interests of the folks who love, use and depend on the Reservoir for their livelihood." He had to, for he recognized more than anyone that his political fate rested in their hands. Left-leaning Gainesville rarely granted its own senator a majority of votes. Thus Kirkpatrick's base of support came from those rural residents who saw him as the lone defender of their way of life. And with the governor and cabinet consistently calling for restoration, he and the people of Putnam County would join forces to fight what they considered an elitist alliance between government bureaucrats and scientific experts, who, at best, ignored them or, at worst, dismissed them as ignorant rednecks.[11]

Ironically, things were coming full circle. In the summer of 1995, a group of Putnam County residents and recreational fishermen organized a group called Save Rodman Reservoir, Inc. to "fight off the wishes of 'those who know better.'" Working within the neopopulist legacy of Ronald Reagan and the conservative revolution, they were determined to protect "their" lake from outsiders, those they considered "paid 'enviro-wonks' [who] pontificated at public hearings about the evil that is Rodman." Relying on strategies strangely similar to the nascent anti-canal movement thirty years earlier, they sought the preservation of Lake Ocklawaha and its new "ecosystem with abundant flora and fauna." "Our band of ragtag supporters had grown into a throng," Kirkpatrick reminisced, "with folks calling and writing from every place imaginable. Weary travelers made the trip to Tallahassee for committee meetings on a weekly basis, sometimes without any plan to speak, but just to be there to make their presence felt. . . . Like a modern day barn raising, they rallied the troops with newsletters, phone calls and faxes. . . . Meanwhile paid consultants and strangers to Rodman pushed the anti-retention agenda." Those very same words could well have described Marjorie Carr's earlier efforts against the Canal Authority and the Army Corps of Engineers. Kirkpatrick's chief legislative aide, Mike Murtha, certainly thought so. "They [FDE] had something they loved back in the Sixties and some bastards came and took it away from them," he exclaimed. "Well, now we have something that we love and some bastards are trying to take it away from us."[12]

Over the next three legislative sessions, Kirkpatrick and his allies did their job well, blocking any effort toward restoration by Governor Chiles and the Department of Environmental Protection. For Marjorie Carr, eighty-two years old and now terminally ill with emphysema, these setbacks must have made it seem as if all her work had been for nought. Rodman Dam—"that obscenity, that ridiculous mistake, that hideous monstrosity"—remained. By the summer of 1997, "feeling lousy," tethered to an oxygen bottle, and forced to move from her cherished Micanopy homestead to a patio home in the middle of Gainesville, Carr plaintively asked, "will I live to see it [the Ocklawaha] run free or not? I don't know." What she did know was that George Kirkpatrick was now the source of all her frustration. Characterizing his defense of Rodman as "an obsession," she added that the senator's success stemmed from the fact that "he is feared and I don't think he cares." Though no longer able to lead the battle for restoration, she still showed signs of her legendary feistiness. She railed against those who failed to see the wisdom of Rodman's removal. She complained that bass fishermen "ought to be ashamed of themselves" for their unyielding support for the reservoir. At the same time, Carr reaffirmed her sentimental attachment to the river, sounding more like Sidney Lanier than a research scientist with a stubborn commitment to the facts. "Once the dam is gone," she reflected, "the manatees will be able to come up there during the winter. What a sight that will be. How lovely that will be."[13]

On October 10, 1997, Marjorie Harris Carr finally succumbed to her illness. Almost immediately, accolades began pouring in for the woman now beatified as "Our Lady of the Rivers." Lawton Chiles commended her as "a true giant in the environmental community. Our state is a truly better place because of her work." Carol Browner, a native Floridian and director of the U.S. Environmental Protection Agency, called her "one of the true pioneers of the movement to preserve what is best about Florida." Bob Graham, who had met with Carr only weeks before her death, said her "name will always be synonymous with conservation." She "served as the environmental conscience for Florida's leaders." Closer to home, her friends and allies within the movement she had created sorrowfully lamented their loss. Her longtime colleague David Anthony reflected on her commitment to the river. Considering she had dedicated nearly forty years of her life to the struggle, he lamented: "it's sad to realize that Marjorie has died without the Ocklawaha running free. It was our dream to have a celebration on its banks." Joe Little, a University of Florida law professor and a veteran of FDE since the 1970s, expressed his "deepest disappointment" that Carr had not lived to see the dam removed. It was "a bitter pill that Marjorie's death leaves us to swallow." Alyson Flournoy, current president of FDE, took Carr's death as a call to action: "Just as she was an inspiration in life . . . [in death] she can only inspire us to continue to work to see that restoration happens. It's the best tribute we can pay to her."[14]

Six days later, almost three hundred people paid their last respects to Marjorie Carr in the crowded sanctuary of downtown Gainesville's First Presbyterian Church. The nearly hour-long service featured eulogies from Joe Little, famed Florida naturalist Al Burt, and Lieutenant Governor Buddy MacKay. Reflecting on Carr's years of activism, MacKay commended her for establishing "the prototype of modern citizen advocacy groups in America." She "redefined our relationship to the environment," he continued, "causing movement from development based on the cash register to an ethic of sustainability." FDE's David Godfrey reinforced the bond between Marjorie Carr and the Ocklawaha with a reading of Sidney Lanier's prosaic ode that had so inspired her to action. Pastor Robert Battles poignantly ended the service by reminding mourners that "Marjorie responded with passionate devotion to the common good. . . . From her, I caught a glimpse of what it means to be a steward of the garden of God." As pallbearers placed the casket in a hearse bound for Carr's final resting place in Gainesville's Evergreen Cemetery, mourners could not help but notice something unusual. A green and white bumper sticker mysteriously appeared on the back window of the big black Cadillac. It read, "Free the Ocklawaha River," a fitting legacy for Marjorie Carr's remarkable life. Her daughter Mimi, who had cared for her in those last difficult years, could only smile as she said, "maybe mother put it there." An editor from the *Gainesville Sun* went a step further. "In death," a headline announced, "she still had last word."[15]

FDE members hoped Carr's demise would signal a change of heart in Tallahassee. Their expectations were buoyed in late May 1998, when the legislature commemorated Carr with the passage of a law that named the Cross Florida Greenway after her. In many respects, it marked the crowning achievement for a woman who had dedicated her life to environmental protection. However, if FDE's membership thought this could provide the political momentum to finally restore the Ocklawaha, they were sadly mistaken. Indeed, the day after the legislature honored Carr with the name change, it also saw fit to memorialize her leading adversary by renaming Rodman Dam after Senator George Kirkpatrick. Calling the senator "an avid bass fisherman, naturalist, and outdoorsman" with a "keen interest in the final disposition of Rodman Dam," the legislature complimented him for leading "the opposition to the removal of the dam throughout his Senate career." It was the worst form of tit-for-tat in an already rancorous debate.[16]

With the turn of a new century, the future of the Ocklawaha still remained unresolved. Even with such federal agencies as the U.S. Forest Service pushing for Rodman's removal, nothing changed. Even with a popular new Republican governor, Jeb Bush, who was publicly committed to restoring the river, nothing changed. Even with Kirkpatrick's forced retirement in 2000, nothing changed. With their nemesis now removed by state-mandated term limits, FDE mistakenly thought they had a chance for success. "Especially with George Kirkpatrick gone," one member asked, "who else is going to be there to champion the dam?" The answer was a bipartisan coalition of north Florida politicians led by Republicans Jim Pickens of Palatka and Jim King of Jacksonville, and Democrat Rod Smith of Gainesville. Smith had not only taken Kirkpatrick's seat, but his passion for the reservoir. Not only would he block restoration efforts, he would even introduce legislation protecting the reservoir as the George Kirkpatrick State Reserve. If such a measure became law, it would make it nearly impossible to remove the dam. Though the legislation was vetoed by Governor Bush in 2003, it remained a legislative perennial, introduced session after session, that demanded FDE's constant vigilance. Even seemingly insignificant issues placed environmental activists on the defensive. Every tax dollar spent on the reservoir's recreational facilities—be they boat ramps, campsites, or bathrooms—reinforced Rodman's permanence. Reservoir supporters argued that after nearly forty years of existence, the artificial lake had become part of the natural environment itself. As one explained: "it's got its own ecology. It's got its own value." The reservoir remained alive, with newspaper headlines as late as the spring of 2007 observing, "Year after Year, It's the Same Dam Debate," and "Ocklawaha Restoration Remains in Limbo."[17]

The dam, now renamed after its staunchest defender, may have endured, but its namesake, former senator George Kirkpatrick, died suddenly on February 5, 2003, at the age of sixty-four. Fittingly, only days beforehand the ex-senator had spent a Sunday improving the fishing in one of his ponds by sinking Christmas trees with concrete blocks. His love of fishing, of course, closely identified him with the long-standing struggle to keep the reservoir intact. Ordinary citizens and politicians alike took a moment at his passing to express just what Kirkpatrick meant to their cause. Ed Taylor, a Palatka resident and president of Save Rodman Reservoir, called the senator "the greatest warrior for the survival of Rodman Reservoir I have ever known. . . . He was also known for stepping on a few toes along the way, no matter who they were attached to. In my dealings with him, the toes he stepped on needed just that." It was a fitting tribute for such a combative man. Representative Joe Pickens recognized both the importance of Kirkpatrick's legislative chicanery and the necessity of continuing that struggle. "We all know," he said, "that I would have no Rodman to protect, no torch to carry, if it were not for Senator Kirkpatrick's lifelong commitment to its preservation."[18]

George Kirkpatrick was no doubt a larger-than-life figure, much like his longtime adversary Marjorie Carr. Just as both were singularly associated with the struggle over the river and the reservoir, they were equally associated with each other. Both were doggedly determined and willing to do whatever necessary to advance their cause. Thus it was only fitting that Kirkpatrick's funeral so eerily paralleled Carr's. His service was not only held in the same downtown Gainesville church, but was officiated by the very same pastor. What must Reverend Robert Battles have thought, knowing he had performed the same ceremony for Marjorie Carr just five years earlier? What did the mourners think when they caught a glimpse of another bumper sticker—this time a blue and white one reading "Save Rodman Reservoir"—attached to the senator's casket? The similarities continued even after the service was completed. Upon leaving the church, Kirkpatrick's funeral procession ended its journey in Gainesville's Evergreen Cemetery, where the senator was laid to rest only yards away from the grave of Marjorie Carr.

Despite the ongoing controversy over the fate of Rodman Dam and the Ocklawaha River, the establishment of the Marjorie Harris Carr Cross Florida Greenway turned the centuries-old boondoggle of a canal into a model conservation project. As Florida representative Bill Grant explained immediately following federal deauthorization, the state now had the chance to "convert an environmental lemon into lemonade for the citizens of Florida." Over the next two decades, Florida's legislature took advantage of an unprecedented opportunity and established a 107-mile greenway dedicated to recreation and natural preservation in a

The land bridge over Interstate 75 represented the completion of the Cross Florida Greenway, a 107-mile linear park along the canal corridor. Courtesy of John Moran and Florida Department of Environmental Protection, Office of Greenways and Trails, Tallahassee.

region undergoing rampant growth and economic development. In 1990, one Marion County resident, excited over the promise of the future, wrote a letter filled with anticipation to the *Ocala Star-Banner*, once the unrivaled voice of procanal boosterism. Recalling the words of Sidney Lanier and William Bartram, Dee Cirino praised "the fruition of our linear park [which] can begin with a system of leisure lanes that lead to a wide oasis of canopied hardwoods, scrub habitat covers, carpets of leaves, and a variety of grasses. We can arrive by car, foot, bicycle, and horseback, until we reach the rivers called Ocklawaha, to the east, and Withlacoochee, to the west. To paddle along shaded waters is to feel the past, understand the present, and be consoled that the future brings hope for 'natural' adventure. For those who think these kind of dreams, this can be a tribute to the future as well."[19]

Marjorie Carr's legacy lies in the Cross Florida Greenway's natural uses so wonderfully captured by Ms. Cirino. Yet canal supporters have their monuments as well. Kirkpatrick Dam and Rodman Reservoir on the eastern end of the park represent an effort to fundamentally tame nature. That the greenway contains elements of both nature and human endeavor points to the ambiguous relationship between human beings and the environment. Politics, science, and economics are deeply embedded in the story of the Cross Florida Barge Canal and its important place

The natural beauty of the free-flowing Ocklawaha River still inspires those who want the last vestiges of the Cross Florida Barge Canal removed. Courtesy of Florida Defenders of the Environment (FDE), Gainesville.

in Florida's history. The story is so compelling because Florida's past is at once its future—as the questions raised by the canal and its legacy continue to persist. After nearly two centuries of extraordinary effort, the unintended consequences of canal development and the unfulfilled dreams of canal boosters created a broad ribbon of undeveloped protected land amid the suburban sprawl that in many respects characterizes modern Florida. The irony of this would no doubt make Marjorie Harris Carr smile.

Guide to Manuscript Collections by Location

ARCHIVAL AND MANUSCRIPT COLLECTIONS

Alexandria, Virginia

U.S. Army Corps of Engineers, Office of History, Humphreys Engineer Center
    Arthur Maass Papers
    U.S. Army Corps of Engineers Papers

Ann Arbor, Michigan

Bentley Library, University of Michigan
    Arthur Vandenberg Papers (microfilm)

Atlanta, Georgia

National Archives and Records Administration, Southeast Region
    Office: Jacksonville District Comprehensive Reports and Studies Files
    Records of the Office of the Chief of Engineers (RG 77)

Austin, Texas

Lyndon Baines Johnson Presidential Library
    Lyndon B. Johnson Papers
    Office Files of Lawrence F. O'Brien
    Presidential Papers, White House Central Files

Bozeman, Montana

Montana State University, Merrill G. Burlingame Special Collections
    William Milner Roberts Papers

Chapel Hill, North Carolina

University of North Carolina, Southern Historical Collection
    Don Shoemaker Papers

Cocoa, Florida

Florida Historical Society
  Miscellaneous materials pertaining to Florida canals and the Intercoastal Waterway

Columbia, South Carolina

South Caroliniana Library, Modern Political Collection, University of South Carolina
  Olin D. Johnston Papers

Gainesville, Florida

Papers in possession of authors
  John Couse Papers
  Mike Murtha Papers
  Bill Partington Papers
P. K. Yonge Library of Florida History, George A. Smathers Library, University of Florida
  Charles Andrews Papers
  Reubin Askew Papers
  Charles Bennett Papers
  Robert Lucas Black Papers
  Napoleon Broward Papers
  Farris Bryant Papers
  Archie Carr Papers
  William Chappell Papers
  John Henry Davis Papers
  Robert "Bert" Dosh Papers
  Florida Defenders of the Environment Papers
  Gleason Family Papers
  Ulysses Gordon Papers
  Bob Graham Papers
  Robert Alexis "Lex" Green Papers
  Theodore Hahn Papers
  Joseph Hendricks Papers
  Albert Sydney Herlong Papers
  Spessard Holland Papers
  IFAS Vice-President for Agricultural Affairs Papers
  Arthur Marshall Papers
  James Hardin Peterson Papers
  Paul Rogers Papers
  George Smathers Papers
  John James Tigert Papers
  Park Trammell Papers
  Wilson Cypress Company Records

Samuel Proctor Oral History Program, Pugh Hall, University of Florida
Reubin Askew interview transcript
Charles E. Bennett interview transcript
Farris Bryant interview transcript
Marjorie Carr interview transcript
Lawton Chiles interview transcript
Thomas LeRoy Collins interview transcript
Don Fuqua interview transcript
Bob Graham interview transcript
R. A. "Lex" Green interview transcript
Claude Kirk interview transcript
Donald R. "Billy" Matthews interview transcript
Claude Pepper interview transcript
Nathaniel Reed interview transcript
George Smathers interview transcript

Hyde Park, New York

Franklin D. Roosevelt Presidential Library
Leland Olds Papers
Franklin Roosevelt Presidential Papers
Samuel Rosenman Papers

Independence, Missouri

Harry S. Truman Presidential Library
Harry S. Truman Presidential Papers

Jacksonville, Florida

Thomas G. Carpenter Library, Special Collections, University of North Florida
John E. Mathews Jr. Papers
Jacksonville Historical Society Archives, Swisher Library, Jacksonville University
Jacksonville Public Library
Vertical files—Cross Florida Barge Canal
Vertical files—Cross Florida Barge Canal, Canals

Norman, Oklahoma

Carl Albert Center, Congressional Archives, University of Oklahoma
Carl Albert Papers
Page H. Belcher Papers
James V. McClintic Papers
Elmer Thomas Papers

Ocala, Florida

Marion County Public Library
  Vertical files—Cross Florida Barge Canal, Silver Springs

Palatka, Florida

Putnam County Public Library, Special Collections
  Chamber of Commerce Papers

Tallahassee, Florida

Florida State Archives
  Tom Adams Papers
  Canal Authority of the State of Florida, Series 127
  Hubbard L. Hart Papers
  State Board of Conservation Papers
Claude Pepper Library, Claude Pepper Center, Florida State University
  Spessard Holland Papers
  Claude Pepper, Official and Personal Papers
Strozier Library, Special Collections, Florida State University
  Malcolm Johnson Papers

Tampa, Florida

Hillsborough County Library Main Library, Tampa, Special Collections
  Vertical files—Cross Florida Barge Canal
University of South Florida Library
  Tom Adams Papers
  Hampton Dunn Papers

Washington, D.C.

Library of Congress, Madison Building
  Geography and Map Division
National Archives and Records Administration, Archives II (College Park, Maryland)
  Central Files, State 1935–1944, Florida
  Richard M. Nixon Presidential Papers
  White House Special Files, Staff Member and Office Files
    John W. Dean III Papers
    Egil Krogh Papers
  White House Central Files, Staff Member and Office Files
    John C. Whitaker Papers
  White House Central Files, Subject Files
    Natural Resources
  Works Progress Administration (WPA) Papers (RG 69)

Winter Park, Florida

Olin Library, Special Collections, Rollins College
    Vertical Files—Florida, Cross Florida Barge Canal
    Gilbert Youngberg Papers
Winter Park Public Library, Winter Park History and Archives Collection
    Paula Hawkins Papers

AUDIOVISUAL MATERIALS

Austin, Texas

Lyndon Baines Johnson Presidential Library, University of Texas
    Recordings of Conversations and Meetings
    Recordings of Telephone Conversations—JFK Series, tape K6312.12
        Conversation with Spessard Holland, 8/23/65
        Conversations with George Smathers and Martin Anderson, 12/20/63
    Recordings of Telephone Conversations—White House Series, tape
    WH6508.09

College Park, Maryland

National Archives and Records Administration, Archives II
Motion Picture Sound and Video Records, Special Media Archives Services Division
    March of Time 1936, "Florida Canal," 4/17/36, newsreel R1 (a)

Gainesville, Florida

Florida Defenders of the Environment Office
    Interview with Marjorie Carr on the Ocklawaha
    Progress, Pork-Barrel, and Pheasant Feathers
    VHS cassettes of documentaries
Interviews conducted by authors and their research assistants (in possession of authors)
    David Anthony
    Margie Bielling
    David Bowman
    David Carr
    Stephen Carr
    Paul Ferguson
    Holly Fisher
    Wayne Little
    Mike Murtha
    William Partington
    Cleveland Powell

Steve Specht
Tom Tyler
JoAnn Valenti
Jim Vearil
John Whitaker
Victor Yannacone

Miami, Florida

Special Collections, Florida International University Library, FAU-FIU Joint Center for Environmental and Urban Problems
  Filmed interviews conducted June 2–4, 1983
  Special Voices: A Tale of Two Women

Tallahassee, Florida

Claude Pepper Library, Claude Pepper Center, Florida State University
  AV A (1197) Tape labeled "Senator Claude Pepper letter of thanks," August 29, 1962
  Claude Pepper interview on Cross State Canal
Department of Environmental Protection, Office of Greenways and Trails
  Crusher video, VHS cassette
  Lyndon B. Johnson, February 1964 Canal Groundbreaking Dedication speech reel-to-reel original, cassette copy

Florida State Archives

Florida Photographic Collection (videos)
  V-16, no. 51, *The Case for the Canal*
  V-11, no. 116, *Florida's Canal, Main Street USA*

Acknowledgments

In the spring of 2001, we were presented with what seemed at the time a simple project—to participate on a grant to develop a master plan for the Marjorie Harris Carr Cross Florida Greenway, a 107-mile linear park administered by the State of Florida's Office of Greenways and Trails. That project would consume eight years of our lives and lead us in directions we could never have imagined. Along the way, we have discovered that, contrary to popular belief, working collaboratively did not fray our friendship; in fact, it made it stronger. This work, however, is not simply a collaboration between two authors; instead, it involved the efforts of literally dozens of people scattered throughout the United States. Without their help, this book would not have been possible. We are deeply indebted to them all.

Les Linscott of the University of Florida's Department of Landscape Architecture gave us the opportunity to work on the Greenway grant by providing a place for historians amid experts in geomatics, recreational studies, and ecological sciences. Gail Hansen and Laura Namm worked closely with us on the project, showing us the value of a true interdisciplinary approach to history and environmental issues. None of this would have been possible without the support of the Florida Department of Environmental Protection's Office of Greenways and Trails (OGT), which not only funded much of our research but also provided unlimited encouragement to our efforts on both the grant and this book. Their open-door policy to our seemingly endless queries was refreshing, especially at a time when many government agencies are restricting access to public records. Special thanks go to OGT Director Jena Brooks and her staff, who allowed us to wander through their Tallahassee offices for more than two weeks, gathering materials and wearing out photocopiers. More locally, Mickey Thomason, OGT's Central Region manager, supported us throughout by providing numerous contacts, locating once-lost materials, and arranging countless visits to various locations on the Greenway itself. On all of those trips, Dave Bowman was our indispensable guide. Having been associated with the Ocklawaha River and its environs for almost forty years, Dave provided a historical and ecological context and an expertise to an issue too often marred by polemics.

Much like the canal itself, our research has taken us across the state of Florida. We particularly would like to thank the directors and staffs at the archives and special collections of Florida State University, Rollins College, University of South Florida, University of North Florida, and Jacksonville University. Special thanks go to the staff of FSU's Claude Pepper Library, who offered us access to an incredible treasure trove of materials. We would also like to acknowledge the help of the Florida State Library and Archives in Tallahassee. Closer to home, we especially appreciate the assistance of the staff of the University of Florida's various libraries, most notably the P. K. Yonge Library of Florida History. At the University of Florida, Jim Cusick, John Nemmers, and Gary Cornwell helped us pursue many leads and avoid at least as many dead ends. Other institutions across the state also aided our work, including public libraries in Gainesville, Jacksonville, Ocala, Orlando, Palatka, Tallahassee, Tampa, and Winter Park. The private research collections of the Florida Historical Society in Cocoa and the Florida Federation of Garden Clubs state headquarters in Winter Park provided gracious hospitality and numerous sources. We also found valuable materials in the libraries of the state's Southwest Florida Water Management District in Brooksville, and the St. Johns River Water Management District in Palatka. Finally, a lot of our sources and inspiration stemmed from informal discussions with editors and staff at the *Daytona Beach News-Journal*, the *Gainesville Sun*, the *Ocala Star-Banner*, and the *Palatka Daily News*.

Since the barge canal was a national issue, much of our work crossed state lines. We are especially indebted to the directors and staff of the Library of Congress in Washington D.C., and the National Archives at College Park, Maryland, and National Archives Southeast Region at Atlanta, Georgia. We would also like to acknowledge the help of the staffs of the following institutions and repositories for their assistance: Lyndon B. Johnson Library, Franklin D. Roosevelt Library, Harry S. Truman Library, the University of Oklahoma's Carl Albert Center's Archives and Collections, and the Arthur Vandenberg Papers at the University of Michigan. Special thanks go to the archivists at the Army Corps of Engineers, IWR Maass-White Reference Room, located on the grounds of Fort Belvoir, Virginia.

The source of much of our material was a series of oral history interviews generously funded by OGT and coordinated under the auspices of the University of Florida's Samuel Proctor Oral History Program. Among those interviewed was a core of individuals who provided us with much more than their memories, notably David Anthony, Margie Bielling, David Carr, Wayne Little, and Bill Partington. Their knowledge, leads, and downright common sense seriously added to the quality of our work and reaffirmed the importance of this project. Special thanks go to Margie Bielling and Bill Partington for kindly sharing their personal files and papers with us. Those materials added greatly to the richness of our story. Over the years, we have shared elements of this story in a variety of academic and public conferences, meetings, and panels. The questions and comments of interested participants and audience members were especially valuable in sharpening our focus and opening new areas of inquiry.

Without the assistance of our institutions, this project would have never come to fruition. Special thanks go to Joe Spillane, chairman of the University of Florida's Department of History, for his constant support and celebration of our efforts. Though above all else a teaching institution, Santa Fe College has also recognized the importance of our research by providing the flexibility for us to pursue this work. Along the way, both graduate and undergraduate students from the University of Florida have aided our efforts as research assistants. Thanks go to Alan Bliss, Evan Nooe, and Kelly Chang. We would also like to thank Peggy McDonald and Leslie Poole, UF graduate students working on similar projects, for sharing their work and insights.

Other people seemingly peripheral to our project contributed much-needed assistance that improved the quality of our work. Joe Knetsch, unofficial dean of Florida historians, helped us navigate the sometimes byzantine world of Florida's bureaucracy. The always gracious Wayne Flynt, emeritus professor of history at Auburn University, assured us that our research was on the right track as he was working on the history of the Ship Canal and Florida politics in the 1930s. Bill Belleville and Karen Ahlers reminded us of the value of the Ocklawaha River at crucial points while we were wading deep in historical documentation.

Holly Fisher shared an insider's view of the 1960s environmental movement that could not be ascertained by archival materials alone. Nick Williams, director of the Florida Defenders of the Environment, provided us with irreplaceable documents, photographs, and film that made this project better. Finally, John Mayfield, chairman of Samford University's Department of History, and Chris Beckmann, history teacher at Oak Hall School in Gainesville, showed us that our investigation of the Cross Florida Barge Canal could be more than a manuscript; it could also provide a valuable teaching tool for students to better appreciate Florida's history and environment.

From the beginning of this long journey, we have greatly benefitted from the example and support of two of Florida's most eminent historians: David Colburn and Gary Mormino. Their faith and enthusiasm pulled us through many a time when we thought this project seemed, like the canal itself, dead in the water. They were always there to answer our questions no matter how arcane or ridiculous. Bertram Wyatt-Brown, for many years the heart and soul of the University of Florida's graduate history program, has been another source of inspiration. More than a mentor, he has been a friend and advocate of our professional development since we were graduate students in the 1980s.

Our positive experience with the University Press of Florida has resulted from the hard work of our editors, Eli Bortz and Meredith Morris-Babb. Their efforts focused our thoughts and tightened our prose. The readers for this work offered invaluable comments, suggestions, and lots of red ink that has greatly improved the final product. Special thanks go to Joy Wallace Dickinson, Fritz Davis, and especially Cynthia Barnett, whose careful editing provided a solid foundation for our revisions.

Finally, our friends and family deserve the last word. They have endured this project as much as we have. Over the last eight years, discussions about the Ocklawaha River have become as much a part of our Friday afternoon basketball games as air balls and blown layups. Our friends suffered through interminable conversations about our research and writing. That the weekly game still persists is a testament to their patience and good temper. The same applies to our families, who had to live through this enterprise the other six days of the week. For Steve, he has seen his children, Jody and Amanda (and recent addition, Jamie McGraw) graduate from high school, go off to college, and strike out on their own. Meanwhile his wife, Beverly, has remained steadfast, maintaining more optimism and tolerance than anyone should be allowed. For Dave, his eight-year-old daughter, Tai Min, has literally grown up with this undertaking. For her, the history of the Cross Florida Barge Canal is synonymous with Sunday-morning writing sessions that often dragged on to late in the afternoon. In the process, Dave's wife, Faith, lost her office, but never her enthusiasm for our work. For all our families, the completion of this endeavor is both exciting and long overdue. It is with great pride that we dedicate this book to them, for we know that it is as much theirs as ours.

Notes

MH	*Miami Herald*
NA	National Archives, College Park, Md.
NARASE	National Archives and Records Administration, Southeast, Atlanta, Ga.
NYT	*New York Times*
OGT	Office of Greenways and Trails, Florida Department of Environmental Protection, Tallahassee
OL	Olin Library, Rollins College, Winter Park, Fla.
OS	*Orlando Sentinel*
OSB	*Ocala Star-Banner*
PBP	*Palm Beach Post*
PDN	*Palatka Daily News*
PKY	P. K. Yonge Library of Florida History, University of Florida, Gainesville
RG	Record Group
RNP	Richard M. Nixon Presidential Papers
SHP	Sydney Albert Herlong Papers
SN	*Stuart News*
SPOHC	Samuel Proctor Oral History Collection, University of Florida, Gainesville
SPT	*St. Petersburg Times*
TD	*Tallahassee Democrat*
TT	*Tampa Tribune*
WCP	William Chappell Papers
WJ	*Waterways Journal*
WP	*Washington Post*
WPP	William Partington Papers
WSJ	*Wall Street Journal*

INTRODUCTION

1. "Return to the River," *OS*, May 5, 2002. Throughout the years, the spelling of the Ocklawaha River has undergone many iterations. We have left those spellings intact in all quotations. In our own narrative, we use the current standard spelling, "Ocklawaha."

2. Marjorie Harris Carr to Marlin Perkins, May 29, 1977, RG 1, Series 1, Box 6, Folder 1977–1, FDE, PK; John Pennekamp, "Canal Project Is Still in Doubt," *MH*, May 19, 1972.

3. "The Case for Restoring the Free-Flowing Ocklawaha River," Florida Defenders of the Environment, www.fladefenders.org/publications/CaseForRestorationSect1. html#canal; "Rodman Reservoir Facts," Save Rodman Reservoir, Inc., www.rodman-reservoir.com/facts.htm; "Another Chapter in the Long Life of Rodman's Darned Dam," *Daytona Beach News-Journal*, March 25, 2007.

4. Cynthia Barnett, "Florida Ports Racing to Handle Giant Cargo Ships," www.flori-datrend.com/article.asp?aID=49254; Deirdre Conner, "Water Wars: Ocklawaha, Rodman Dam Are Flash Points, *FTU*, May 18, 2008.

5. "Canal Less Harmful Than Disney," *GS*, January 25, 1971.

6. Nancy Lee Rogers to Canal Lands Advisory Committee, July 13, 1992, Accession II, RG 1, Series 1, Box 33, Folder CLAC Correspondence 1991–1992, FDE, PKY; Carr and Anthony to Commissioners of Alachua, Citrus, Clay et al., July 13, 1965, Adverse

File, ACE, OGT; "Return to the River," *OS*, May 5, 2002; "Career 'Troublemaker' Loves His Job," *TT*, May 13, 1990; Marjorie Harris Carr, "A Florida Scandal," November 1996, www.fladefenders.org/publications/FloridaScandal.html.

CHAPTER 1. SURVEYS AND STEAMBOATS

1. Barber, "The History of the Florida Cross-State Canal," 10.

2. "Roads and Canals—Report of the Committee on so much of the President's Message as Relates to Roads and Canals, February 7, 1817," *Niles' Weekly Register* 11, no. 286 (February 22, 1817): 423–27.

3. "Memorial of James Gadsden and E. R. Gibson, Commissioners Appointed by the Legislative Council of Florida, to Examine into the Expediency of Opening a Canal through the Peninsula of Florida, January 5, 1826," FHSL.

4. "Memorial to Congress by the Legislative Council, December 28, 1824," in *Territorial Papers of the United States*, ed. Clarence Carter (Washington, D.C.: National Archives, 1958), 23: 136; "Internal Improvements, House of Representatives, February 26, 1825, Report of the Committee on Roads and Canals, upon the Subject of Internal Improvements, accompanied by a bill concerning 'Internal Improvements,'" *Niles' Weekly Register* 28, no. 708 (April 9, 1825): 91.

5. "John C. Calhoun, Documents from the War Department, The Secretary of War to the President of the U. States, Department of War, December 3, 1824," *Niles' Weekly Register* 27, no. 693 (December 25, 1824): 265, 267; Richard Call to House Committee on Canals and Roads, February 24, 1824, included in "Internal Improvements, House of Representatives, February 26, 1825, Report of the Committee on Roads and Canals, upon the Subject of Internal Improvements, accompanied by a bill concerning Internal Improvements," *Niles' Weekly Register* 28, no. 708 (April 9, 1825): 95; "An act Appointing Commissioners to report on the expediency of opening a canal from the Gulf of Mexico to the Atlantic Ocean, December 8, 1825," Florida Historical Legal Documents, http://fulltext10.fcla.edu/cgi/t/text/pageviewer-idx?c.

6. Simon Bernard to John Quincy Adams, July 7, 1827, in *Territorial Papers*, ed. Carter, 23: 855 n. 17.

7. Anonymous, "Florida," *American Quarterly Review* 2, no. 3 (September–December 1827): 236. There is some information that suggests the article was written by Achille Murat, a French émigré to Florida, nephew of Napoleon and a major land and plantation owner in Leon County, but this is unsubstantiated. Joseph White to Peter Porter, July 8, 1828, in *Territorial Papers*, ed. Carter, 24: 33–34; "Atlantic and the Gulf of Mexico," *Niles' Weekly Register* 36, no. 922 (May 16, 1829): 189.

8. John Pickell, *Report of a survey for a canal across the peninsula of Florida to connect the waters of the Atlantic with those of the Gulf of Mexico, March 6, 1832*, House of Representatives Document 185, Jacksonville Public Library.

9. "Memorial to Congress and Resolution by the Legislative Council, February 16, 1834," in *Territorial Papers*, ed. Carter, 24: 976–77.

10. M. L. Smith, "Report on Survey for a Ship Canal across the Peninsula of Florida, May 1, 1855," Special Collections, Tampa–Hillsborough County Public Library, 25, 32.

11. "Florida—The Key to the Gulf," *DeBow's Review and Industrial Resources* 1, no. 3 (September 1856): 283–86.

12. Ibid., 286; "Ship Canal across the Peninsula of Florida," *DeBow's Review and Industrial Resources* 14, no. 2 (February 1853): 122–25.

13. *Tallahassee Floridian*, April 14, 1860, in "The Florida Railroad Story," www.flarr.com/frrstory.htm.

14. George Barbour, *Florida for Tourists, Invalids, and Settlers* (New York: D. Appleton and Company, 1884), 126–27.

15. Marjorie Carr, "The Oklawaha River Wilderness, *FN* 38 (August 1965): 3.

16. Mitchell, *Paddle-Wheel Inboard*, 6; Martin, *Eternal Spring*, 169; Joseph Siry, "What Was the Ocklawaha Once Like?" http://web.rollins.edu/~jsiry/OklawahaRiver.html#Overview.

17. Martin, *Eternal Spring*, 107; George Bancroft to his wife, March 17, 1855, in "'A Tale to Tell from Paradise Itself': George Bancroft's Letters from Florida, March 1866," ed. Patricia Clark, *FHQ* 48, no. 2 (January 1970): 272.

18. Mitchell, *Paddle-Wheel Inboard*, 16.

19. Edward Mueller, *Ocklawaha River Steamboats* (Jacksonville, Fla.: Mendelson Printing, 1997), 7.

20. Ibid., 6; Lanier, *Florida*, 20.

21. Barbour, *Florida for Tourists*, 127; Lanier, *Florida*, 18; Glicksberg, "Letters of William Cullen Bryant from Florida," *FHQ* 14 (April 1936): 268–69; Harriet Beecher Stowe, *Palmetto Leaves* (1873; facsimile ed., Gainesville: University Press of Florida, 1999), 261–62.

22. Stowe, "Up the Okalawaha," 393; Glicksberg, "Letters of William Cullen Bryant from Florida," *FHQ*, 268–69.

23. "Picturesque America," 582–83; Stowe, "Up the Okalawaha," 393.

24. *Minutes of the Board of Trustees, Internal Improvement Fund of the State of Florida*, July 22, 1881, vol. 3, *1881–1888* (Tallahassee, Fla.: J. B. Hilson, 1904), 18; Hubbard Hart to Board of Trustees, Internal Improvement Fund, July 22, 1881, ibid., 19–21.

25. *Charter and Letters Patent of the Florida Atlantic and Gulf Ship-Canal Company* (New York: Evening Post Steam Presses, 1881), 4, 70, FHSL.

26. Jackson, "Sidney Lanier in Florida," 120; Lanier, *Florida*, 21, 38.

27. Carr, "Oklawaha River Wilderness," 1; Lanier, *Florida*, 20.

28. "Picturesque America," 582; Stowe, "Up the Okalawaha," 393; Martha Holmes, "Log Book of the *Okahumkee* [sic] from Jacksonville up the Ocklawaha and back to Palatka," handwritten manuscript, 6, PKY.

29. "Picturesque America," 583.

30. *Raymond & Whitcomb's Tour Brochure for the Season of 1904*, 9, 33–35, PKY.

31. "The Great Florida Phosphate Boom," *Cosmopolitan* 14, no. 6 (April 1893): 706–14; *New York Republic*, April 9, 1890, quoted in J. Lester Dinkins, *Dunnellon: Boomtown of the 1890s* (St. Petersburg, Fla.: Great Outdoors, 1969), 97.

32. "Great Florida Phosphate Boom," 708; *Florida Times-Union*, quoted in Dinkins, *Dunnellon*, 111; L. L. Aiken, "A Tribute to a Benefactor," *Ocala Banner*, n.d., 1899.

33. A. Dix Stephens, *Withlacoochee Notes* (Trenton, Fla.: Lulu Press, 2008).

34. "10-Foot Channel Is Endorsed," *Marion County Advocate*, August 7, 1914.

35. "Death of Col. Hart—Killed by an Electric Car at Atlanta Thursday Morning," *Palatka Times-Herald*, December 13, 1895.

36. "1904 Hart Line Brochure," Series M81–007, Box 2, Folder Publicity Materials,

Hubbard Hart Papers, FSA; *FTU*, March 20, 1903; *Palatka News and Advertiser*, March 20, 1903.

37. Shaffer and Greenwood, *Maud Powell*, 328; Maud Powell, "Up the Ocklawaha (An Impression)," *San Francisco Examiner*, December 22, 1912.

38. "1904 Hart Line Brochure"; Mueller, *Ocklawaha River Steamboats*, 18.

39. Marion County Board of Trade to Major W. B. Ladue, August 25, 1915, "Proposed Canal Silver Springs to Ocala, Florida," Box 03317, Folder 31, ACE, Jacksonville District Papers, NARASE; W. W. Fineren to District Engineer Officer, Jacksonville, Florida, December 12, 1916, "Proposed Canal Silver Springs to Ocala, Florida," Box 03317, Folder 38, ibid.; "Notice of Unfavorable Report, February 28, 1917," "Proposed Canal Silver Springs to Ocala, Florida," Box 03317, Folder 38, ibid.

40. *Raymond & Whitcomb's Tour Brochure*, 32.

41. *Joint Memorial of the Legislature of Alabama Relating to Ship Communication between the waters of the Gulf of Mexico and the Atlantic Ocean by ship-canal through the Florida Peninsula, March 11, 1872*, 42nd Cong., 2nd sess., March 11, 1872, H. Misc. Doc. 110; Josiah T. Walls, "Speech of Hon. Josiah T. Walls, M.C., of Florida before the Transportation Committee of the United States Senate, January 28, 1874," Florida Collection, Jacksonville Public Library; *Resolution of Mr. Jones of Florida*, 47th Cong., 1st sess., December 14, 1881, S. Misc. Doc. 16.

42. William Gleason to Frank Sherwin, June 25, 1878, Box 1, Folder 1878–1881, Gleason Family Papers, PKY.

43. *Second and Third Biennial Supplement to Johnson's New Universal Cyclopedia: A Scientific and Popular Treasury of Useful Knowledge*, ed. Frederick Barnard and Arnold Guyot (New York: A. J. Johnson and Co., 1885), 245. Much of the older scholarship maintains that Disston killed himself over his financial problems. Recent work tends to emphasize natural causes. See Michael Grunwald, *The Swamp: The Everglades, Florida, and the Politics of Paradise* (New York: Simon and Schuster, 2006), 96; and Army Corps of Engineers, "Draining the Swamp: Development and the Beginning of Flood Control in South Florida, 1845–1947," www.evergladesplan.org/docs/river_interest/river_int_chap_01.pdf, esp. n. 18.

44. *Papers & Map to Accompany Memorial of Major Robert Gamble of Tallahassee Showing the Advantages and Value of a Barge Canal*, 53rd Cong., 3rd sess., December 1894, S. Misc. Doc. 37, 3. For his letter, see "Florida Ship Canal," *DeBow's Review* 1, no. 4 (October 1852): 419–20.

45. Buker, *Sun, Sand and Water*, 75.

46. Baldwin, "St. John's Bar," 340.

47. McGee, "Our Inland Waterways," 291–92.

48. Marion County Board of Trade to Major W. B. Ladue, August 25, 1915, Box 03317, Folder 31, ACE, Jacksonville District Papers, NARASE.

49. Lynn Alperin, History of the Gulf Intercoastal Waterway, www.usace.army.mil/usace-docs/misc/nws83-9/; G. D. Luetscher, "Atlantic Coastwise Canals: Their History and Present Status," *Annals of the American Academy of Political and Social Science* 31 (January 1908): 99.

50. "Proposed St. Mary's–St. Mark's Canal, Urged by Governor Dorsey Will Link Inland Waterway System Covering Half of America," *Atlanta Constitution*, June 9, 1918.

51. "The Inland Waterways Commission," 556–57.

52. Editorial, *Ocala Banner*, March 20, 1903.

1. Phillips, *Transportation in the Eastern Cotton Belt to 1860*, 361.

2. Duncan Fletcher, "The Florida Trans-Peninsular Ship Canal—It's [*sic*] Meaning," *Pensacola Journal*, March 20, 1910.

3. *Why You Should Be a Member of or Subscribe to the Mississippi to Atlantic Inland Waterway Association*, pamphlet, n.d. (ca. 1909), Series 2, Box 10, Folder Miscellaneous Publications, Napoleon Broward Papers, PKY; "By-Laws and Resolutions of the Mississippi to Atlantic Waterway Association," n.d. (ca. 1909), Series 1, Box 9, Folder September 1909, ibid.; *Why You Should Be a Member of or Subscribe*.

4. Duncan Fletcher to Napoleon Broward, September 14, 1909, Series 1, Box 9, Folder September 1909, Napoleon Broward Papers, PKY; Jacksonville Board of Trade to Officers of the Atlantic Deeper Waterways Association, August 27, 1910, Series 1, Box 9, Folder 1910, ibid.; "Projects of the Atlantic Deeper Waterways Association," Philadelphia 1909, Series 4, Box 12, Folder Atlantic Deeper Waterways Association, ibid.

5. *Intracoastal Waterway across Florida Section*, 63rd Cong., 1st sess., September 11, 1913, H. Doc. 233, 10–11.

6. "Proposed St. Mary's—St. Mark's Canal, Urged by Governor Dorsey, Will Link Inland Waterway System Covering Half of America," *Atlanta Constitution*, June 9, 1918.

7. Chapter 8578 (No. 183), June 24, 1921, *Laws of Florida*, 1921 (Tallahassee, Fla.: T. J. Appleyard Printers, n.d.), 1: 389; *Help to Secure the All-American Canal*, pamphlet published by the Florida State Canal Commission, 1926, 9, Jacksonville and Florida Collections, Jacksonville Public Library, Jacksonville.

8. Joseph Arnold, *The Evolution of the 1936 Flood Control Act* (U.S. Army Corps of Engineers, 1988), 11, www.usace.army.mil/publications/eng-pamphlets/ep870-1-29/.

9. Buckman, *Documentary History*, 1; *Inland and Coastal Waterways of Florida. Report of the Florida Inland and Coastal Waterways Association*, in Buckman, 1.

10. Sinclair Chiles, *A Florida Cross-State Ship Canal* (Bethlehem, Pa.: Bethlehem Printing, 1928), Box 8, Folder 2, GYP, OL.

11. Gilbert Youngberg, "The Gulf-Atlantic Ship Canal across Florida," *Florida Engineering Society Bulletin*, no. 3 (August 1935): 3

12. Buckman, *Documentary History*, 7; Gilbert Youngberg, "The Gulf-Atlantic Ship Canal across Florida: An Economic Study," Supplement B, Box 5, Folder 2, GYP, OL.

13. Youngberg, "The Proposed Gulf-Atlantic Ship Canal Across Florida," 136; Youngberg, "Jacksonville in Relation to the Gulf-Atlantic Ship Canal Across Florida—Remarks Made at a Meeting of Life Insurance Underwriters' Association at Jacksonville, March 21, 1932," Box 5, Folder 2, GYP, OL.

14. Proceedings of the 24th Convention of the National Rivers and Harbors Congress, December 5–6, 1928, 132, Folder 466 ACE, Office of History, Humphreys Engineer Center, Fort Belvoir, Alexandria, Va.; "Waterways Serving Florida," *Tampa Daily Times*, January 24, 1930; A. F. Knotts, *The Across Florida Canal*, undated pamphlet, FHSL; F.R.S. Phillips, for the Florida State Canal Commission, to A. F. Knotts, September 7, 1929, Box 03321, Folder 67, ACE, NARASE.

15. "Resolution by the National Gulf-Atlantic Ship Canal Association, Adopted at New Orleans, March 31, 1932, in Buckman, *Documentary History*, 27; "Jacksonville's Growth as Port," *WSJ*, December 14, 1931.

16. "$175,000,000 U.S. Aid Asked to Cut Cross-Florida Canal," *WP*, July 19, 1932.

17. Buckman, *Documentary History*, 30–31.

18. "Ship Canal Loan Not Favored," *WSJ*, October 22, 1932.

19. *1933 Laws of Florida—General Laws*, vol. 1 (n.p., n.d.), 714–15.

20. Syllabus, U.S. Army Corps of Engineers Report, December 30, 1933, Box 03321, Folder 67, ACE, NARASE; *Atlantic-Gulf Ship Canal, Fla.* 75th Cong., 1st sess., April 5, 1937, H. Doc. 194, 2.

21. Morris Sheppard to Franklin Roosevelt, March 21, 1933, Box 635, Folder Florida Ship Canal—1933, FDR; Lex Green to Bert Dosh, March 22, 1933, Box 1, Folder Canal 1930–39, BDP, PKY; Governor David Sholtz to Franklin Roosevelt, March 18, 1933, in Buckman, *Documentary History*, 80; Joint Memorial of the Senate and House of Representatives of the State of Florida, May 27, 1933, ibid., 82–83.

22. Report of the President's Board of Review, June 28, 1934, in Buckman, *Documentary History*, 105–10; Supplemental Report of Board of Review to the President, September 15, 1934, 3, Office Files, Box 635, Folder Florida Ship Canal—1934, FDR; Major General E. M. Markham to Franklin Roosevelt, January 19, 1935, Office Files, Box 635, Folder Florida Ship Canal—1935, ibid.

23. Franklin Roosevelt memorandum for Secretary of War, January 21, 1935, Office Files, Box 635, Folder Florida Ship Canal—1935, FDR; "Cash Promised for Cross-Florida Canal," *Reporter Star* [OS], April 15, 1935.

24. "Canal Group Plans New Activity," *GS*, August 14, 1935; "FDR Calls for Canal Data," *GS*, August 14, 1935.

25. John Alsop to Duncan Fletcher, August 21, 1935, Office Files, Box 635, Folder Florida Ship Canal—1935, FDR; "Sore Thumb," *Time*, February 17, 1936, 11–12; George Hills to Marvin McIntyre, August 22, 1935, Office Files, Box 635, Folder Florida Ship Canal—1935, FDR; Sumter Lowry to Duncan Fletcher, August 25, 1935, ibid.

26. Hills to McIntyre, August 22, 1935, Office Files, Box 635, Folder Florida Ship Canal—1935, FDR.

27. Ibid.; "Works Board Prepares Fund for Canal Jobs," *TT*, August 20, 1935.

28. "Concerning the Canal," *Tampa Morning Tribune*, June 20, 1935.

29. "Canal Will Avoid Hazards Says Summerall," *GS*, September 6, 1935; Duncan Fletcher to Franklin Roosevelt, September 3, 1935, Office Files, Box 635, Folder Florida Ship Canal—1935, FDR; "Disaster: Heroism and Tragedy in Florida's 'Mild Hurricane,'" *Newsweek*, September 14, 1935, 12; "Sore Thumb," *Time*, February 17, 1936, 11–12.

30. Tentative Program for the Construction of the Atlantic Gulf Ship Canal Across Florida, War Department, Corps of Engineers (Washington, D.C.) 1934, Box 635, Folder Florida Ship Canal—1935, FDR; *Atlantic-Gulf Ship Canal*, 75th Cong., 1st sess., April 5, 1937, H. Doc. 194, 46.

31. Address by Walter F. Coachman Jr. before the National Rivers and Harbors Congress, December 9, 1931, in Buckman, *Documentary History*, 25.

32. "Engineers in Charge of Canal Operations Here to Start Construction," *Ocala Evening Star*, September 5, 1935.

33. Frank Parker Stockbridge and John Holliday Perry, *So This Is Florida* (Jacksonville: John H. Perry Publishing, 1938), 191; "Canal Digging Started," *TT*, September 15, 1935; CAM, July 2, 1936, 1: 204, OGT.

34. "Canal Digging Started," *TT*, September 15, 1935.

35. "Roosevelt Pushes Key to Start Canal Construction in 1935," *OSB*, May 5, 1996.

36. "146,000,000 Canal Cash Held Assured," *OS*, September 20, 1935; "President Sets Off Explosion to Start Canal," *TT*, September 20, 1935.

37. "President Sets Off Explosion to Start Canal," *Tampa Morning Tribune*, September 20, 1935.

38. Duncan Fletcher to Senator Elmer Thomas, November 21, 1935, including enclosures of letter from Fletcher to Franklin Roosevelt, November 7, 1935, and description of commemorative tray presented to the president, Legislative Series, Box 18, Folder 57, Elmer Thomas Papers.

39. "Beginning the Florida Canal," *ENR*, April 2, 1936, 479, 483.

40. *NYT*, November 24, 1935; "Boom Awakens Sedate O-Can-Ala," *OS*, October 13, 1935; "Canal Digging Started," *TT*, September 15, 1935; editorial, *SPT*, September 20, 1935.

41. "Canal Digging Started," *TT*, September 15, 1935; John Snell to J. Hardin Peterson, September 18, 1935, Box 50, Folder Cross-State Canal 1935 (1), HPP, PKY; Peterson to Snell, September 21, 1935, ibid.; Sidney Brown to J. Hardin Peterson, September 5, 1935, ibid.

42. R.E.L. Chancey to Harry Hopkins, August 20, 1935, WPA Central Files, RG 69, State of Florida 1935–1939, Box 1092, Folder Florida State Canal 1935–1938, NA.

43. Kenneth Van der Hulse, *A Report on Conditions in Marion County Resulting from the Gulf-Atlantic Ship Canal Project*, 1, 4, Box 30, Florida Ship Canal Papers, PKY; "Canal Digging Started," *TT*, September 15, 1935.

44. Van der Hulse, *Report on Conditions*, 12, 5.

45. "Nailed to a Cross, with Lips Sewed," *NYT*, March 19, 1936, 3.

46. *Report of Proceedings of Hearings Held in Jacksonville, Florida by the Special Board of Army Engineers*, February 10, 1933, in Buckman, *Documentary History*, 74; James M. Howe to Franklin Roosevelt, June 12, 1933, Office Files, Box 635, Folder Florida Ship Canal—1933, FDR.

47. Communication from Harry Slattery, Personal Assistant to the Secretary of the Interior, to Representative J. Hardin Peterson, in Buckman, *Documentary History*, 154.

48. Resolution by the Board of County Commissioners, Hillsborough County, June 24, 1935, in Buckman, *Documentary History*, 149; Address by Marvin H. Walker, editor of the *Florida Grower*, "The Stygian Canal," at Winter Haven, June 12, 1935, in Buckman, *Documentary History*, 148, 144.

49. *Report of the Special Board of Geologists and Engineers Appointed by the District Engineer of Ocala, Florida*, December 18, 1935, in Buckman, *Documentary History*, 159–60; *NYT*, December 27, 1935.

50. *Report of Proceedings of Hearings Held in Jacksonville, Florida by the Special Board of Army Engineers*, February 10, 1933, in Buckman, *Documentary History*, 72; "Current Events," *Literary Digest* 120 (November 2, 1935): 18; L.H.H. Calhoun to Harold Ickes, September 17, 1935, Office Files, Box 635, Folder Florida Ship Canal—1935, FDR; Dudley V. Haddock to unknown, September 21, 1935, ibid.

51. "Memorandum—Proposed Canal between Atlantic & Gulf, December 6, 1933, Growers and Shippers League of Florida, Orlando, Florida," Box 25, Folder Cross-Florida Canal, Spessard Holland Papers, PKY; Frank Anderson, *A Brief Against the Construction of the Proposed Florida Cross-State Ship Canal on Behalf of the Central and South*

Florida Water Conservation Committee (Central and South Florida Water Conservation Committee, February 3, 1936).

52. "Peterson Acts to Halt Canal," *Tampa Daily Times*, August 24, 1935.

53. "Resolution of the Anti-Canal Committee of the Winter Park Garden Club, February 11, 1937," Florida Federation of Garden Clubs Board Reports, 1930–1936, Florida Federation of Garden Clubs Library, Winter Park, Box 52, Folder 1937, HPP, PKY; J. Hardin Peterson to Emily Herron, May 24, 1937, ibid.

54. *NYT*, December 18, 1935; Ickes, *The Secret Diary of Harold L. Ickes*, 1: 488–89.

55. Arthur Vandenberg to Richard H. Mahard, January 28, 1936, Roll 2, AVP; "Big Michigander," *Time*, October 2, 1939, 16. Vandenberg also graced the magazine's cover that week. Arthur Vandenberg to Frank B. Shutts, March 17, 1936, Roll 2, AVP.

56. Arthur Vandenberg to Duncan Fletcher, January 2, 1936, Roll 2, AVP; Vandenberg to James Peterson, January 4, 1936, ibid.; Hearings before a Subcommittee of the Committee on Commerce, United States Senate (74th Cong., 2nd sess.), on Senate Resolution 210, in Buckman, *Documentary History*, 193–97.

57. "Moon Unharnessed from Recovery," *Literary Digest* 121 (February 22, 1936): 8.

58. "Division on Canal Persists in Florida," *NYT*, March 22, 1936; "War Department Appropriation Bill for the Fiscal Year Ending June 30, 1937," 74th Cong., 2nd sess., H.R. 11035, Amendment Proposed in the Senate by Senator Fletcher, *Congressional Record*, March 16, 1936, in Buckman, *Documentary History*, 312; *Congressional Record*, March 17, 1936, in Buckman, *Documentary History*, 338.

59. Arthur Vandenberg to Edwin P. Thomas, May 16, 1936, Roll 2, AVP; Frank Kay Anderson to Marvin H. McIntyre, May 24, 1936, Office Files, Box 635, Folder Florida Ship Canal—1936, FDR.

60. Henry Buckman, "The Florida Canal—Correspondent Summarizes Arguments Favoring Relief Project," *WP*, April 5, 1936; Arthur Vandenberg, "The Post Has Nothing to Take Back," *WP*, April 8, 1936; Buckman to Vandenberg, April 10, 1936, Roll 2, AVP; Vandenberg to Buckman, April 11, 1936, ibid.; Central and South Florida Water Conservation Committee to Buckman, April 11, 1936, ibid.

61. David Sholtz to Marvin McIntyre, March 7, 1936, Office Files, Box 635, Folder Florida Ship Canal—1936, FDR; G. B. Pillsbury to Marvin McIntyre, March 11, 1936, ibid.

62. "What's Ahead," *WSJ*, April 17, 1936; "Two States Ponder Pet Projects' Fate," *NYT*, April 19, 1936.

63. "Senate in Race for Adjournment, Hurdles Relief Bill but Stumbles over Tax Measure," *Newsweek*, June 6, 1936, 8–9; Frank Parker Stockbridge, "The Chamber of Commerce of Jacksonville, Florida Presents the Truth about the Florida Ship Canal," May 1936, Florida Canals Folder, Florida Vertical File, OL.

64. "Speech of Representative Green in the House of Representatives, June 15, 1936," in Buckman, *Documentary History*, 461; "Ship Canal 'Death' Is Seen in Florida," *NYT*, June 28, 1936; "Senator Duncan U. Fletcher Dies in His Washington Home from Effects of Heart Attack," *FTU*, June 18, 1936.

CHAPTER 3. THE DREAM DEFERRED

1. "Abandoned Camp of Florida Canal Finds Partial Use," *CSM*, November 16, 1936; "Ship Canal 'Death' Is Seen in Florida," *NYT*, June 28, 1936; Blake, *Land into Water—*

Water into Land, 161–62; Paul Mathis, "Memo on Camp Roosevelt," n.d. (ca. 1946), University of Florida Archives, Public Records Collection, Series P7b, Office of the President, John James Tigert General Correspondence, 1929–1957, Box 9, Folder 1946 (3), PKY.

2. Lowry, "Gulf-Atlantic Ship Canal," 7; "Army Engineers Back Completion of Florida Canal," *CSM*, November 17, 1936; "Hope Stirs for Canal and for Quoddy Also," *NYT*, November 22, 1936.

3. "Is It Worth the Gamble?" *DeLand Sun News*, December 9, 1936; "Florida Canal Plan Is Threatened by Water Supply Issue," *NYT*, December 11, 1936; "Railroads Assail Canal for Florida," *NYT*, December 18, 1936; "Split Is Widening on Florida Canal," *NYT*, December 27, 1936; "The Atlantic States and the Florida Canal, A Statement by the Honorable John H. Small, Vice-President, The Atlantic Deeper Waterways Association before the Board of Engineers for Rivers and Harbors, U.S.A., Washington, D.C.," December 16, 1936, 5, Folder Florida Canals, Florida Vertical File, OL; "Renews Warfare on Florida Canal," *NYT*, December 31, 1936.

4. "Report of the Board of Engineers for Rivers and Harbors, February 24, 1937," Summary of House Document No. 194, *Atlantic-Gulf Ship Canal, Fla.*, April 5, 1937, Box 7, Folder History, OGT.

5. "Examination and Survey of the Florida Ship Canal, April 1, 1937," Box 7, Folder History, OGT; "Boost Florida Canal," *Business Week*, April 10, 1937, 32; "The Florida Canal," *CT*, January 17, 1937.

6. Charles Summerall to Ship Canal Authority, June 4, 1937, CAM, 2: 304, OGT.

7. "Pepper v. Sholtz v. Wilcox," *Time*, May 2, 1938, 9–10; Alexander Stoesen, "Claude Pepper and the Florida Canal Controversy," *FHQ* 50, no. 3 (January 1972): 238; "Pepper v. Sholtz v. Wilcox," *Time*, May 8, 1938, 12; letter to the editor, Thomas Swanson, *MH*, n.d., clipping in RG 201, S 200, Box 53, Folder 13, CPP.

8. Claude Pepper diary entry, April 4, 1937, RG 400, Series 439, Box 1, Folder 1, CPP; letter to the editor, Thomas Swanson, *MH*, n.d., clipping in RG 201, S 200, Box 53, Folder 13, CPP.

9. "Florida Canal Passage Seen," *CSM*, December 14, 1937.

10. Memorandum from Claude Pepper to James Roosevelt, May 24, 1938, Box 635, Folder Florida Ship Canal—1938, FDR; memorandum from Henry Buckman to Claude Pepper, June 21, 1938, RG 201, S 200, Box 53, Folder 10, CPP.

11. *Report of the National Resources Committee to President Franklin Roosevelt, Frederic Delano, Chairman*, January 16, 1939, Box 635, Folder Florida Ship Canal—1939, FDR.

12. CAM, 2: 356–58, OGT.

13. "Schedule of Official Inspection Tour of the Atlantic-Gulf Ship Canal, January 18–23, 1939," Box 635, Folder Florida Ship Canal—1939, FDR; "Schedule of Official Inspection Tour of the Atlantic-Gulf Ship Canal, March 2–6, 1939," RG 200, Series 201, Box 53, Folder 14, CPP; Claude Pepper diary, March 11, 1939, RG 400, Series 439, Box 1, Folder 3, ibid.; Franklin Roosevelt to Josiah Bailey and Joseph Mansfield, January 16, 1939, Box 635, Folder Florida Ship Canal—1939, FDR; "Schedule of Official Inspection Tour of the Atlantic-Gulf Ship Canal, January 18–23, 1939," ibid.

14. Mrs. W. T. Walters to Roosevelt, January 19, 1939, Box 635, Folder Florida Ship Canal—1939, FDR; Fred P. Cone to Roosevelt, January 21, 1939, ibid.

15. Miami Beach Chamber of Commerce to Roosevelt, January 19, 1939, Box 635,

Folder Florida Ship Canal—1939, FDR; editorial, *MH*, January 18, 1939; editorial, *Miami Daily News*, January 18, 1939; "In Today's News," *MH*, January 20, 1939.

16. W. E. Briggs to Claude Pepper, January 20, 1939, RG 201, Series 200, Box 53, Folder 11, CPP; W. H. Frasse to Pepper, April 6, 1939, RG 200, Series 201, Box 53, Folder 15, CPP; Coral Gables Chamber of Commerce to Arthur Vandenberg, January 13, 1938, Roll 2, AVP; Resolution of Greater Melbourne Chamber of Commerce, January 3, 1938, ibid.; "Vandenberg and Pepper Clash on Ship Canal for Florida," *WP*, January 20, 1939.

17. H.R. 3223, introduced January 25, 1939; Series 1100, introduced February 1, 1939, 76th Cong., 1st sess.; House Committee on Rivers and Harbors, *Authorizing the Completion of the Construction of the Atlantic-Ship Canal Across Florida*, 76th Cong., 1st sess., April 26, 1939, H. Rep. 509, 2; Testimony of H. V. Borges before the House Committee on Rivers and Harbors, n.d., 1, Series 201, Box 53, Folder 15, CPP.

18. House Committee on Rivers and Harbors, *Authorizing the Completion of the Construction of the Atlantic-Ship Canal Across Florida*, 76th Cong., 1st sess., April 26, 1939, H. Rep. 509, 9–11.

19. *Congressional Record*, 76th Cong., 1st sess., 1939, vol. 84, pt. 5, p. 5583; "Again the Florida Canal," *WSJ*, May 3, 1939; Ickes, *Secret Diary of Harold Ickes*, 2: 592.

20. "Florida Canal Waste Blasted by Vandenberg," *CT*, May 17, 1939; *Congressional Record*, 76th Cong., 1st sess., 1939, vol. 84, pt. 5, p. 5647.

21. Randall Chase to J. Hardin Peterson, May 19, 1939, Box 52, Folder Cross State Canal 1939 (1), HPP, PKY; M. L. Cullum, Sales Manager Chase & Company to Gentlemen, May 19, 1939, ibid.; J. Hardin Peterson to J. K. Singletary, May 23, 1939, Box 52, Folder Cross State Canal 1939 (2), ibid.

22. Miami Beach Chamber of Commerce to Arthur Vandenberg, June 1, 1939, Roll 2, AVP; "Senate Kills New Deal Florida Canal," *CT*, May 18, 1939.

23. Buckman to Pepper, June 21, 1939, in Stoessen, "Claude Pepper and the Florida Canal Controversy," 246; "Notice of Public Hearing on the Provision of an Intracoastal Waterway from Anclote River (Tarpon Springs) to St. Marks, Fla., and a Barge Canal across the State of Florida to Connect with the St. Johns River," U.S. Engineers Office, Jacksonville, Fla., April 5, 1940, ACE, OGT.

24. "Plans for Canal Project Are Outlined," *FTU*, August 2, 1939.

25. Henry Buckman, *Defense Coordination of the Panama and Florida Canals—A Preliminary Study*, 76th Cong., 3rd sess., April 30, 1940, S. Doc. 198.

26. "Remarks by Senator Pepper in the Senate Wednesday, April 10, 1940," RG 200, Series 201, Box 54, Folder 8, CPP; "Pepper Urges Construction of Canal," *Ocala Banner*, April 12, 1940; "Canal Urged as Important for Defense," *FTU*, April 11, 1940; "Pepper Tells Senate Florida Canal Is Vital to Defense of Nation," *JJ*, April 11, 1940; "Why Dig This Ditch," *Orlando Morning Sentinel*, April 11, 1940.

27. "Fin" to Claude Pepper, August 15, 1940, RG 200, Series 201, Box 54, Folder 8, CPP; C. Lyman Spencer to Claude Pepper, September 2, 1940, ibid.

28. "Again the Florida Canal," *WSJ*, May 3, 1939.

29. "Senate Kills New Deal Florida Canal: Roosevelt Rebuffed on Pet Scheme," *CT*, May 18, 1939; *The Truth about Taxes*, 1940 Republican party film, www.archive.org/details/TruthAbo1940; H. C. Kopf letter to the editor, *WSJ*, August 25, 1941.

30. Henry Buckman to Bert Dosh, March 14, 1941, Box 1, Folder Canal 1940–1941, BDP, PKY.

31. "Mr. Chase on the Canal," *OS*, February 18, 1941.

32. "Florida Canal Unnecessary, Survey Shows," *MH*, December 31, 1941; "Canal Out for Duration, *TT*, February 5, 1942; "Florida Barge Canal Urged to Offset U-Boats," *TT*, February 10, 1942.

33. Walter Coachman to Members of Executive Committee of Canal Counties, January 24, 1942, Box 1, Folder Canal 1940–49, BDP, PKY; Walter Coachman to Members of Executive Committee of Canal Counties, February 2, 1942, Box 1, Folder General Correspondence, 1940–49, ibid.; M. C. "Doc" Izlar, February 20, 1942, Box 1, Folder Canal 1940–49, ibid.; Coachman to John Campbell et. al, April 23, 1942, Box 1, Folder General Correspondence, 1940–49, ibid.; "Perry Urges Barge Canal to Win War," *JJ*, April 7, 1942; telegram from M. C. "Doc" Izlar, Marion County Chamber of Commerce to Joe Hendricks, January 25, 1942, Box 1, Folder Cross Florida Canal June 1940–June 1942, JHP, PKY.

34. Arthur Vandenberg to Harold Ickes, January 23, 1942, Roll 3, AVP; "Help Fight the Canal and You Will Help Lick the Axis!" news bulletin, Florida Citrus Producers Trade Association, April 4, 1942, Box 53, Folder Cross State Canal 1942, JHP, PKY.

35. "Screen for the Big Ditch," *MH*, May 29, 1942.

36. William Pittenger to Franklin Roosevelt, April 24, 1942, Box 635, Folder Florida Ship Canal—1942, FDR; Eugene Reybold to Joseph Mansfield, June 12, 1942, *Waterway across Northern Florida for Barge Traffic*, 79th Cong., 1st sess., H. Doc. 109; H.R. 6999, 77th Cong., 2nd sess., April 27, 1942.

37. Franklin Roosevelt to Joseph Mansfield, March 25, 1942, in *Enlargement and Extension of the Gulf Intracoastal Waterway, Including the Construction of a Barge Channel and Pipeline across Northern Florida—Hearings before the Committee on Rivers and Harbors House of Representatives, 75th Congress, 2nd Session on H.R. 6999*, 13; Lex Green, oral history interview, August 11, 1970, 52, SPOHC; Arthur Vandenberg to Boake Carter, June 12, 1942, Roll 3, AVP; Carrie Anderson to "Pete" Peterson, June 2, 1942, Box 53, Folder Cross State Canal 1942, HPP, PKY; telegram from Randall Chase to Peterson, June 2, 1942, ibid.; Agnes Purnell to Peterson, June 2, 1942, ibid.

38. Charles Summerall to Carl Hayden, June 14, 1942, Box 1, Folder Cross Florida Canal June 1940–June 1942, JHP, PKY; "Predicts New Vote on Florida Pipeline," *NYT*, June 9, 1942; Lex Green to "Colleagues," June 15, 1942, Box 1, Folder Cross Florida Canal June 1940–June 1942, JHP, PKY; "House Passes Bill for Florida Canal, Pipeline for Oil," *NYT*, June 18, 1942.

39. Dave Curtis to "Pete" Peterson, June 18, 1942, Box 53, Folder Cross State Canal 1942, HPP, PKY; Peterson to Curtis, June 22, 1942, ibid.; Joe Hendricks to Claude Pepper, June 25, 1942, Box 1, Folder Cross Florida Canal, June 1940–June 1942, JHP, PKY.

40. "Vandenberg Hits Bill for Pipeline," *NYT*, July 6, 1942; "Florida Canal Bill Assailed in Senate," *NYT*, July 17, 1942; "Canal Bill Passed," *WSJ*, July 18, 1942; "Signs Florida Canal Bill," *NYT*, July 24, 1942.

41. Statement on canal and pipeline bill by Senator Pepper of Florida, July 18, 1942, Box 53, Folder Cross State Canal 1942, HPP, PKY; telegram from Joe Hendricks to Jane Harris, July 17, 1942, Box 1, Folder Cross Florida Canal July 1942–December 1942, JHP, PKY; telegram from Wilson Turnipseed to Joe Hendricks, July 17, 1942, ibid.

42. Memorandum from Franklin Roosevelt to Director of Budget, August 22, 1942, Box 635, Folder Florida Ship Canal— 1942, FDR; "Canal 'Dead'—Again," *TT*, October 23, 1942.

43. "Florida Canal Would Relieve Gas Shortage," *WP*, January 18, 1943; "Forty Four Million Asked by Hendricks for Barge Canal," *Orlando Sentinel-Star*, January 20, 1943; Claude Pepper diary entry, September 26, 1942, Series 439, Box 1, Folder 7, CPP; "Here Is How We Can Get OIL," *WP*, February 24, 1943; *Congressional Record*, 78th Cong., 1st sess., vol. 89, pt. 3, p. 4071; Charles Summerall to Raymond Baldwin, January 16, 1943, Box 1, Folder Cross State Canal 1943, JHP, PKY; H.R. 1353, January 20, 1943, ibid.; "Funds for Florida Canal Voted Down by 21 to 19," *CSM*, March 31, 1943; "Ditch Resurrected," *Time*, April 19, 1943, 23.

44. Ralph Davies to George Dondero, March 20, 1943, Box 53, Folder Cross State Canal 1943, HPP, PKY; Joseph Hendricks to Harold Ickes, May 15, 1943, Box 1, Folder Cross Florida Canal 1943, JHP, PKY; Ickes to Hendricks, June 4, 1943, ibid.; Hendricks to Franklin Roosevelt, July 2, 1943, ibid.; Roosevelt to Hendricks, July 21 1943, ibid.

45. Franklin Roosevelt to Joseph Mansfield, July 1, 1943, Box 635, Folder Florida Ship Canal—1943, FDR; "Definite Project Report—Cross-Florida Barge Canal," U.S. Engineer Office, Jacksonville, Florida, December 1943, ACE, OGT; "Premature Tom-Toms," *TT*, July 12, 1943.

46. Buckman to Hendricks, February 12, 1945, Box 1, Folder Cross Florida Canal, 1944–46, JHP, PKY.

47. Resolution of the Board of County Commissioners of Marion County, August 30, 1945, Box 1, Folder Cross Florida Canal, 1944–46, JHP, PKY; Resolution of the Board of County Commissioners of Putnam County, August 28, 1945, ibid.; John Perry to Abe Hurwitz, August 14, 1945, Box 1, Folder Canal 1940–1945, BDP, PKY.

48. Hal P. Phillips to J. W. Campbell, October 7, 1947, Official File, HST; Harry S. Truman to Charles Andrews, March 16, 1946, ibid.; Harry S. Truman to Margaret Chase Smith, September 24, 1947, ibid.; Memo from Harry S. Truman to Robert A. Lovett, Deputy Secretary of Defense, May 18, 1951, ibid.

49. Henry Buckman to John Perry, January 25, 1951, Box 1, Folder Canal 1950–1959, BDP, PKY; Robert Lovett to Harry S. Truman, May 29, 1951, Official File, HST.

50. Henry Buckman to John Perry, January 25, 1951, Box 1, Folder Canal 1950–1959, BDP, PKY; Robert Dow to Bert Dosh, January 3, 1951, ibid.; Minutes of Meeting of Board of Directors, July 18, 1951, 895, CAM, OGT; Minutes of Meeting of Board of Directors, June 30, 1953, 977, ibid.

51. "Know Your Waterways, Already Two Wars Late," advertisement, *WP*, June 22, 1951, and April 21, 1952.

52. "Freight Boom on the Inland Waterways," *Business Week*, October 3, 1953, 27; "The Intracoastal Waterway," *Time*, October, 1, 1956, 82.

53. Frank Norris to Major General Samuel D. Sturgis, Jr., August 9, 1954, cover letter for brief prepared by Jacksonville Chamber of Commerce on the Cross-Florida Barge Canal, folder 1, Vertical Files, JHS.

54. *Traffic Analysis and Estimated Tonnage Prospectus of the Cross-State Florida Barge Canal*, prepared for Florida Geological Survey and the Ship Canal Authority of the State of Florida, Gee & Jensen, West Palm Beach Florida, November 1956, 52, Florida Vertical File, Folder Cross Florida Barge Canal II, JHS.

55. *Economic Restudy of Cross-Florida Barge Canal*, January 10, 1958, U.S. Army Engineer District, Jacksonville, Corps of Engineers, Jacksonville, Fla., Box 19, Folder 4, ACE, Office of History, Humphreys Engineer Center, Fort Belvoir, Alexandria, Va.

56. "Century's Dream Near at Hand," *OS*, April 9, 1958; "Governor Collins Gives

Okay to Cross-Florida Barge Canal, *JS*, May 1958, 5; "Senator Holland Displays Leadership in Barge Canal Budget Hearing," *JS*, October 1959, 2–3; Allen Ellender to Spessard Holland, June 25, 1959, Box 1, Folder Canal 1950–1959, BDP, PKY.

57. "Truman Tells Bennett of Canal Views," *JS*, May 1959, 11.

58. "Kennedy Backs Florida Canal," *OSB*, October 26, 1960.

59. "Kennedy Asks Money for Barge Canal: Link between Atlantic and Gulf Waterways," *JS*, May 1961, 4; "Governor Bryant Makes Strong Plea for Cross-State Canal, *JS*, July 1961, 1; "Tom Adams Sparks Cross State Canal," *OS*, June 11, 1961.

60. "Cross-Florida Canal Figures in U.S. Defense," *GS*, March 25, 1958; Dale Miller, "The Intracoastal Waterway," address before the National Rivers and Harbors Conference, May 15, 1959, in Barber, "The History of the Florida Cross-State Canal," 271; "Action This Year," *WJ*, July 13, 1963, 5.

61. "Blast Starts Florida Canal," *CSM*, February 28, 1964; The Cross Florida Barge Canal as an Element of National Defense, June 20, 1963, typescript summary of House document, Series Legislative, Box 70, Folder 88, CAP, Carl Albert Congressional Archives, University of Oklahoma, Norman.

62. "Barge Canal Study Funds Win Approval of Senate," *OS*, October 12, 1962; Charles Bennett to Carl Albert, September 24, 1962, Series Legislative, Box 53, Folder 20, CAP; "Green Light at Last for Canal," *OS*, October 14, 1962; Luther Carter, *The Florida Experience: Land and Water Policy in a Growth State* (Baltimore: Johns Hopkins University Press), 277-78.

63. "Bringing Cape Canaveral into Georgia," *Atlanta Journal and Constitution*, March 10, 1963; Henry Buckman to Bert Dosh, March 1, 1963, Box 1, Folder Canal 1960–65, BDP, PKY; Dosh to John Bailey, March 5, 1963, ibid.; Report of the Chairman of the Canal Authority to the State of Florida, Covering Period March 18 to June 5, 1963, CAM 7: 1520, OGT.

64. "Dreams Do Come True," *OS*, November 20, 1963; Bert Dosh to John Bailey, December 28, 1963, Box 1, Folder Canal 1960–65, BDP, PKY.

CHAPTER 4. GROUNDBREAKING, BREAKING GROUND

1. Editorial, *PDN*, February 27, 1964; U.S. Army Corps of Engineers news release, n.d., Cross Florida Canal Project Files, Series 1727, Box 10, Folder 4, FSA.

2. "Barbecue Pits Burned All Night to Cook Chicken for Ranch Lunch," *PDN*, February 27, 1964.

3. "Floridians Wait in Rain to Get View of Johnson," *FTU*, February 28, 1964; "The Rains Came, But So Did President Johnson," *PDN*, February 28, 1964.

4. "Florida Barges Ahead, *OSB*, February 28, 1964; McCree to A. Sydney Herlong, March 2, 1964, Box 57, Folder Waterways—Cross-Florida Barge Canal 1964, SHP, PKY.

5. "Definite Project Report—Cross-Florida Barge Canal, U.S. Engineer Office, Jacksonville, Florida, December 1943," 1, 6, 22, ACE, OGT.

6. *A Review and Appraisal of the Cross Florida Barge Canal* (Bureau of Sports Fisheries and Wildlife, Fish and Wildlife Service, Department of Interior, June 1970), 4; Cross-Florida Barge Canal, Report of Meeting Arranged by Senator Holland with the Bureau of the Budget, Washington, D.C., August 19, 1959, Series 100B, Box 6, Folder Cross Florida Barge Canal, IFAS Vice-President for Agricultural Affairs, Administrative Policy Records, 1949–1962, PKY.

7. "Statement of Robert O. Vernon, State Geologist of Florida on the Geology and Hydrology of Florida as They Relate to the Proposed Florida Barge Canal," given at hearing before Bureau of Budget, Washington, D.C. August 19, 1959, Accordion File, ACE, OGT; Theodore Hahn to editors, *Sailing South*, June 2, 1950, Box 1, Theodore Hahn Papers, PKY.

8. Marjorie Carr, oral history interview, 1984, 5, 15, SPOHC.

9. "Is the Cross-State Barge Canal Worthwhile?" *Alachua Audubon Society Bulletin* 4, no. 1 (November 1962), RG 1, Series 1, Box 1, Folder (1962–64), FDE, PKY; Richard Bowles to John Couse, November 26, 1962, RG 1, Series 1, Box 1, Folder 1962–1964, FDE, PKY.

10. www.fladefenders.org/marjorie/quotes.html; Marjorie Carr, "The Fight to Save the Ocklawaha River," paper presented to the Twelfth Biennial Sierra Club Wilderness Conference, September 1971, Box 1, Folder (1971), FDE, PKY.

11. R. Malcolm Fortson to Willard Fifield, March 21, 1960, Series 100b, Box 6, Folder Cross-Florida Barge Canal, IFAS Vice-President for Agricultural Affairs, Administrative Policy Records, PKY; Fifield to Fortson, March 25, 1960, ibid.

12. "Carl Radder to Marjorie Carr, January 20, 1965, RG 1, Series 1, Box 1, Folder 1965–1, FDE, PKY; "I Just Got Sick of All of the Destruction," *PBP*, September 13, 1970.

13. "A Brief Summary of the Destruction of Natural Resources That Will Result from the Construction of the Cross Florida Barge Canal," Citizens for the Conservation of Florida's Natural and Economic Resources, Inc., n.d., JCP.

14. *A Brief Outline of the Inadequate Economic Justification of the Cross Florida Barge Canal Project*, Citizens for the Conservation of Florida's Natural and Economic Resources, Inc., Special Bulletin No. 2, March 1965, JCP.

15. "River Killers at Work: Now It's the Ocklawaha," *Stuart News*, February 4, 1965; "Brief Outline of the Inadequate Economic Justification," JCP.

16. "Editors Notebook," *Citrus County Chronicle—Dunnellon Press*, Special Supplement, April 29, 1965; "Governor, General Help Dig Ditch at Inglis Lock, Bridge Dedication," *TT*, May 2, 1965.

17. Jim Hughes to Sydney Herlong, January 11, 1965; Box 65, Folder Waterways—CFBC 1965, SHP, PKY; Herlong to Hughes, January 18, 1965, ibid.; "Canal Value Proclaimed," *OS*, May 27, 1965.

18. "Authority Chairman Says Claims 'False,'" *GS*, May 13, 1965; A. Sydney Herlong to James G. Hughes, August 3, 1965, Civil Works Account #07796–0039, Box 7, Folder August–October 1965, ACE, NARASE; D. R. Matthews to James G. Hughes, July 28, 1965, ibid.

19. "Society Asks Saving of Oklawaha Wild River," *FN* 38 (July 1965): 99; Marjorie Carr to Mrs. Forrest, February 12, 1965, RG 1, Series 1, Folder 1965, FDE, PKY; "McCree Blasts Canal Foes, Col. Hodge Doesn't Mind the Label," *GS*, May 13, 1965; "Alachua Pleads for Wild River," *FN*, 38, July 1965, 103.

20. "The Fate of the Oklawaha," *NYT*, July 21, 1965; "Misstatements about Canal," *OSB*, July 25, 1965; Randolph Hodges to John Oakes, August 23, 1965, Adverse File, ACE, OGT.

21. Hodges to Oakes, August 23, 1965, Adverse File, ACE, OGT; Dosh to Koperski, August 4, 1965, NARASE.

22. Joseph Livingston to "Editorial Friends," August 31, 1965, Series 301, Box 36,

Folder 1, CPP; Report of the Chairman to the Canal Authority, June 9, 1965, CAM, 9: 1, OGT; "Cross-State Barge Canal on Schedule," *FT*, 8, December 1965, 5.

23. Minutes for February 28, 1965, CAM, 9–10: 1569, OGT; Edwin Eden to Chief of Engineering Division, United States Army Corps of Engineers Memorandum, January 14, 1965, Adverse File, ACE, OGT; Sydney Herlong to Randolph Hodges, August 16, 1965, Box 65, Folder Waterways CFBC 1965, SHP, PKY.

24. Memo for Chief, Engineering Division, January 14, 1965, Adverse File, ACE, OGT; Brown to Koperski, memorandum, January 13, 1965, ibid.

25. "We Love Nature Too," *GS*, August 20, 1965; Joseph Livingston to "Editorial Friends," August 31, 1965, Series 301, Box 36, Folder 1, CPP.

26. Buchheister to Carr and Anthony, June 25, 1965, Gen NR 7–1, Box 24, Folder CFBC, LBJ; "Testimony of Dr. Spencer M. Smith, Jr.," *FN* 38 (August 1965); statement found in letter from Michael McCloskey, conservation director of the Sierra Club, to A. Sydney Herlong, August 3, 1966, Box 73, Folder Waterways—CFBC 1966, SHP, PKY.

27. Marjorie Carr to Lee White, June 26, 1965, Elmer Staats reply to Carr, July 21, 1965, Gen NR 7–1, Box 24, Folder CFBC, LBJ; Carr to Herlong, September 9, 1965, Box 65, Folder Waterways—CFBC 1965, SHP, PKY; Herlong to Carr, September 12, 1965, ibid.

28. Marjorie Carr to Governor Hayden Burns, July 13, 1965, letter printed in *FN* 38 (August 1965).

29. *Brief Summary of the Destruction of Natural Resources*, JCP.

30. Leopold, *Sand County Almanac*, viii.

31. Simpson, *In Lower Florida Wilds*, 140–41.

32. Kline, *First Along the River*, 77; Rome, "Give Earth a Chance," 536.

33. Carr to Burns, July 13, 1965, *FN* 38 (August 1965).

34. Marjorie Carr, "What Do Users Want? Wilderness!!" speech to the Southeastern Section, American Society of Foresters, January 12–13, 1967, Orlando, Florida, Series 4, Box 3, Folder 7, Archie Carr Papers, PKY.

35. Marjorie Carr to Claude Pepper, December 4, 1965, Series 301, Box 36, Folder 1, CPP.

36. "Barge Canal Delaying Action Attacked by Commissioners" *GS*, July 26, 1965; "Misstatements about Canal," *OSB*, July 25, 1965.

37. Managers Report, Period 4 October–15 November 1965, CAM, 9–10: 1654, OGT; "Floridians Being Alerted to Loss of Oklawaha," *GS*, December 9, 1965; William Partington to LC Ringhaver, November 16, 1965, JCP; "River vs. Canal," *GS*, December 8, 1965; Smathers' Press Secretary Jerry Blizin to Ed Johnson, November 23, 1963, Box 106, Folder Barge Canal, Smathers Papers, PKY; "Floridians Being Alerted to Loss of Oklawaha," *GS*, December 9, 1965; Carr to Claude Pepper, December 4, 1965, Series 301, Box 36, Folder 1, CPP.

38. "'Hearing' on Canal Called," *GS*, December 8, 1965; "Canal Board to Take Up Hearing," *OSB*, December 9, 1965.

39. Annual Report of the Chief of Engineers, U.S. Army on Civil Works Activities, 1966, 1: 435–36, OGT; "Cross Florida Barge Canal Locks Started," *Avon Park Sun*, December 2, 1965.

40. Information Release, Canal Authority, January 6, 1966, 3, OGT; Managers Report, Period 20 November–10 December 1965, CAM, 9–10: 1658, OGT.

41. "River vs. Canal," *GS*, December 8, 1965.

42. Ibid.

43. Carr and Anthony to Mrs. Lyndon Johnson, September 30, 1965, Series 301, Box 36, Folder 1, CPP.

44. "Pirates at Work?" *GS*, December 12, 1965; Giles Evans to Billy Matthews, January 3, 1966, Series 1727, Box 3, Folder 8, Cross Florida Canal Project Files, FSA; "Only Washington Can Stop the Canal Now," *Fort Myers News-Press*, December 31, 1965.

45. Address of Tom Adams, Englewood Board of Realtors, December 8, 1965, Box 7, Tom Adams Speeches, USF; "Adams Sees Canal as Economic Boost," *FTU*, December 31, 1965; Giles Evans to Billy Matthews, January 3, 1966, Series 1727, Box 3, Folder 8, Cross Florida Canal Project Files, FSA.

46. "Many Express Faith in Canal Future," *DP*, December 23, 1965; "McCree Defends the Canal," *OSB*, January 2, 1966; "Barge Canal Forges Ahead Notwithstanding Opposition," *DP*, December 30, 1965; "Editors Outlook," *CCC*, December 30, 1965.

47. L. E. Snagg Holmes to State Legislators, January 13, 1966, JCP.

48. U. S. Green, "Prayer Offered on the Banks of the Oklawaha, January 23, 1966," Box 60, Cross-Florida Barge Canal Project Folder, 6A, No. 1, GSP, PKY.

49. "Blocking Florida's Ditch," *Newsweek*, February 1, 1971, 55.

50. "Board of Conservation to Study All Canal Reports," *GS*, January 27, 1966.

51. "Oklawaha River Backers Speak Up at Hearing," *GS*, January 26, 1966; "Says Canal Hearings 'Rigged,'" *GS*, January 27, 1966; letter to the editor, "Protecting Oklawaha," *Cocoa Tribune*, January 31, 1966; "Says Canal Hearings 'Rigged,'" *GS*, January 27, 1966; "Barge Canal Route Protests Blunted," *GS*, January 26, 1966; "Says Canal Hearings 'Rigged,'" *GS*, January 27, 1966; "Barge Canal Route Protests Blunted," *GS*, January 26, 1966.

52. "Impressions of Hearing on Canal Construction," *OSB*, January 27, 1966.

53. Carr to George Smathers, July 25, 1966, Box 60, Folder CFBC Project, 6A, No. 1 GSP, PKY.

54. "Rallying Time," *Sarasota Herald-Tribune*, January 29, 1966; L. E. Snagg Holmes to State Legislators, January 13, 1966, JCP; David Anthony, interview by Steven Noll and David Tegeder, June 14, 2005; Margie Bielling, interview by Steven Noll, June 9, 2005; Anthony, interview.

CHAPTER 5. NEW VISIONS, NEW STRATEGIES

1. Barnetta Couse to Rose Allegato, July 1, 1966, RG 1, Series 1, Box 1, Folder 1966, FDE, PKY; Marjorie Carr to Sydney Herlong, May 30, 1966, Box 73, Folder Waterways—Cross State Barge Canal, SHP, PKY; Herlong to Carr, June 3, 1966, ibid.

2. R. T. Fisher to Robert High, June 3, 1966, RG 1, Series 1, Box 1, Folder 1966, FDE, PKY.

3. Mrs. J. W. Patterson to High, July 20, 1966, RG 1, Series 1, Box 1, Folder 1966, FDE, PKY.

4. Robert King High to Jim Redford, August 9, 1966, RG 1, Series 1, Box 1, Folder 1966, FDE, PKY.

5. John Couse to Jack Poorbaugh, May 25, 1967, RG 1, Series 1, Box 1, Folder 1967, FDE, PKY.

6. Information release, December 29, 1966, 1–3, Canal Authority Public Information Releases, 1966–1970, OGT; Minutes, February 28, 1966, CAM, 10: 1693, OGT; infor-

mation release, December 29, 1966, 1–3, Canal Authority Public Information Releases, 1966–1970, OGT.

7. Managers Report, November 15, 1965, CAM, 10: 1657, OGT.

8. "'Paul Bunyan's Bull' Adds a Touch of Devastation Here," *JS*, June 1967, 15; "New Type 'Monster' at Work on Canal Project," *OSB*, April 12, 1967; "Paul Bunyan's Bull," 14.

9. "Machine Tramples 85-Ft. Trees," *JJ*, April 12, 1967; "'Monster' Clears Land," *OS*, April 13, 1967; "300-Ton Masher Clears Reservoir Site," *ENR*, June 1967, 25; "'Monster' Clears Land," *OS*, April 13, 1967; "New Type 'Monster' at Work on Canal Project," *OSB*, April 12, 1967.

10. "'Monster' Clears Land," *OS*, April 13, 1967; "'Paul Bunyan's Bull,'" 15; "'Monster' Clears Land," *OS*, April 13, 1967; "Machine Tramples 85-Ft. Trees," *JJ*, April 12, 1967.

11. "The Man Who Was Building the Barge Canal," *Florida Accent*, Sunday supplement to *TT*, February 7, 1971, 12.

12. Tom Adams speech, March 17, 1966, Box 7, Tom Adams Speeches Collection, University of South Florida, Florida History Collection, Tampa.

13. Ibid; "'Missing Link' in Waterway Approved by Engineer Corps," *TT*, September 30, 1967; "The Missing Waterway Link," *SPT*, October 4, 1967.

14. *Ocala–Marion County Chamber of Commerce Progress Bulletin* 3, no. 1 (1965), Civil Works, accession number 077960039, Box 7, Folder August–October 1965, ACE, NARASE; "Waterway to Prosperity," *SPT*, October 15, 1967.

15. Information release, "Canal Physical Progress Excellent But Funding Disturbingly Inadequate," December 23, 1967, CAAR 1967, OGT; "Pessimistic Note Sounded in Canal Report," *OS*, December 29, 1967; "Canal Physical Progress Excellent," December 23, 1967, CAAR 1967, OGT.

16. Charles Bennett, "That Florida Canal," correspondence to *Florida Field Report*, March 17, 1966, Series 131, Box 8, Folder 6, HBP, FSA; information release, "Canal Physical Progress Excellent," December 23, 1967, CAAR 1967, OGT.

17. Charles Bennett, "That Florida Canal," March 17, 1966, HBP, FSA, Series 131, Box 8, Folder 6; "War Slows Progress of Cross Florida Canal," *SPT*, December 30, 1967; Manager's Report, January 26, 1967, CAM, 11: 1801, OGT; "War Slows Progress of Cross Florida Canal," *SPT*, December 30, 1967.

18. "Canal Physical Progress Excellent," December 23, 1967, CAAR, 1967, OGT.

19. Herlong to James Hughes, August 3, 1965, Civil Works, accession number 077960039, Box 7, Folder August–October 1965, ACE, NARASE; Herlong to Ringhaver, June 21, 1967, Box 86, Folder Waterways—Cross State Barge Canal, 1967, SHP, PKY.

20. "Remarks Prepared for Ed Hensley," May 25, 1965, Tom Adams press files, Series 697, Box 5, Folder Cross Florida Barge Canal II, Tom Adams Papers, FSA.

21. *Developing Florida Through Its Waterways*, promotional pamphlet, Florida Waterways Committee, September 1962, JHS.

22. Florida Board of Conservation Resolution, March 1, 1966, Box 2 of 2, PD-E 1735, ACE, OGT; welcoming remarks, Colonel R. P. Tabb, Meeting with Conservation Interests, Rodman Pool, Cross Florida Barge Canal, Jacksonville, September 22, 1966, Box 7, Folder General History II, ACE, OGT; presentation, Ocala–Marion Chamber of Commerce, "Meeting with Conservation Interests," internal summary of Army Corps of Engineers meeting with conservationists, Rodman Pool, Cross Florida Barge Canal, Jacksonville, September 22, 1966, ibid.

23. *Recreational and Sports Potentials Almost Limitless Upon Completion of the Cross Florida Barge Canal*, Canal Authority, n.d., 1–6, Box 610, Folder 37A, Cross Florida Barge Canal, Malcolm Johnson Papers, Strozier Library Special Collections, Florida State University, Tallahassee; *The Cross-Florida Barge Canal Can Provide Virtually Limitless Facilities*, pamphlet, Canal Authority, n.d., ibid.; Florida Board of Conservation news release, 1–3, January 1966, Series 697, Box 5, Folder Central Florida Barge Canal 2, State Board of Conservation Papers, FSA; *Recreational and Sports Potentials Almost Limitless*, 6; "29,000 Miles of Boating," *OS*, September 25, 1967.

24. *The Cross-Florida Barge Canal Can Provide Virtually Limitless Facilities*, pamphlet, Canal Authority, n.d., Box 610, Folder 37A, Cross Florida Barge Canal, Malcolm Johnson Papers; "Progress Report—Barge Canal Pushes Recreation Potential," *TT*, September 19, 1968; "Barge Canal Opens Recreation Areas," *Dade City Banner*, September 26, 1968.

25. Marjorie Carr, "What Do Users Want?—Wilderness!!" speech to Southeastern Section, Society of American Foresters, January 12–13, 1967, Orlando, Fla., Series 4, Box 3, Folder 7, Archie Carr Papers, PKY.

26. F. Browne Gregg, with Margie Sloan, *Progress through Innovation* (privately printed, n.d.), 94–95; Dave Bowman, interview by Steven Noll and David Tegeder, March 23, 2003.

27. E. Jane Maiden to J. J. Koperski, April 24, 1967, Miscellaneous Adverse File, General 1962–1968, ACE, OGT.

28. "New Waterway Called Passkey to Great Gains," *OS*, February 28, 1964; www.lbjlib.utexas.edu/johnson/archives.hom/speeches.hom/650407.asp.

29. "Conservationists Seek to Stop Construction of Cross-Florida Canal," *WSJ*, August 24, 1970.

30. Couse to Spessard Holland, November 30, 1962, JCP.

31. Couse to William F. Buckley, November 4, 1965, RG 1, Series 1, Box 1, Folder 1965 (1), FDE, PKY; William Schulz, "Mike Kirwan's Big Ditch," *Reader's Digest*, June 1967, 59; Couse to Spessard Holland, December 14, 1962, JCP.

32. David Carr, interview by Steven Noll and David Tegeder, November 22, 2005; Marjorie Carr, "The Fight to Save the Ocklawaha," paper presented at the Twelfth Biennial Sierra Club Wilderness Conference, September 25, 1971, 4, JCP.

33. Quote from Marjorie Carr, "The Fight to Save the Oklawaha," paper presented at the Twelfth Biennial Sierra Club Wilderness Conference, September 21, 1971, 10–11, JCP.

34. "St. Marks Channel Seen Shipping Boom If Built," *TD*, March 23, 1966; "Dear Friend of Florida" announcement, January 24, 1968, RG 1, Series 1, Box 1, Folder 1968, FDE, PKY; Carr to Randolph Hodges, January 26, 1968, ibid.

35. "St. Marks to Tampa Link Raises Some Dissension," *TD*, January 30,1968; Col. R. P. Tabb to William Partington, February 27, 1968, RG 1, Series 1, Box 1, Folder 1968, FDE, PKY; "St. Marks to Tampa Link Raises Some Dissension," *TD*, January 30, 1968.

36. Florida Defenders of the Environment, *Environmental Impact of the Cross-Florida Barge Canal with Special Emphasis on the Oklawaha Regional Ecosystem*, March 1970, 2, JCP.

37. "Viet to Sap Waterway Projects $," *OS*, June 7, 1968; "Ocala Interests Ponder Waning Canal Support," *TT*, March 23, 1968; "Canal Cuts Are Deplored," *SPT*, January 31, 1968; "Canal Authority Chief Criticizes Budget Cut," *JJ*, February 1, 1968; informa-

tion release, March 11, 1968, Canal Authority Public Information Releases, 1966–1970, OGT.

38. "Delegation Should Seek Hike in Canal Allocation," *OSB*, February 1, 1968; C. C. Leiby to Sydney Herlong, March 13, 1968, Box 88, Folder Waterways CFBC, SHP, PKY; Sydney Herlong to Thomas Brownlee, April 18, 1968, ibid.

39. "An Honor for Harry Buckman," *OSB*, March 21, 1968.

40. "Canal Progress Prospect Dim," *JS*, October 1968, 14.

41. "St. Johns Lock Is Born," *PDN*, September 24, 1968; "Canal Reality," *PDN*, September 15, 1968.

42. "Rep. Boland Pledges Support to Finish Canal by Mid-1973," *JS*, January 1969, 8–9, 20.

CHAPTER 6. FLOATING LOGS, DYING TREES, AND CLOGGING WEEDS

1. www.lawtonchiles.org/walkin19.htm; "Sailing Down the River Gives Chiles Sick Feeling," *SPT*, June 26, 1970.

2. "300-Ton Masher Clears Reservoir Site," *ENR*, June 1, 1967, 25; Partington to Michael Frome, December 17, 1968, RG 1, Series 1, Box 1, Folder 1968, FDE, PKY.

3. Bill Partington to Michael Frome, December 17, 1968, RG 1, Series 1, Box 1, Folder 1968, FDE, PKY; Summary of Environmental Consideration Involved in the Recommendation for Termination of Construction of the Cross Florida Barge Canal, Council on Environment Quality, 14, RG 300, Series 301, Box 746, Folder 9, CPP.

4. "Report Answers Critics of Florida Barge Canal," *WJ* 83, no. 2 (April 12, 1969): 4; "Backers of Barge Canal Called to Action," *WJ* 83, no. 7 (May 17, 1969): 5; Florida Board of Conservation Comments on Progress in Constructing the Cross-Florida Barge Canal, March 28, 1969, 1, Box 9, Folder 4, Cross Florida Barge Canal, WCP, PKY.

5. Florida Board of Conservation Comments on Progress in Constructing the Cross-Florida Barge Canal, March 28, 1969, 2–3, Box 9, Folder 4, Cross Florida Barge Canal, WCP, PKY.

6. Ibid., 4.

7. Gilbert Rogin, "All He Wants to Do Is Save the World," *Sports Illustrated*, February 3, 1969, 24–29.

8. "Charles F. Wurster Environmental Defender," www.newsday.com/community/guide/lihistory/ny-century_of_science_dishist,0,4320247.story?coll=ny-lihistory-navigation.

9. Rogin, "All He Wants to Do Is Save the World," 24.

10. Ibid., 25.

11. Ibid., 24.

12. Joseph Hassett to Carr, March 26, 1969, RG 1, Series 1, Box 1, Folder 1969, FDE, PKY.

13. Middleton, "Cutting Through Paradise," 179.

14. Articles of Incorporation of Florida Defenders of the Environment, Inc., May 31, 1970, RG 1, Series 1, Box 10, Folder Trustees Correspondence 1, FDE, PKY.

15. Melissa Shepard Carver, "Florida Defenders of the Environment: A Case Study of a Volunteer Organization's Media Utilization" (master's thesis, University of Florida, 1973), 43; Victor Yannacone, interview by Steven Noll and David Tegeder, February 24, 2006.

16. "New Canal Carries First Cargo—With Truck's Help," *National Observer*, July 7, 1969; "1st Commercial Cargo Shipped Via Cross-Florida Barge Canal," *JS*, August 1969, 10–11; "New Canal Carries First Cargo—With Truck's Help," *National Observer*, July 7, 1969.

17. "Western End of Canal Used by Palatka Firm," *OSB*, June 25, 1969; "Cross Florida Barge Canal Wins 'Beauty Contest,'" *Florida Vacation Fun Times*, October 1969; Conservation of Natural Beauty Award, Rodman Reservoir, n.d., Box 6, Folder January 69–March 70, ACE, NARASE.

18. Gene Brown to Chief of Engineers, August 5, 1969, Box 6, Folder January 69–March 70, ACE, NARASE; Statement of Giles Evans to Florida Senate Committee on Natural Resources and Conservation, March 16, 1970, Cross Florida Barge Canal Vertical File, Ocala Public Library.

19. "Biologists Describe Canal 'A Sickening Sight to View,'" *GS*, November 2, 1969; "'No,' Say Conservationists Armed for Battle, The Canal Is a 'Crime Against Nature,'" *OS*, November 2, 1969.

20. "Who's Bankrolling Anti-Canal Group?" *JS*, July 1970, 3–4.

21. Ibid., 4.

22. David Anthony to Colonel John McElhenny, September 5, 1969, in Carver, "Florida Defenders of the Environment," 47; "Barge Canal 'Blow-Up' Urged," *Tampa Tribune-Times*, September 14, 1969; Anthony interview, by Steven Noll and David Tegeder, June 14, 2005; Poole, "Florida: Paradise Redefined," 119.

23. Anthony interview; Fund Drive Launched to Halt Florida Barge Canal, October 14, 1969, RG 1, Series 1, Box 1, Folder 1969, FDE, PKY; George Remington to John Couse, October 15, 1969, ibid.

24. Carver, "Florida Defenders of the Environment," 77–78.

25. *EDF v. Corps of Engineers*, Plaintiff's Brief, RG 1, Series 1, Folder 1969, FDE, PKY; Unique Environmental Suit Filed on Barge Canal, FDE news release, September 20, 1969, 1–3, ibid.

26. *EDF v. Corps of Engineers*, Plaintiff's Brief, RG 1, Series 1, Folder 1969, FDE, PKY.

27. Ibid.

28. Ibid.

29. "EDF Attorney Asks Governor Kirk to Intervene in Barge Canal Suit," FDE news release, September 23, 1969, RG 1, Series 1, Folder 1969, FDE, PKY; "We Can Have Barge Canal and Conservation Too," *OSB*, November 26, 1969.

30. "Notorious Hearsay," http://info-pollution.com/unquote.htm.

31. Poole, "Florida: Paradise Redefined," 120–21.

32. "A Sickening Sight to View," *GS*, November 4, 1969. Newspapers throughout the state reprinted the Funk article. Other examples include: "Florida Canal Attacked," *PBP*, November 2, 1969; "New Battle Erupts over Barge Canal," *MH*, November 2, 1969; "The Big Ditch Battle Rages," *Sarasota Herald-Tribune*, November 2, 1969; and "War Over the Barge Canal Escalates," *SPT*, November 2, 1969.

33. "State Plans Full-Scale Restudy of Barge Canal," *GS*, November 11, 1969; *A Brief Assessment of the Ecological Impact of the Cross Florida Barge Canal*, November 1969, Florida Game and Fresh Water Fish Commission, 1–4, Folder Eureka, 1966–69, ACE, OGT.

34. "Seafaring with Dave Howard," *JS*, September 1969, 18–19; "Barge Canal Is

Praised," *SPT*, November 25, 1969; "Adams Charges 'Sly-Lie' Used to Block Canal," *OS*, September 21, 1969; "'Yes,' Says Secretary of State Tom Adams, 'Conservation Means Planned Management,'" *Florida Magazine, OS*, November 2, 1969.

35. "Ecology: The New Jeremiahs," *Time*, August 15, 1969, 37.

36. Train, "The Environmental Record of the Nixon Administration," 185; Gordon, "The American Environment," 45.

37. www.let.rug.nl/usa/P/rn37/speeches/rn_1970.htm.

38. James Nathan Miller to John Couse, September 10, 1969, JCP; Charles Pintchman to John Couse, December 9, 1969, RG 1, Box 1, Folder 1969, FDE, PKY.

39. James Nathan Miller, "Rape on the Oklawaha," *Reader's Digest*, January 1970, 54–60.

40. Ibid.

41. Partington to Charles Pintchman, December 13, 1969, JCP; Couse to Charles Pintchman, December 26, 1969, JCP.

42. Memorandum for file, subject: *Reader's Digest* visit in Jacksonville District, July 17, 1969, Cross Florida Canal project files, Series 1727, Box 6, Folder 24, Rape on the Ocklawaha, FSA; William McCree to Hobart Lewis, January 9, 1970, ibid.; Robert Shortle to Paul Thompson, January 19, 1970; Shortle to Theodore Braaten, August 6, 1970, ibid.

43. Giles Evans to John Eden, January 30, 1974, Cross Florida Canal project files, Series 1727, Box 6, Folder 24, Rape on the Ocklawaha, FSA.

44. "Canal Defended at Lock Ceremony" *OSB*, January 11, 1970.

45. Kenneth Morrison to William Partington, December 26, 1969, WPP; Partington to Morrison, December 30, 1969, WPP.

46. Open Letter to Richard Nixon, January 27, 1970, WPP.

47. http://www.jacksbromeliads.com/thetamiamitrail.htm

48. United States Department of the Interior, "Environmental Impact of the Big Cypress Swamp Jetport," September 1969, 150, 151.

49. Florida Defenders of the Environment, *Environmental Impact of the Cross-Florida Barge Canal with Special Emphasis on the Oklawaha Regional Ecosystem*, March 1970, 1, JCP.

50. Ibid., 73, 77.

51. Ibid., 43, 95, 90, 95.

52. Ibid., 49.

53. Ibid., 104.

54. Ibid., 105, 111.

55. "Cross-Florida Barge Canal Shrouded in Controversy," Transcript of CBS Evening News Segment, March 7, 1970, Box 6, Folder 2, ACE, NARASE; Gerald A. Lewis, "Death on the Oklawaha," *Florida and Tropic Sportsman*, December–January 1969–70, 18, 20.

56. Warren Henderson to Members of Congress, January 1970, Series Legislative, Box 123, Folder 10, CAP, University of Oklahoma Special Collections, Norman.

57. Memorandum from Russell Train to John Ehrlichman, November 3, 1969, White House Files, Egil Krogh files, Box 63, Folder 485, RNP, NA.

58. Flippen, *Nixon and the Environment*, 42.

59. Memorandum from John C. Whitaker to Russell Train, March 16, 1970, White House Files, Egil Krogh files, Box 63, Folder 485, RNP, NA.

60. *A Review and Appraisal of the Cross Florida Barge Canal*, United States Depart-

ment of the Interior, Fish and Wildlife Service, Bureau of Sport Fisheries and Wildlife, Atlanta, Ga., March 30, 1970, 1, 8, 12, WPP.

61. Don Fuqua to Richard Nixon, June 11, 1970, Series Natural Resources, Box 20, Folder 3, RNP, NA; Spessard Holland to Walter Hickel, June 15, 1970, WPP.

62. Powers, *Martin Andersen*, 108.

63. Refutation and Summary of Findings by Florida Waterways Association, n.d., 5, WPP.

64. Statement of Giles L. Evans, Jr., Manager, Canal Authority of the State of Florida to Florida Senate Committee on Natural Resources and Conservation, March 16, 1970, 1, 8, Marion County Public Library, Ocala Branch, Cross-Florida Barge Canal Vertical File.

65. Giles Evans to Dyersburg Chamber of Commerce, April 13, 1970, JCP; Joseph Boyd to Evans, April 20, 1970, JCP; Joseph Boyd to National Audubon Society, April 23, 1970, JCP.

66. "Canal 'Forerunner' Is Comic Hangup," *St. Petersburg Independent*, July 7, 1970.

67. Anthony interview.

68. William Cook Advertising Agency press release, July 1, 1970, WPP; "6,000-Ton Shipment Gets Stuck in Canal," *GS*, June 30, 1970; "Col. Evans Says Big Barge Didn't Get Stuck in Canal," *TT*, August 19, 1970; "Canal 'Forerunner' Is Comic Hangup," *St. Petersburg Independent*, July 7, 1970.

69. Manager's Report, July 17, 1970, CAM, 13: 2259, OGT; S. D. Dornbirer, letter to the editor, *TT*, August 26, 1970.

70. L.C. Ringhaver, "The Real Purpose of Canal Is to Relieve Industry from Burden of High Freight Rates," *JS*, April 1970, 4; "Canal Official Rips Critics," *Melbourne Times*, April 7, 1970; "Ecological Hysteria," *WJ* 84, no. 1 (April 4, 1970): 4; "The Good Earth," *WJ*, 84, no. 3 (April 18, 1970): 4.

71. *Eureka Alternative Alignment Cross Florida Barge Canal, July 1970*, Army Corps of Engineers Report, Box 20, Folder 4, ACE, NARASE.

72. "The Big Ditch," *GS*, July 14, 1970; "Let's Decide If Canal Needed before Debating a New Route," *OS*, July 20, 1970.

73. "Environment Veto Threatens Florida Canal," *CSM*, June 29, 1970; "The Politics Behind the Cross Florida Barge Canal Project," *SPT*, August 16, 1970.

74. Lyman Rogers to Richard Nixon, July 9, 1970, Series 5301, Box 791, Folder 18, CPP, Candidates' Questionnaire, distributed by FDE, Summer and Fall 1970, WPP. FDE sent these questionnaires to all statewide and congressional candidates and compiled the responses of 123 candidates.

75. Candidates' Questionnaire, distributed by FDE, Summer and Fall 1970, WPP.

76. FDE press release, September 4, 1970, WPP.

77. "But, Mr. President, It's Your Canal," *FTU*, October 29, 1970. For other versions of this now-apocryphal story, see Carter, *The Florida Experience*, 295; Blake, *Land into Water-Water into Land*, 209; Kallina, *Claude Kirk and the Politics of Confrontation*, 165; and "Canal Foes Believe Election Strengthened Their Hand," *TT*, November 7, 1970.

78. Edward Lee Rogers to Marjorie Carr, January 28, 1970, WPP; both First and Second Amended Complaints to the original EDF lawsuit in WPP. The First Complaint was filed on February 16, 1970; the Second Complaint was filed on April 5, 1970.

79. Claude Kirk to Walter Hickel, May 14, 1970, WPP.

80. John Whitaker to John Ehrlichman, memorandum, July 10, 1970, White

House Central Files, Box 44, Cross Florida Barge Canal (1970–1971), Folder 2, JWP, RNP, NA; John Whitaker to John Ehrlichman, memorandum, August 6, 1970, ibid.

81. John Whitaker to John Ehrlichman, memorandum, November 4, 1970, White House Central Files, Box 44, Cross Florida Barge Canal (1970–1971), Folder 2, JWP, RNP, NA; "Who's Building the Canal?" *SPT*, October 31, 1970; Donald Walden, *Early Warning Report, Subject: Possible Presidential Action to Modify the Cross-Florida Barge Canal for Environmental Considerations*, November 16, 1970, White House Central Files, Box 44, Cross Florida Barge Canal (1970–1971), Folder 2, JWP, RNP, NA.

82. Russell Train to Mr. Whitaker, memorandum, December 1, 1970, White House Central Files, Box 44, Cross Florida Barge Canal (1970–1971), Folder 2, JWP, RNP, NA; *Summary of Environmental Considerations Involved in the Recommendation for Termination of Construction of the Cross Florida Barge Canal*, Council on Environmental Quality, December 1970, 11, ibid.

83. Whitaker to the President, memorandum, December 23, 1970, White House Central Files, Box 44, Cross Florida Barge Canal (1970–1971), Folder 2, JWP, RNP, NA.

84. Draft Presidential Statements on Termination of the Cross-Florida Barge Canal, n.d., White House Central Files, Box 44, Cross Florida Barge Canal (1970–1971), Folder 2, JWP, RNP, NA.

85. Statement about Halting Construction of the Cross Florida Barge Canal, January 19, 1971, www.nixonlibrary.org/clientuploads/directory/archive/1971_pdf_files/1971_0020.pdf?PHPSESSID=ab209e3323828a4705c424a689c4e914.

86. John Whitaker, telephone interview by Steven Noll, June 10, 2006.

87. Stephen Carr, telephone interview by Steven Noll, June 12, 2006.

CHAPTER 7. DEATH OF A DREAM

1. "Yankeetown Citizens Bitter over Stoppage of 'Big Ditch,'" *FTU*, January 25, 1971.

2. "Nixon Nixes Canal Dream," *PDN*, January 20, 1971; "President Makes a Serious Error in Judgment!!!" *PDN*, January 20, 1971; William McCree to Richard Nixon, January 20, 1971, Series 301, Box 746, Folder 8, CPP; Dewey Dye to Richard Nixon, January 21, 1971, ibid.; "Florida Canal Authority Caught by Surprise," *GS*, January 24, 1971; Managers Report, February 8, 1971, CAM, 13: n.p., OGT.

3. John Eden, "An Open Letter to the People of the United States of America," *CCC*, February 11, 1971; "Boosters Keep Your Cool," advertisement, *FTU*, February 17, 1971; F. M. Simms to Colonel A. S. Fullerton, March 24, 1971, Box 6, Folder 1971, ACE, NARASE.

4. Charles Bennett, oral history interview, May 15, 1995, 239, SPOHC; statement of Congressman Charles E. Bennett before Subcommittee on Separation of Powers, Senate Judiciary Committee, March 23, 1971, Box 6, Folder 1971, ACE, NARASE.

5. Transcript of Tom Adams speech, May 1, 1971, Series 961, Box 1, Folder 17, Tom Adams Papers, FSA.

6. Telegram from Tropical Audubon Society to Claude Pepper, January 21, 1971, Series 301, Box 746, Folder 8, CPP; telegram from Tom Pankowski to Richard Nixon, n.d., ibid.; telegram from Oklahoma Wildlife Federation to Richard Nixon, January 27, 1971, Box 163, Folder 23b, Page Belcher Papers, Carl Albert Congressional Archives,

University of Oklahoma, Norman; Arthur Godfrey to John Whitaker, April, 27, 1971, White House Central Files, Box 44, Cross Florida Barge Canal (1970–71) Folder 4, JWP, RNP, NA.

7. Noble Cunningham to John Ehrlichman, memorandum, March 1, 1971; John Whitaker to Richard Nixon, memorandum, February 18, 1971, White House Central Files, Box 44, Cross Florida Barge Canal (1970–71) Folder 4, JWP, RNP, NA.

8. "Halting the Barge Canal, A Timely and Proper Move," editorial, *MH*, January 21, 1971; *Stuart Florida News* and *OS* clips in John Whitaker to Richard Nixon, February 18, 1971, White House Central Files, Box 44, Cross Florida Barge Canal (1970–71) Folder 4, JWP, RNP, NA; "Death of a Boondoggle," editorial, *NYT*, January 22, 1971; "Environmental Wins and Losses," *WP*, January 23, 1971; "Blocking Florida's Ditch," *Newsweek*, February 1, 1971, 55.

9. Nathaniel Reed, oral history interview, November 2, 2000, 145, SPOHC.

10. Stine, *Mixing the Waters*, 102, 104.

11. Joe Browder, "Decision-Making in the White House," in Rathlesberger, ed., *Nixon and the Environment*, 106.

12. "Barrington Parker Dies at 77, Judge in Many Notable Trials," *WP*, June 4, 1993; "Environmental Defense Fund v. Corps of Engineers—Civil Action No. 2655–69," January 27, 1971," RG1, Series 1, Box 3, Folder 1971–1974, FDE, PKY.

13. Dennis Puleston to Marjorie Carr, January 26, 1971, RG 1, Series 1, Box 3, Folder 1971 (2), FDE, PKY.

14. "Dale Miller Will Direct National Effort for Canal," *JS*, March 1971, 8–9.

15. Jack Lucas, President of Florida Waterways Association, editorial letter, March 26, 1971, *JS*, April 1971, 4; "The Cross Florida Barge Canal and Its Environs, A Presentation to the Council on Environmental Quality," Charles Bennett, February 1, 1971, Series 301, Box 746, Folder 9, CPP.

16. "No Alternate Canal Route Acceptable to Conservationists," *GS*, February 3, 1971; Paul Roberts, "Cross Florida Barge Canal: A Current Appraisal," *Business and Economic Dimensions* 7, no. 6 (June 1971): 7.

17. "Lawmaker Urges Alternate Route," *MH*, February 2, 1971; "Nixon's Order Slams Door on Alternate Canal Routes," *GS*, February 12, 1971; "Area Officials and Citizens Talk Canal with Dr. Talbot," *DP*, February 25, 1971; "Editor's Notebook," *DP*, February 25, 1971.

18. Reubin Askew to Richard Nixon, January 27, 1971, Series 301, Box 746, Folder 9, CPP; "Askew Asks for Evidence That Brought Canal Halt," *GS*, February 8, 1971.

19. CAM, April 22, 1971, 13: n.p. OGT; "'Florida Looks Ridiculous' in Canal Suit—Shevin," *GS*, February 16, 1971.

20. Statement by Jack W. Lucas before the Governor and Cabinet, March 1, 1971, Tallahassee, Florida, Series 301, Box 746, Folder 8, CPP; David Anthony, Summary of Recommendation for Cross-Florida Barge Canal Lands and Waters, RG 1, Series 1, Box 3, Folder 1971 (4), FDE, PKY.

21. O. E. Frye, Jr. to Reubin Askew, March 3, 1971, Box 2 of 2, pd-e 1735, ACE, OGT; Recommendations for Alteration and Future Use of Cross Florida Barge Canal Project Facilities by Florida Game and Fresh Water Fish Commission, March 1, 1971, ibid.

22. Secretary of State Richard Stone news release, March 1, 1971, Series 301, Box 791, Folder 18, CPP; "Rail-Barge Proposal Presented by Stone," *FTU*, March 2, 1971; Richard "Dick" Stone to Richard Nixon, March 1, 1971, RG 1, Series 1, Box 3, Folder 1971(3),

FDE, PKY; "Secretary Stone's Unique Solution to Canal Issue," *JS* 20, no. 4 (April 1971): 22–24.

23. *Complaint Canal Authority of the State of Florida vs. Stanley R. Resor*, United States District Court, Middle District of Florida, Jacksonville Division, February 12, 1971, 7–10, OGT.

24. "Block Florida's Ditch," *Newsweek*, February 1, 1971, 56.

25. *Canal Authority of the State of Florida v. Howard H. Callaway, Secretary of the Army*, Opinion and Judgment, January 31, 1974, 29, Box 1, Folder U. S. District Court, Middle District 1974, John Henry Davis Papers, PKY.

26. Ibid, 10–11.

27. John Whitaker, memorandum to John Ehrlichman, June 9, 1971, White House Central Files, Box 44, Cross Florida Barge Canal (1970–71) Folder 4, JWP, RNP, NA; Arthur Godfrey to John Whitaker, April 27, 1971, ibid.

28. David Anthony to Reubin Askew, May 3, 1971, Box 1, U.S. District Court, District of Columbia Folder, John Henry Davis Papers, PKY; Robert Shevin to Art Marshall, June 9, 1971, MS Group 73, Box 1, Folder 46—Shevin Correspondence, Arthur Marshall Papers, PKY.

29. Marjorie Carr to Reubin Askew, June 3, 1971, WPP; testimony of Dr. John H. Kaufman, Chairman, Scientist Advisory Committee, Florida Defenders of the Environment, before the Florida State Cabinet, June 15, 1971, RG 1, Series 1, Box 3, Folder 1971 (1), FDE, PKY; interview with Jack Kaufman, Working with the FDE, "A Tale of Two Women," http://everglades.fiu.edu/two/transcripts/SPC935_3.htm; Reubin Askew to David Anthony, June 16, 1971, RG 1, Series 1, Box 3, Folder 1971 (1), FDE, PKY.

30. Avery Fullerton to Reubin Askew, March 29, 1971, White House Central Files, Box 44, Cross Florida Barge Canal (1970–71) Folder 4, JWP, RNP, NA; Donald Chapman to Reubin Askew, May 25, 1971, Box 2, Folder Cross Florida Barge Canal, John Henry Davis Papers, PKY; Floyd Christian to Art Marshall, June 24, 1971, Manuscript Group 73, Box 1, Folder 46, Arthur Marshall Papers, PKY.

31. *Canal Authority of the State of Florida v. Howard H. Callaway, Secretary of the Army*, United States Court of Appeals, Fifth Circuit, February 15, 1974, 1067.

32. "U.S. Proposes Scenic River for Fla. Canal Nixon Halted," *WP*, May 19, 1972; press release, Council on Environmental Quality, June 21, 1972, Box 20, Folder 6, ACE, Office of History, Humphreys Engineer Center, Fort Belvoir, Alexandria, Va.

33. "Visit by Critics of Canal Wrapped," *OSB*, May 2, 1972; "CEQ Representative on Scene 'to Map Another Attack' on Ocklawaha Level," *PDN*, May 2, 1972.

34. *Canal Authority of the State of Florida v. Howard H. Callaway, Secretary of the Army*, United States Court of Appeals, Fifth Circuit, February 15, 1974, 1069; "Major Barge Canal Defeat: Lowering Rodman Pool Is Ordered by U.S. Judge," *JJ*, July 24, 1972.

35. "As Rodman Debate Goes On, Water Goes Down," *SPT*, August 23, 1972.

36. *Canal Authority of the State of Florida v. Howard H. Callaway, Secretary of the Army*, United States Court of Appeals, Fifth Circuit, February 15, 1974, 1069, 1075.

37. "Death by Degrees," *OS*, January 24, 1973; "Canal or No Canal, Raise Rodman and Leave it Alone," *OSB*, January 21, 1973; "Rodman Ruling Viewed with Optimism," *OS*, January 17, 1973.

38. *Canal Authority of the State of Florida v. Howard H. Callaway, Secretary of the Army*, Opinion and Judgment, January 31, 1974, pp. 14–15, 38, 34, 29, John Henry Davis Papers, Box 1, Folder U.S. District Court, Middle District 1974, PKY.

39. "Judge Revives 'Dead' Canal," *OSB*, February 4, 1974; "Florida Water Interests Applaud Canal Ruling," *WJ*, 87, 46 (February 16, 1974): 4; "Canal Ruling Seen Ending 'Hysteria,'" *FTU*, February 5, 1974; Giles Evans, Memorandum, February 8, 1974, Box 2, Folder Canal Authority, John Henry Davis Papers, PKY.

40. "Canal Ruling Seen Ending 'Hysteria,'" *FTU*, February 5, 1974; "'Dear Judge . . .,'" *St. Petersburg Independent*, August 20, 1974; Marjorie Carr to Bill Partington, April 25, 1971, WPP.

41. "Press Release by L. C. Ringhaver, Chairman, Canal Authority of the State of Florida, June 21, 1972," Cross Florida Canal Project Files, Series 1727, Box 7, Folder 8, FSA; "Canal-Ruling Reaction Varied," *SPT*, February 6, 1974.

42. Cabinet Resolution, August 1, 1972, *ENFO Newsletter*, 1974, WPP; "Remarks by Tom Adams, Lieutenant Governor of Florida, at the Dedication Cross-Florida Barge Canal Bridge (S.R. 40 and Oklawaha River), May 11, 1972," Tom Adams Speeches, Box 17, University of South Florida, Florida History Collection, Tampa.

43. "Congressman Paul Rogers Report on Congress, 92nd Congress, 1st session, November–December 1972," WPP; *Congressional Record*, June 26, 1972, H6063; House Bill H.R. 9115, submitted June 29, 1973, RG 1, Series 1, Box 4, Folder 1973 (2), FDE, PKY.

44. Lee Ogden, memorandum of July 27, 1972, WPP.

45. "Cross Florida Barge Canal Restudy Report: The Highest and Best Uses of the Oklawaha River Basin and Lake Rousseau for the Economy and the Environment," Institute of Natural Resources, the University of Georgia, Athens, July 1976, 79, Box 2, Folder Cross Florida Barge Canal, John Henry Davis Papers, PKY; "Statement Prepared by Florida Defenders of the Environment for a Series of Public Meetings on the Draft Environmental Impact Statement of the Cross-Florida Barge Canal," September 17–27, 1976, Box 5, Folder 15, Art Marshall Papers, PKY.

46. Claude Pepper to D. Richard Mead, August 26, 1976, Series 301, Box 499, Folder 5, CPP; "Barge Canal Is a Loser," transcript of a WJXT Television editorial, September 24, 1976, RG 1, Series 1, Box 5, Folder 1976 (1), FDE, PKY; John Campbell to Reubin Askew, September 16, 1976, ibid.; "Marion Chamber Directors Oppose Cross-Florida Canal Building," *OS*, October 1, 1976; "Barge Canal Would Run $350 Million, Corps Says," *MH*, August 21, 1976.

47. "Cabinet Votes 6–1 Against Completion of Barge Canal," *SPT*, December 18, 1976; "Bryant to Askew, Facts on Canal Now Apparent," *Cross Florida Canal Counties Association* 2, no. 3 (September 1976), Series 301, Box 499, Folder 5, CPP; "Canal Backers Give Plea," *OSB*, December 17, 1976; Recommendations for Future of Canal—Marjorie Carr's Testimony, December 16, 1976, RG 1, Series 1, Box 5, Folder 1976 (2), FDE, PKY.

48. "Corps Deals Barge Canal Major Blow," *TT*, December 17, 1976; "State's Top Environmentalist Pleads with Cabinet to Kill Canal," *SPT*, December 17, 1976.

49. "Cabinet Votes Canal Death," *TT*, December 18, 1976.

50. Marjorie Carr to John Bustered, December 30, 1976, RG 1, Series 1, Box 5, Folder 1976 (2), FDE, PKY; "Cabinet Votes 6–1 Against Completion of Barge Canal," *SPT*, December 18, 1976; "Cabinet Votes Against Canal," *FTU*, December 18, 1976; "Cabinet Minutes of the Canal Authority," January 31, 1977, OGT; Marjory Stoneman Douglas to Marjorie Carr, December 30, 1976, RG 1, Series 1, Box 5, Folder 1976 (1), FDE, PKY.

51. Lieutenant General John W. Morris, Memorandum for Washington Policy Group, January 11, 1977, Box 20, Folder 5, ACE, Office of History, Humphreys Engineer

Center, Fort Belvoir, Alexandria, Va.; Nathaniel Reed et. al., Memorandum for Lieutenant General John W. Morris, January 12, 1977, Box 20, Folder 4, ibid.

52. Edward Greene, news release, "Chief of Army Engineers Recommends Cross-Florida Barge Canal Project be Terminated," February 24, 1977, Box 20, Folder 6, ACE, Office of History, Humphreys Engineer Center, Fort Belvoir, Alexandria, Va.; "Terminate Canal, Army Engineers Say," *OSB*, February 25, 1977; Nathaniel Reed to John Bustered, December 9, 1976, Box 20, Folder 4, ACE, Office of History, Humphreys Engineer Center, Fort Belvoir, Alexandria, Va.

53. Ernest Graves, "Engineer Memoirs," 168, www.usace.army.mil/publications/eng-pamphlets/ep870-1-52/part1.pdf.

54. "End Florida Canal Project Carter Urges Congress," *SPT*, May 24, 1977.

55. "End Florida Barge Canal: President," *FTU*, May 24, 1977; "Marion Commissioners' Canal Reaction Mixed," *OSB*, May 24, 1977; "Congress Asked to put 'RIP' on Barge Canal," *OS*, May 24, 1977; "Carter Urges Death for Canal," *OSB*, May 24, 1977.

56. "Carter Urges Death for Canal," *OSB*, May 24, 1977; "Close Ledger on Useless Project," *OS*, February 19, 1977; "Florida House Votes to Kill Barge Canal," *PDN*, April 5, 1979; "Barge Canal May Face Watery Grave," *Independent Florida Alligator*, April 20, 1983; "Environmentalist Wants Cemetery to Mean Death for Canal," *Independent Florida Alligator*, January 25, 1983; Jim Smith to Paula Hawkins, May 3, 1983, Series VI, Box 3, Folder Cross Florida Barge Canal, Paula Hawkins Papers, Winter Park Public Library, Winter Park, Fla.

57. "Florida House Votes to Kill Barge Canal," *PDN*, April 5, 1979.

58. "Feud over Florida Barge Canal Reheats," *MH*, June 10, 1985.

59. "Why Cross Florida Barge Canal Refuses to Die," *SPT*, May 26, 1985; Bill Chappell to Marsha Chance, April 18, 1983, RG 1, Series 1, Box 7, Folder 1983 (1), FDE, PKY; "Cross-Florida Barge Canal Shows Signs of Life," *Independent Florida Alligator*, September 21, 1982.

60. LeRoy Collins to Claude Pepper, July 14, 1983, Series 301, Box 472A, Folder 3, CPP; Resolution, Ocala Board of Realtors, Inc., April 7, 1983, RG 2, Box 4, Folder CFBC Easy Reference, FDE, PKY.

61. Andy Johnson to Friends of the Cross Florida Barge Canal, October 19, 1981, Box 9, Folder 3, WCP, PKY; "Barge Canal Opponents and Proponents Line Up on Either Side of Project, *Daytona Beach Morning Journal*, February 9, 1983; "Feud over Florida Barge Canal Reheats," *MH*, June 10, 1985; David Gencarelli, memorandum to Senator Hawkins, April 21, 1983, Series VI, Box 3, Folder Cross Florida Barge Canal, Paula Hawkins Papers, Winter Park Public Library, Winter Park, Fla.

62. "Barge Canal Is Still Hotly Debated," *GS*, June 17, 1985.

63. Paula Hawkins press release, undated, Series VI, Box 3, Folder Cross Florida Barge Canal, Paula Hawkins Papers, Winter Park Public Library, Winter Park, Fla.

64. Water Resources Development Act of 1986, PL 99–662, November 7, 1986, www.fws.gov/habitatconservation/Omnibus/WRDA1986.pdf; "A Monster Slain," *MH*, November 20, 1986; Section 1114, PL 99–662, www.fws.gov/habitatconservation/Omnibus/WRDA1986.pdf.

65. Charles Bennett press release, January 20, 1990, RG 1, Series 1, Box 7, Folder 1990s (2), FDE, PKY; "Bill Would Kill Canal, Return Land to State," *OSB*, March 9, 1990; FDE press release, March 11, 1990, RG 1, Series 1, Box 7, Folder 1990s (2), FDE, PKY; "Land for Florida Barge Canal May Become a Ribbon of Parks," *MH*, May 10, 1990; "Gra-

ham, Stearns: Give Canal to State," *GS*, March 9, 1990; "Land for Florida Barge Canal May Become a Ribbon of Parks," *MH*, May 10, 1990.

66. "Land for Florida Barge Canal May Become a Ribbon of Parks, *MH*, May 10, 1990; Water Resources Development Act of 1990, PL 101–640, November 28, 1990, www.nab.usace.army.mil/whatwedo/civwks/wrda90.pdf.

EPILOGUE: FROM CANAL TO GREENWAY

1. For a listing of CLAC members, see "Minutes of Meeting of Canal Lands Advisory Committee, April 9, 1992," Accession II, Box 19, Cross Florida Greenbelt Plan Folder, FDE, PK; "Barge Canal's Fate in Dispute," *SPT*, October 29, 1992; "One Strip of Land Is the Focus of Many Different Desires, *SPT*, September 4, 1991; Richard Coleman, "The Joys of Fishing," letter to the editor, *GS*, May 24, 1998; David Bruderly, "Fishing Would Be Great," op-ed column, *GS*, June 1, 1998.

2. "Panel: Study Rodman 3 Years," *GS*, September 18, 1982.

3. "Ex-Senator Kirkpatrick Dead at 64," *GS*, February 6, 2003; "Kirkpatrick Created Legacy of Action, Controversy," *GS*, February 6, 2003.

4. Daniel Canfield, Eric Schulz, Mark Hoyer, "'To Be or Not To Be'—The Rodman Reservoir Controversy, A Review of Available Data, February 1993," v, viii, St. Johns Water Management District Library, Palatka, Florida.

5. George Kirkpatrick, "Musings on Rodman," n.d., 3, George Kirkpatrick Papers, Mike Murtha Collection, Gainesville, Fla.; "Rodman Decision Relayed," *PDN*, September 18, 1992; "Rodman's Fate Put on Hold, *OSB*, September 18, 1992; "Panel: Study Rodman 3 Years," *GS*, September 18, 1982.

6. "Cabinet: Pull Rodman Plug," *FTU*, December 16, 1992; "Cabinet Urges End to Dam," *GS*, December 16, 1992; Motion by Commissioner Betty Castor, December 15, 1992, Historic Documents Pertaining to Ocklawaha River Restoration, FDE, 2000 (blue folder), FDE Headquarters, Gainesville, Fla.; "Cabinet: Pull Rodman Plug," *FTU*, December 16, 1992.

7. "Fishermen Worried About Another One Getting Away," *FTU*, December 16, 1992; "Cabinet: Pull Rodman Plug," *FTU*, December 16, 1992; "Cabinet Votes to Restore River," *OSB*, December 16, 1992; "Cabinet Urges End to Dam," *GS*, December 16, 1992.

8. "Cabinet Urges End to Dam," *GS*, December 16, 1992; "Cabinet Votes to Restore River," *OSB*, December 16, 1992.

9. *Environmental Studies Concerning Four Alternatives for Rodman Reservoir and the Lower Ocklawaha River*, vol. 1, *Executive Summary*, St. Johns River Water Management District, 22, St. Johns Water Management District Library, Palatka, Fla.; George Kirkpatrick, "Musings on Rodman," n.d., 5, George Kirkpatrick Papers, Mike Murtha Collection, Gainesville, Fla.

10. George Kirkpatrick to Virginia Wetherell, April 4, 1995, George Kirkpatrick Papers, Mike Murtha Collection, Gainesville, Fla.

11. George Kirkpatrick, Memorandum to Members of the Board of Trustees of the Internal Improvement Trust Fund, n.d., 11, George Kirkpatrick Papers, Mike Murtha Collection, Gainesville, Fla.

12. George Kirkpatrick, "Musings on Rodman," n.d., 3, 4, 6, George Kirkpatrick Papers, Mike Murtha Collection, Gainesville, Fla.; "The Battle of the Dams," *Smithsonian* 28 (November 1998): 52.

13. "Barge Canal's Nemesis Takes on the Dam," *MH*, February 4, 1990; "Activist Recalls Fight to Let the River Run," *GS*, June 13, 1997.

14. "Gainesville's Marjorie Carr, Environmentalist, Dies at 82," *GS*, October 11, 1997; "'Steward of God's Garden' Laid to Rest," *GS*, October 17, 1997.

15. Ibid.

16. http://election.dos.state.fl.us/laws/98laws/ch_98–398.pdf.

17. "Should the River Run Free?" *FTU*, March 13, 2000; "Ocklawaha Restoration Remains in Limbo," *GS*, April 26, 2007; "Year after Year, It's the Same Dam Debate," *OSB*, April 26, 2007.

18. "Kirkpatrick Hailed for Work on Putnam's Behalf," *PDN*, February 7, 2003.

19. "Bill Would Kill Canal, Return Land to State," *OSB*, March 9, 1990; Dee Cirino, "Conserving Land Where Bridge Was to Cross Canal," letter to editor, *OSB*, March 8, 1990.

Bibliography

ARCHIVAL AND MANUSCRIPT COLLECTIONS

Adams, Tom. Papers. Florida State Archives, Tallahassee.

———. Papers. University of South Florida Library, Special Collections, Tampa.

Albert, Carl. Papers. Carl Albert Center, Congressional Archives, University of Oklahoma, Norman.

Andrews, Charles. Papers. PKY.

Askew, Reuben. Interview transcript. SPOHC.

———. Papers. PKY.

Belcher, Page H., Papers. Carl Albert Center, Congressional Archives, University of Oklahoma, Norman.

Bennett, Charles. Interview transcript. SPOHC.

———. Papers. PKY.

Black, Robert Lucas. Papers. PKY.

Bryant, Farris. Interview transcript. SPOHC.

———. Papers. PKY.

Canal Authority of the State of Florida. Records, Series 127. FSA.

Carr, Archie. Papers. PKY.

Carr, Marjorie. Interview transcript. SPOHC.

Chappell, William. Papers. PKY.

Chiles, Lawton. Interview transcript. SPOHC.

Collins, Thomas LeRoy. Interview transcript. SPOHC.

Couse, John. Papers. In possession of authors, Gainesville, Fla.

Davis, John Henry. Papers. PKY.

Dean, John, III. Papers. White House Special Files, Staff Member and Office Files, Richard Nixon Presidential Papers, NA.

Dosh, Robert "Bert." Papers. PKY.

Dunn, Hampton. Papers. University of South Florida Library, Special Collections, Tampa.

Florida Defenders of the Environment. Papers. PKY.

Fuqua, Don. Interview transcript. SPOHC.

Gleason Family. Papers. PKY.

Gordon, Ulysses. Papers. PKY.

Graham, Bob. Interview transcript. SPOHC.

———. Papers. PKY.

Green, Robert Alexis "Lex." Interview transcript. SPOHC.

————. Papers. PKY.

Hahn, Theodore. Papers. PKY.

Hart, Hubbard H., Papers. FSA.

Hawkins, Paula. Papers. Winter Park Public Library, Winter Park History and Archives Collection, Winter Park, Fla.

Hendricks, Joseph. Papers. PKY.

Herlong, Albert Sydney. Papers. PKY.

Holland, Spessard. Papers. Claude Pepper Library, Claude Pepper Center, Florida State University, Tallahassee.

————. Papers. PKY.

Jacksonville District Comprehensive Reports and Studies Files. Records of the Office of the Chief of Engineers (RG 77), NARASE.

Johnson, Lyndon B., Papers. Lyndon Baines Johnson Presidential Library, Austin, Tex.

————. Recordings of Conversations and Meetings. Recordings of Telephone Conversations, JFK Series, tape K6312.12, Conversations with George Smathers and Martin Anderson, 12/20/63, Lyndon Baines Johnson Presidential Library, Austin, Tex.

————. Recordings of Conversations and Meetings. Recordings of Telephone Conversations, Recordings of Telephone Conversations—White House Series, tape WH6508.09, Conversation with Spessard Holland, 8/23/65. Lyndon Baines Johnson Presidential Library, Austin, Tex.

Johnson, Malcolm. Papers. Strozier Library, Special Collections, Florida State University, Tallahassee.

Johnston, Olin D. Papers. South Caroliniana Library, Modern Political Collection, University of South Carolina, Columbia.

Kirk, Claude. Interview transcript. SPOHC.

Krogh, Egil. Papers. White House Special Files, Staff Member and Office Files, Richard Nixon Presidential Papers, NA.

Maas, Arthur. Papers. U.S. Army Corps of Engineers, Office of History, Humphreys Engineer Center, U.S. Army Corps of Engineers Papers, Alexandria, Va.

Marshall, Arthur. Papers. PKY.

Mathews, John E. Jr., Papers. Thomas G. Carpenter Library, Special Collections, University of North Florida.

Matthews, Donald R. "Billy." Interview transcript. SPOHC.

McClintic, James V. Papers. Carl Albert Center, Congressional Archives, University of Oklahoma, Norman.

Murtha, Mike. Papers. In possession of authors, Gainesville, Fla.

Nixon, Richard. Presidential Papers. White House Central Files, Subject Files, Natural Resources, NA.

O'Brien, Lawrence F. Office Files. Lyndon Baines Johnson Presidential Library, Austin, Tex.

Olds, Leland. Papers. Franklin D. Roosevelt Presidential Library, Hyde Park, N.Y.

Partington, Bill. Papers. In possession of authors, Gainesville, Fla.

Pepper, Claude. Interview transcript. SPOHC.

————. Official and Personal Papers. Claude Pepper Library, Claude Pepper Center, Florida State University, Tallahassee.

Peterson, James Hardin. Papers. PKY.

Reed, Nathaniel. Interview transcript. SPOHC.

Roberts, William Milner. Papers. Montana State University, Merrill G. Burlingame Special Collections, Bozeman.

Rogers, Paul. Papers. PKY.

Roosevelt, Franklin. Presidential Papers. Franklin D. Roosevelt Presidential Library, Hyde Park, N.Y.

Rosenman, Samuel. Papers. Franklin D. Roosevelt Presidential Library, Hyde Park, N.Y.

Shoemaker, Don. Papers. University of North Carolina, Southern Historical Collection, Chapel Hill, N.C.

Smathers, George. Interview transcript. SPOHC.

————. Papers. PKY.

State of Florida, Board of Conservation. Papers. FSA.

Thomas, Elmer. Papers. Carl Albert Center, Congressional Archives, University of Oklahoma, Norman.

Tigert, John James. Papers. PKY.

Trammell, Park. Papers. PKY.

Truman, Harry S. Presidential Papers. Harry S. Truman Presidential Library, Independence, Mo.

University of Florida Archives, Public Records Collection, IFAS Vice-President for Agricultural Affairs. Papers. PKY.

Vandenberg, Arthur. Papers (microfilm). Bentley Library, University of Michigan, Ann Arbor.

Whitaker, John. Papers. White House Central Files, Staff Member and Office Files, Richard Nixon Presidential Papers, NA.

White House Central Files, Presidential Papers. Lyndon Baines Johnson Presidential Library, Austin, Tex.

Wilson Cypress Company. Records. PKY.

Works Progress Administration (WPA), Central Files, State 1935–1944, Florida. Papers (RG 69). NA.

Youngberg, Gilbert. Papers. Olin Library, Special Collections, Rollins College, Winter Park, Fla.

GOVERNMENT DOCUMENTS

Canal Authority of the State of Florida. Annual reports, 1963–1975.

Fernald, Edward A. *An Optimum Land Use Model for a Delimited Area Contiguous to the Cross Florida Barge Canal.* Study prepared for the Canal Authority of the State of Florida. 1967.

Final Environmental Statement, Proposal for Oklawaha River; Ocala National Forest, Florida. Tallahassee, Fla.: United States Forest Service, 1973. 3 vols. (vol. 1, Final Environmental Statement; vol. 2, Appendices 1–14; vol. 3, Appendices 15–44).

U.S. Army, Corps of Engineers. *Definite Project Report, Cross-Florida Barge Canal, Addendum No. 1.* Jacksonville U.S. Engineer Office, March, 1945.

————. *Economic Restudy of Cross-Florida Barge Canal.* Jacksonville, January 10, 1958.

————. *Economic Evaluation of Cross-Florida Barge Canal Project.* Jacksonville, March 13, 1962.

————. *Enclosure to Economic Evaluation of Cross-Florida Barge Canal Project*. Report prepared by Arthur D. Little, Inc., March 1962.

U.S. Army, Department of the Army. *Annual Report of the Chief of Engineers, U.S. Army, on Civil Works Activities*. See annual reports starting in 1935 through 1973 for complete data on construction.

U.S. House of Representatives Documents

U.S. Congress. House. *Memorial of the Legislative Council of the Territory of Florida*. 18th Cong., 2nd sess., December 28, 1824. H. Doc. 36.

————. Committee on Roads and Canals. *Report of the Committee on Roads and Canals, upon the subject of Internal Improvements, Accompanied by a Bill Concerning Internal Improvements*. 18th Cong., 2nd sess., February 26, 1825. H. Rep. 83.

————. *Daniel Webster, Resolution Relative to a Canal in Florida*. H.R. 6. 19th Cong., 1st sess., December 19, 1825.

————. *Letter from the Secretary of War, Transmitting an Estimate of the Cost of Completing a Survey, etc., of a Canal to Connect the Waters of the Atlantic and the Gulf of Mexico*. 21st Cong., 1st sess., January 27, 1830. H. Doc. 41.

————. *Letter from the Secretary of War, Transmitting a Report of Lieutenant John Pickell Respecting the Practicability of a Canal across Florida*. 22nd Cong., 1st sess., March 26, 1832. H. Exec. Doc. 185.

————. *Joint Memorial of the Legislature of Alabama, Relating to Ship Communication between the Waters of the Gulf of Mexico and the Atlantic Ocean by Ship canal through the Florida Peninsula*. 42nd Cong., 2nd sess., March 11, 1872. H. Misc. Doc. 110.

————. *Letter from the Secretary of War, Transmitting the Report of the Chief of Engineers on the Practicability and Cost of Inland Water Communication between the Mississippi River and the Atlantic Ocean, etc.* 44th Cong., 1st sess., April 6, 1876. H. Exec. Doc. 157.

————. Committee on Rivers and Harbors. *Atlantic-Gulf Ship Canal. Hearings on H.R. 6150, a Bill for the Completion of the Construction of the Atlantic-Gulf Ship Canal across Florida*. 75th Cong., 1st sess., 1937.

————. *Atlantic-Gulf Ship Canal, Florida. Letter from the Secretary of the Army transmitting to the House of Representatives a letter from the Chief of Engineers of the United States Army, dated April 1, 1937. Submitting a Report, together with Accompanying Papers and Illustrations on a Preliminary Examination and Survey of Various Routes for a Waterway Across Southern Georgia or Northern Florida to Connect the Atlantic Ocean with the Gulf of Mexico, Authorized by the Rivers and Harbors Act Approved January 21, 1927 and July 3, 1930*. 75th Cong., 1st sess., April 5, 1937. H. Doc. 194.

————. Committee on Rivers and Harbors. *Report of the Committee on Rivers and Harbors: Atlantic-Gulf Ship Canal, Florida*. 75th Cong., 1st sess., June 8, 1937. H. Doc. 950.

————. *Atlantic-Gulf Ship Canal across Florida*. 76th Cong., 1st sess., 1938. H. Doc. 853.

————. *Atlantic-Gulf Ship Canal across Florida, Hearings before the Committee on Rivers and Harbors, House of Representatives, on H.R. 3222 and H.R. 3223*. 76th Cong., 1st sess., 1939.

————. *Authorizing the Completion of the Construction of the Atlantic-Gulf Ship Canal across Florida. Report to Accompany H.R. 3223*. 76th Cong., 1st sess., April 26, 1939. H. Rep. 509.

————. *Enlargement and Extension of the Gulf Intracoastal Waterway, Including the Con-*

struction of a Barge Canal and Pipe Line across Northern Florida, Hearings on H.R. 6999. 77th Cong., 2nd sess., 1942.

———. *War Department Civil Functions Appropriation Bill for 1944, Hearings.* 78th Cong., 1st sess., 1943.

———. *War Department Civil Functions Appropriation Bill for 1945, Hearings.* 78th Cong., 2nd sess., 1944.

———. *War Department Civil Functions Appropriation Bill for 1946, Hearings.* 79th Cong., 1st sess., 1945.

———. Committee on Printing, Atlantic-Gulf Ship Canal, Florida. *Report to Accompany House Resolution 301, House Report No. 2.* 79th Cong., 1st sess., January 8, 1945.

———. *War Department Civil Functions Appropriation Bill for 1947, Hearings.* 78th Cong., 2nd sess., 1946.

———. *War Department Civil Functions Appropriation Bill for 1948, Hearings.* 80th Cong., 1st sess., 1947.

———. Special Committee on the Panama Canal of the National Rivers and Harbors Congress. *The Panama Canal, The Sea-level Project and National Security.* 84th Cong., 2nd sess., 1956. H. Doc. 446.

———. *Public Works Appropriations for 1963, Hearings.* 87th Cong., 2nd sess., 1962.

———. *Public Works Appropriations for 1964, Hearings.* 88th Cong., 1st sess., 1963.

———. *Public Works Appropriations for 1965, Hearings.* 88th Cong., 2nd sess., 1964.

———. Subcommittee of the Committee on Appropriations. *Public Works for Water and Power Development and Energy Research Appropriation Bill, 1978.* 95th Cong., 1st sess., April 21, 1977.

———. Subcommittee on Water Resources of the Committee on Public Works and Transportation. *Cross-Florida Barge Canal.* 99th Cong., 1st sess., June 10, 1985, Palatka, Fla.

U.S. Senate Documents

U.S. Congress. Senate. *Memorial of James Gadsden and E. R. Gibson, Commissioners appointed by the Legislative Council of Florida, to Examine into the Expediency of Opening a Canal through the Peninsula of Florida.* 19th Cong., 1st sess., January 5, 1826. S. Doc. 15.

———. *Report of the Secretary of War, Made in Compliance with a Resolution of the Senate, Calling for Information in Relation to the Construction of a Canal across the Peninsula of Florida.* 32nd Cong., 1st sess., June 23, 1852. S. Exec. Doc. 85.

———. *Letter from the Secretary of War, Transmitting, in compliance with a Senate Resolution of January 9, 1877, a copy of a Report of Lieutenant Colonel Q.A. Gillmore, Corps of Engineers, on a Waterline of Transportation from the Mouth of St. Mary's River to the Gulf of Mexico.* 44th Cong., 2nd sess., January 22, 1877. S. Doc. 22.

———. *Letter from the Secretary of War, Transmitting Report of the Results of an Examination made of the Peninsula of Florida, with a View to the Construction of a Ship Canal from St. Mary's River to the Gulf of Mexico.* 46th Cong., 2nd sess., August 22, 1880. S. Doc. 154.

———. *Exposition of the Advantages and Value of a Barge Canal Connecting the Waters of the Mississippi River, through the State of Florida, with the Waters of the Atlantic Seaboard, by Major Robert Gamble, Tallahassee, Florida.* 53rd Cong., 2nd sess., March 13, 1894. Sen. Misc. Doc. 118.

———. *Florida Ship Canal, Hearings before the Committee on Commerce, United States Senate, 76th Congress, 1st Session, on S. 1100.* 76th Cong., 1st sess., 1939.

———. *Defense Coordination of the Panama and Florida Canals, A Preliminary Study, by Henry Holland Buckman.* 76th Cong., 3rd sess., April 30, 1940. S. Doc. 198.

———. *War Department Civil Functions Appropriation Bill, Fiscal Year 1947.* Report to Accompany H.R. 5400. 79th Cong., 2nd sess., March 18, 1946. S. Rep. 1067.

———. *Public Works Appropriations, 1960, Hearings on H.R. 7509.* 86th Cong., 1st sess., 1959.

———. *Public Works Appropriations, 1961, Hearings on H.R. 12326.* 86th Cong., 2nd sess., 1960.

———. *Public Works Appropriations, 1962. Hearings on H.R. 9076.* 87th Cong., 1st sess., 1961.

———. *Public Works Appropriation Bill, 1962.* Report to Accompany H.R. 9076. 87th Cong., 1st sess., September 20, 1961. S. Rep. 1097.

———. *Public Works Appropriation Bill, 1963.* Report to Accompany H.R. 12900. 87th Cong., 2nd sess., September 28, 1962. S. Rep. 2178.

———. *Public Works Appropriations, 1963. Hearings on H.R. 12900.* 87th Cong., 2nd sess. 1962.

———. *Public Works Appropriation Bill, 1964.* Report to Accompany H.R. 9140. 88th Cong., 1st sess., December 5, 1963. S. Rep. 746.

———. *Public Works Appropriations, 1964. Hearings on H.R. 9140.* 88th Cong., 1st sess., 1963.

———. *Public Works Appropriations, 1965. Hearings on H.R. 11579.* 88th Cong., 2nd sess., 1964.

———. Committee on Agriculture, Nutrition, and Forestry and the Committee on Environment and Public Works. *Termination of the Cross-Florida Barge Canal Project and Extension of the Boundaries of the Ocala National Forest* [microform]. Report to accompany S. Doc. 3337. Washington, D.C.: U.S. Government Printing Office, 1978.

———. Committee on the Environment and Public Works. *De-authorization of the Cross-Florida Barge Canal: Hearing Before the Subcommittee on Water Resources.* 95th Cong., 2nd sess., July 26, 1978.

OTHER SOURCES

"Action This Year." Editorial. *Waterways Journal* 77, no. 15 (July 13, 1963): 5.

"Alachua Pleads for Wild River." *Florida Naturalist* 38, no. 3, (July 1965): 103.

Anderson, Frank Kay. *A Brief Against the Construction of the Proposed Florida Cross-State Ship Canal on Behalf of the Central and South Florida Water Conservation Committee.* Sanford: Central and South Florida Water Conservation Committee, February 3, 1936.

Anderson, Frederick, with the assistance of Robert Daniels. *NEPA in the Courts: A Legal Analysis of the National Environmental Policy Act.* Baltimore: Johns Hopkins University Press, 1973.

Angevine, Robert. *The Railroad and the State: War, Politics, and Technology in Nineteenth-Century America.* Stanford, Calif.: Stanford University Press, 2004.

"Backers of Barge Canal Called to Action." *Waterways Journal* 83, no. 7 (May 17, 1969): 5.

Baldwin, Abel S. "St. John's Bar." *Semi-Tropical* 2 (June 1876): 340.

Barber, H. E. "The History of the Florida Cross-State Canal." Ph.D. diss., University of Georgia, 1969.

"The Battle of the Dams," *Smithsonian*, November 1998, 52.

"Beginning the Florida Canal." *Engineering News-Record*, April 2, 1936, 479–83.

Belleville, Bill. *River of Lakes: A Journey on Florida's St. Johns River*. Athens: University of Georgia Press, 2000.

Bennett, Charles E. "Early History of the Cross-Florida Barge Canal." *Florida Historical Quarterly* (October 1966): 32–143.

Berry, Wendell, and Gene Meatyard. *The Unforeseen Wilderness: An Essay on Kentucky's Red River Gorge*. Lexington: University of Kentucky Press, 1971.

"Big Michigander." *Time*, October 2, 1939, 13–14.

Blake, Nelson. M. *Land into Water—Water into Land: A History of Water Management in Florida*. Tallahassee: University Presses of Florida, 1980.

"Blocking Florida's Ditch." *Newsweek*, February 1, 1971, 56.

"Boost Florida Canal." *Business Week*, April 10, 1937, 52.

Bradbury, Robert W. "Cross Florida Barge Canal, Part II—A Proponent's Viewpoint." *Business and Economic Dimensions: Journal of the Graduate Faculty, College of Business Administration, University of Florida* 1, no. 7 (October 1965): 7–8.

Breton, Mary Joy. *Women Pioneers for the Environment*. Boston: Northeastern University Press, 1998.

A Brief Assessment of the Ecological Impact of the Cross-Florida Barge Canal, Addendum to November 1969 Report, Florida Game and Fresh Water Fish Commission, March 1970; with Comments by U.S. Army Corps of Engineers. U.S. Army Corps of Engineers, 1971.

Brinton, Daniel. *Notes on the Floridian Peninsula: Literary History, Indian Tribes and Antiquities*. Philadelphia: Joseph Sabin, 1859.

Buckman, Henry H., ed. *Documentary History of the Florida Canal: Ten-Year Period, January 1927 to June 1936*. Washington: U.S. Government Printing Office, 1936.

Buker, George. *Jacksonville: Riverport-Seaport*. Columbia: University of South Carolina Press, 1992.

———. *Sun, Sand and Water: A History of the Jacksonville District U.S. Army Corps of Engineers, 1821–1975*. Fort Belvoir, Va.: U.S. Army Corps of Engineers, 1981.

Byerley, M. H. "Cross-Florida Barge Canal: An Annotated Bibliography." Master's thesis, Florida State University, 1961.

"Canal Projected for Northern Florida." *Science*, May 27, 1932, Supp. 12, p. 14.

Carr, Marjorie. "The Oklawaha River Wilderness," *Florida Naturalist* 38, no. 3–A (August 1965): 1.

Carter, Luther. J. *The Florida Experience: Land and Water Policy in a Growth State*. Baltimore: Johns Hopkins University Press, 1974.

Carver, Melissa. "Florida Defenders of the Environment: A Case Study of a Volunteer Organization's Media Utilization." Master's thesis, University of Florida, 1973.

Catlin, Yates. "By Barge—River Transportation's Confluence." In *Transportation Century*, edited by George Fox Mott, 114–21. Baton Rouge: Louisiana State University, 1966.

"Caught in the Courts." *Time*, May 15, 1972, 63.

Cerrato, Cynthia L. "Davenport: A Prehistoric Village and Mound Site in the Ocala National Forest." Master's thesis, University of South Florida, 1994.

Chazal, L. H. *Ocala, Marion County, Florida*. Ocala: Marion County Board of Trade, ca. 1920.

"Cloudy Sunshine State." *Time*, April 13, 1970, 48–49.

Colburn, David, and Lance deHaven-Smith. *Government in the Sunshine State: Florida since Statehood*. Gainesville: University Press of Florida, 1999.

"Congress Says 'No' to Quoddy and Florida Canal, but President Says 'Maybe, Sometime.'" *Newsweek*, April 25, 1936, 9–10.

"Congress, Senate, in Race for Adjournment, Hurdles Relief Bill but Stumbles over Tax Measure." *Newsweek*, June 6, 1936, 8–9.

"Congress to Get New Plea on Cross-Florida Canal." *Engineering News Record*, December 25, 1958, 25.

"Connection of the Atlantic with the Gulf." *DeBow's Review*, June 1853, 567–72.

"Cross Florida Barge Canal." *Living Wilderness*, Summer 1965, 39.

"Cross Florida Barge Canal Wins 'Beauty Contest.'" *Florida Vacation Fun Times*, October 1969.

"Cross-State Barge Canal on Schedule." *Florida Trend*, December 1965, 5.

"Current Events." *Literary Digest* 120, no. 18 (November 2, 1935): 29.

Cusick, Joyce E. *City of Dunnellon Historic District Survey: "Boomtown" of the Hardrock Phosphate Mining Industry*. Dunnellon, Fla.: n.p., 1987.

Crum, Lou Jean. "The Ocklawaha River." Master's thesis, Florida State University, 1954.

"Dale Miller Will Direct National Effort for Canal." *Jacksonville Seafarer* 20, no. 3 (March 1971): 8–9.

"Dam Ditched: Ditch Damned." *Time*, April 27, 1936, 9–19.

Danese, Tracy E. *Claude Pepper and Ed Ball: Politics, Purpose, and Power*. Gainesville: University Press of Florida, 2000.

Dasmann, Raymond F. *No Further Retreat: The Fight to Save Florida*. New York: Macmillan, 1971.

Davenport, Walter. "Splitting Florida." *Collier's*, December 14, 1935, 7–8.

Davis, Frederick. "'Get the Facts—and Then Act': How Marjorie H. Carr and the Florida Defenders of the Environment Fought to Save the Ocklawaha River." *Florida Historical Quarterly* 83, no. 1 (Summer 2004): 46–69.

Derr, Mark. *Some Kind of Paradise: A Chronicle of Man and the Land in Florida*. Gainesville: University Press of Florida, 1998.

Dinkins, J. Lester. *Dunnellon—Boomtown of the 1890s: The Story of Rainbow Springs and Dunnellon*. St. Petersburg: Great Outdoors, 1969.

"Disaster: Heroism and Tragedy in Florida's 'Mild Hurricane.'" *Newsweek*, September 14, 1935, 12–14.

"Ditch Resurrected." *Time*, April 19, 1943, 23–24.

Dodd, Dorothy. "Florida in 1845." *Florida Historical Quarterly* 24 (July 1945): 3–27.

———. "The Wrecking Business on the Florida Reef, 1822–1860." *Florida Historical Quarterly* 22 (April 1944): 171–99.

Dosh, R.N. "Cross-State Canal: Completing the Proposed Canal Project Would Benefit Commerce, Industry and Agriculture in Florida." *All Florida Magazine*, August 26, 1956, 1–2.

"Double Death." *Time*, June 29, 1936, 15.

Douglas, Marjorie Stoneman. *Florida: The Long Frontier*. New York: Harper and Row, 1967.

Dunn, Hampton. *Back Home: A History of Citrus County, Florida*. Clearwater: Citrus County Bicentennial Steering Committee, 1977.

Dworsky, Leonard, David Allee, and Sandor Csallany, eds. *Social and Economic Aspects of Water Resources Development*. Urbana, Ill.: American Water Resources Association, 1971.

"Ecology: The New Jeremiahs." *Time*, August 15, 1969, 37.

"Ecological Hysteria." Editorial. *Waterways Journal* 84, no. 1 (April 4, 1970): 4.

Ellerson, Amy. "The Cross-Florida Barge Canal: Ideologies in Conflict." Master's thesis, Florida State University, 1995.

"End of the Barge Canal." *Time*, February 1, 1971, 43.

Environmental Impact of the Cross-Florida Barge Canal with Special Emphasis on the Oklawaha Regional Ecosystem. Gainesville: Florida Defenders of the Environment, 1970.

Eriksson, Barbara Dunlap. "The History of the Florida Ship Canal from its Earliest Beginnings to its Temporary Abandonment in 1942." Master's thesis, Stetson University, 1964.

"Exit 'White House Pet.'" *Literary Digest*, March 28, 1936, 6–7.

Federal Writers' Project of the Works Progress Administration for the State of Florida. *The WPA Guide to Florida: The Federal Writers' Project Guide to 1930s Florida*. New York: Pantheon Books, 1939.

Ferejohn, John. *Pork Barrel Politics: Rivers and Harbors Legislation, 1947–1968*. Stanford: Stanford University Press, 1974.

"1st Commercial Cargo Shipped Via Cross-Florida Barge Canal." *Jacksonville Seafarer* 18, no. 8 (August 1969): 10–11.

Flippen, J. Brooks. *Conservative Conservationist: Russell E. Train and the Emergence of American Environmentalism*. Baton Rouge: Louisiana State University Press, 2006.

———. *Nixon and the Environment*. Albuquerque: University of New Mexico Press, 2000.

"Florida: Another Step toward Resurrecting of Canal Project." *Newsweek*, December 26, 1936, 15.

"Florida Canal, Oliver Twists and Uncle Sam." *Literary Digest*, October 12, 1935, 34.

"Florida—The Key to the Gulf." *DeBow's Review and Industrial Resources*, September 1856, 283–86.

"Florida Canal and Quoddy Must Get Approval of Congress." *Engineering News-Record*, April 16, 1936, 576.

"Florida Canal Funds Voted Down." *Engineering News-Record*, March 19, 1936, 434.

"Florida Readies for Drive for Barge Canal." *Engineering News-Record*, January 18, 1951, 31.

"Florida Ship Canal." *DeBow's Review*, October 1852, 419–20.

"The Florida Ship Canal." *Scientific American*, January 18, 1890, 41.

"Florida Water Interests Applaud Canal Ruling." *Waterways Journal* 87, no. 46 (February 16, 1974): 4.

"The Florida Wreckers." *DeBow's Review*, February 1859, 181.

Flynt, Wayne. *Duncan Upchurch Fletcher: Dixie's Reluctant Progressive*. Tallahassee: Florida State University Press, 1971.

Fogelsong, Robert. *Married to the Mouse: Walt Disney World and Orlando*. New Haven: Yale University Press, 2001.

"Freight Boom on the Inland Waterways." *Business Week*, October 3, 1953, 27.

Gannon, Michael V., ed. *The New History of Florida*. Gainesville: University Press of Florida, 1996.

Glicksberg, Charles. "Letters of William Cullen Bryant from Florida." *Florida Historical Quarterly* 14 (April 1936): 268–69.

"The Good Earth." Editorial. *Waterways Journal* 84, no. 3 (April 18, 1970): 4.

Gordon, John. "The American Environment: The Big Picture Is More Heartening Than All the Little Ones." *American Heritage* 44, no. 6 (October 1993): 30–51.

"Governor Bryant Makes Strong Plea for Cross-State Canal." *Jacksonville Seafarer* 10, no. 7 (July 1961): 1.

"Governor Collins Gives Okay to Cross-Florida Barge Canal." *Jacksonville Seafarer* 7, no. 5 (May 1958): 5.

"The Great Florida Phosphate Boom" *Cosmopolitan* 14, no. 6 (April 1893): 706–14.

Gregg, F. Browne, with Margie Sloan. *Progress through Innovation*. Privately printed, n.d.

Grossman, Elizabeth. *Watershed: The Undamming of America*. New York: Counterpoint, 2002.

Grove, William G. "Some of the Bridge Problems in Connection with the Atlantic-Gulf Canal." *Florida Engineering Society Bulletin* 11 (August 1936): 11.

Hines, Lawrence. *Environmental Issues: Population, Pollution, and Economics*. New York: Norton, 1973.

Hoffman, Paul E. *Florida's Frontiers*. Bloomington: Indiana University Press, 2002.

Ickes, Harold L. *The Secret Diary of Harold L. Ickes: The First Thousand Days*. New York: Simon and Schuster, 1953.

An Industrial Survey of Ocala and Marion County Florida. Ocala: Marion County Chamber of Commerce, 1928.

"The Inland Waterways Commission." *Science*, April 5, 1907, 556–57.

"In Race for Adjournment, Senate Stumbles over Tax Measure." *Newsweek*, June 6, 1936, 34.

"Instead of a Canal." *Survey*, November 1937, 350.

"Insurance Tax upon Gulf Commerce of the Southwest." *DeBow's Review*, November 1853, 516–18.

"The Intracoastal Way." *Time*, October 1, 1956, 61.

Irby, Lee. "A Passion for Wild Things: Marjorie Harris Carr and the Fight to Free a River." In *Making Waves: Female Activists in Twentieth-Century Florida*, edited by Jack E. Davis and Kari A. Frederickson, 375–97. Gainesville: University Press of Florida, 2003.

Jackson, Lena, "Sidney Lanier in Florida." *Florida Historical Quarterly* 15 (October 1936): 120–28.

Jones, S. A. "The Florida Inland Canal." *Scientific American*, December 12, 1908, 431.

Judge, Joseph. "Florida's Becoming-and Beleaguered Heartland." *National Geographic*, November 1973, 585–621.

Kallina, Edmund F., Jr. *Claude Kirk and the Politics of Confrontation*. Gainesville: University Press of Florida, 1993.

Kennedy, David M. *Freedom from Fear: The American People in Depression and War, 1929–1945*. New York: Oxford University Press, 1999.

"Kennedy Asks Money for Barge Canal: Link between Atlantic and Gulf Waterways." *Jacksonville Seafarer* 10, no. 5 (May 1961): 4.

Keyser, C. F. "Cross-Florida Barge Canal (Selected References)." 1961. General manuscripts, P. K. Yonge Library of Florida History, Dept. of Special Collections, Smathers Libraries, University of Florida.

The Kingdom of the Sun: Ocala, Marion County, Florida; Region of Great Natural Wonders. Ocala: Marion County Chamber of Commerce, 1939.

Kline, Benjamin. *First Along the River: A Brief History of the U.S. Environmental Movement.* Lanham, Md.: Rowman and Littlefield, 2007.

Knetsch, Joe. *Timber Agents, the Law and Lumbering in Historic Citrus County, Florida: The Early Decades.* Inverness, Fla.: Citrus County Historical Society, 1999.

LaMoreaux, P. E. *The Impact of the Cross-Florida Barge Canal on the Ground Water of Florida.* Tuscaloosa: Canal Authority of the State of Florida, 1973.

Lanier, Sidney. *Florida: Its Scenery, Climate, and History with an Account of Charleston, Savannah, Augusta, and Aiken and a Chapter for Consumptives; Being a Complete Handbook and Guide.* Introduction and index by Jerrell H. Shofner. Bicentennial Floridiana Facsimile Series. Gainesville: University of Florida Press, 1973.

Leopold, Aldo. *A Sand County Almanac and Sketches Here and There.* New York: Oxford University Press, 1949.

Lewis, Gerald A. "Death on the Oklawaha." *Florida and Tropic Sportsman* 1, no. 2 (December–January 1969–70): 18–20.

Life in Citrus County, Florida. Inverness, Fla.: Citrus County Chamber of Commerce, 1967.

"Limited Effect on Water Supplies Expected from Florida Canal." *Engineering News-Record,* January 9, 1936, 59–61.

Long, John E. "The Cross-Florida Canal." *American Forests,* June 1964, 17–19, 48–49.

Lowry, Sumter. "A Canal across Florida." *Review of Reviews,* May 1934, 40, 55.

———. "A Florida Ship Canal." *Review of Reviews,* August 1932, 41–42.

———. "Gulf-Atlantic Ship Canal." Special canal edition, *Southeastern Waterways* 1, no. 7 (July 1936): 7.

Luetscher, G. D. "Atlantic Coastwise Canals: Their History and Present Status." *Annals of the American Academy of Political and Social Science* 31 (January 1908): 99.

Lyon, Eugene. "Pedro Menendez's Plan for Settling La Florida." In *First Encounters: Spanish Explorations in the Caribbean and the United States, 1492–1570,* edited by Jerald T. Milanich and Susan Milbrath, 150–66. Gainesville: University of Florida Press, 1989.

Marion County, Florida, The Kingdom of the Sun: Ocala, Silver Springs. Ocala, Fla.: Marion County Chamber of Commerce, 1927.

Martin, Richard A. *Eternal Spring: Man's 10,000 Years of History at Florida's Silver Springs.* St. Petersburg: Great Outdoors, 1966.

McGee, W. J. "Our Inland Waterways." *Popular Science Monthly,* April 1908, 291–92.

Michaels, Brian E. *The River Flows North: A History of Putnam County, Florida.* Palatka: Putnam County Archives and History Commission, 1976.

Middleton, Sallie R. "Cutting through Paradise: A Political History of the Cross-Florida Barge Canal." Ph.D. diss., Florida International University, 2001.

Miller, James J. *An Environmental History of Northeast Florida.* Gainesville: University Press of Florida, 1998.

Miller, James Nathan. "Rape on the Oklawaha." *Reader's Digest,* January 1970, 54–60.

Mitchell, C. Bradford. *Paddle-Wheel Inboard: Some History of Ocklawaha River Steamboating and of the Hart Line.* Providence, R.I.: Steamship Society of America, 1983.

"Moon Unharnessed from Recovery." *Literary Digest,* February 22, 1936, 9.

Morgan, Arthur E. *Dams and Other Disasters: A Century of the Army Corps of Engineers in Civil Works*. Boston: Porter Sargent, 1971.

Noll, Steven. "Steamboats, Cypress, and Tourism: An Ecological History of the Ocklawaha Valley in the Late Nineteenth Century." *Florida Historical Quarterly* 83, no. 1 (Summer 2004): 6–23.

Odum, Howard, Elisabeth Odum, and Mark Brown. *Environment and Society in Florida*. Boca Raton: Fla.: Lewis, 1998.

Ohl, John Kennedy. *Supplying the Troops: General Somervell and American Logistics in WWII*. DeKalb: Northern Illinois Press, 1994.

Ott, Eloise Robinson, and Louise Hickman Chazal. *Ocali Country: Kingdom of the Sun: A History of Marion County, Florida*. Ocala: Greene's Printing, 1974.

Partington, William. "A Concern for the Quality of Life." *Reader's Digest*, July 1971, 45.

Patterson, James. *Grand Expectations: The United States, 1945–1974*. New York: Oxford University Press, 1996.

"Pepper v. Sholtz v. Wilcox." *Time*, May 8, 1938, 26.

Phillips, P. "The Gulf Trade." *DeBow's Review*, April 1852, 399–402.

Phillips, Ulrich B. 1908. *A History of Transportation in the Eastern Cotton Belt to 1860*. Reprint New York: Octagon Books, 1968.

"Picturesque America: The St. Johns and Ocklawaha Rivers, Florida." *Appleton's Journal of Literature, Science, and Arts*, November 12, 1870, 582–83.

Poole, Leslie Kemp, "Florida: Paradise Redefined—the Rise of Environmentalism in a State of Growth," MLS thesis, Rollins College, 1991.

Porter, Kenneth W. *The Black Seminoles: A History of a Freedom-Seeking People*. Gainesville: University Press of Florida, 1996.

Powers, Ormund. *Martin Andersen: Editor, Publisher, Galley Boy*. Chicago: Contemporary Books, 1996.

"Public Works Bill to Be Submitted to Next Congress." *Engineering News-Record*, December 19, 1935, 864.

"Quoddy and Florida." *Newsweek*, January 30, 1939, 16.

Rathlesberger, James, ed. *Nixon and the Environment: The Politics of Devastation*. New York: Village Voice, 1972.

Rawlings, Marjorie Kinnan. *South Moon Under*. New York: Charles Scribner's Sons, 1933.

A Report for Industry Concerning Ocala, Marion County, Florida. Ocala: Ocala Florida Committee of One Hundred, 1956.

"Rep. Boland Pledges Support to Finish Canal by Mid-1973." *Jacksonville Seafarer* 18, no. 1 (January 1969): 8–9, 20.

"Report Answers Critics of Florida Barge Canal." *Waterways Journal* 83, no. 2 (April 12, 1969): 4.

Reynolds, Charles. *The Standard Guide to Florida*. New York: Foster and Reynolds, 1871.

Ringhaver, L. C. "The Real Purpose of Canal Is to Relieve Industry from Burden of High Freight Rates." *Jacksonville Seafarer* 19, no. 4 (April 1970): 4.

Roberts, Paul. "Cross Florida Barge Canal: A Current Appraisal." *Business and Economic Dimensions, College of Business Administration, University of Florida* 7, no. 6 (June 1971): 7.

Rogers, Benjamin F. "The Florida Ship Canal Project." *Florida Historical Quarterly* 36 (July 1957): 14–23.

Rogin, Gilbert, "All He Wants to Save Is the World." *Sports Illustrated*, February 3, 1969, 24–29.

Rome, Adam. "'Give Earth a Chance': The Environmental Movement and the Sixties." *Journal of American History* 90, no. 2 (September 2003): 525–54.

Roosevelt, Theodore. "Our National Inland Waterway Policy." *Annals of the American Academy of Political and Social Science* 31 (January–June 1908): 1–11.

Rothman, Hal. *The Greening of a Nation? Environmentalism in the United States since 1945.* Belmont, Calif.: Thomson/Wadsworth, 1998.

Schmidt, Lewis G. *The Civil War in Florida: A Military History.* Allentown: L. G. Schmidt, 1989.

Schulz, William. "Mike Kirwan's Big Ditch." *Reader's Digest*, June 1967, 59–64.

"Seafaring with Dave Howard." *Jacksonville Seafarer* 18, no. 9 (September 1969): 18–19.

"Secretary Stone's Unique Solution to Canal Issue." *Jacksonville Seafarer* 20, no. 4 (April 1971): 22–24.

"Senator Holland Displays Leadership in Barge Canal Budget Hearing." *Jacksonville Seafarer* 8, no. 10 (October 1959): 2–3.

Sewell, J. Richard. "Cross-Florida Barge Canal, 1927–1968." *Florida Historical Quarterly* 46 (April 1968): 369–75.

Shaffer, Karen, and Neva Garner Greenwood. *Maud Powell, Pioneer American Violinist.* Ames: Iowa State University Press, 1988. Distributed for the Maud Powell Foundation.

"Ship Canal across the Peninsula of Florida." *DeBow's Review*, February 1853, 122–25.

"Ship Canals Projected and in Progress." *Scientific American*, February 9, 1895, 90.

"Shipwreck, Storm and Florida's New Canal." *Literary Digest*, September 14, 1935, 25.

Shores, Venila Lovina. "Canal Projects of Territorial Florida." *Tallahassee Historical Society Annual* (1935): 12–16.

Silver Springs Transportation Company, Ocala, Fla. *Silver Springs Daylight Route, [Oklawaha River], Florida.* Ocala: Silver Springs Transportation Co., 1917.

Simpson, Charles Torrey. *In Lower Florida Wilds.* New York: G. P. Putnam's Sons, 1920.

"Society Asks Saving of Oklawaha Wild River." *Florida Naturalist* 38, no. 3 (July 1965): 99.

Somervell, Brehon. *Digest of Reports on the Atlantic-Gulf Ship Canal.* Ocala: U.S. Engineer Office, 1936.

"Sore Thumb." *Time*, February 17, 1936, 11–12.

Steinberg, Ted. *Down to Earth: Nature's Role in American History.* New York: Oxford University Press, 2002.

Stephens, A. Dix. *Withlacoochee Notes.* Trenton, Fla.: Lulu Press, 2008.

Stevens, John P. "The Relation of Railways to Canals." *Engineering Magazine*, February 1909, 844–46.

Stine, Jeffrey K. *Mixing the Waters: Environment, Politics, and the Building of the Tennessee Tombigbee Waterway.* Akron: University of Akron Press, 1993.

Stockbridge, Frank Parker, and John Holliday Perry. *So This Is Florida.* Jacksonville: John H. Perry, 1938.

Stoessen, Alexander. "Claude Pepper and the Florida Canal Controversy, 1939–1943." *Florida Historical Quarterly* 50 (January 1972): 235–51.

Stoll, Steven. *U.S. Environmentalism since 1945: A Brief History with Documents.* Boston: Bedford/St. Martin's, 2007.

Stowe, Harriet Beecher. "Up the Okalawaha: A Sail into Fairy-Land." *Christian Union*, May 14, 1873, 393.

"'A Tale to Tell from Paradise Itself': George Bancroft's Letters from Florida, March 1866." Edited by Patricia Clark. *Florida Historical Quarterly* 48, no. 2 (January 1970): 272.

Tebeau, Charlton W. *A History of Florida.* Coral Gables: University of Miami Press, 1971.

Tegeder, Michael David. "Economic Boom or Political Boondoggle? Florida's Atlantic Gulf Ship Canal in the 1930s." *FHQ* 83, no. 1 (Summer 2004): 24–45.

"300-Ton Masher Clears Reservoir Site." *Engineering News-Record*, June 1, 1967, 25.

Train, Russell, E. "The Environmental Record of the Nixon Administration." *Presidential Studies Quarterly* 26, no. 1 (1996): 185–96.

"Truman Tells Bennett of Canal Views." *Jacksonville Seafarer* 8, no. 5 (May 1959): 11.

Trumbull, Stephen. "The River Spoilers." *Audubon Magazine*, March 1966, 102–10.

"U.S. Starts Anew on an Old Canal." *Engineering News-Record*, February 27, 1964, 21.

Van Der Hulse, Kenneth. "A Report on Conditions in Marion County Resulting from the Gulf-Atlantic Ship Canal Project." Typescript, March 17, 1936. P. K. Yonge Library of Florida History, Dept. of Special Collections, Smathers Libraries, University of Florida.

Ware, John D. *George Gauld: Surveyor and Cartographer of the Gulf Coast.* Revised and completed by Robert R. Rea. Gainesville: University Press of Florida, 1982.

Whitaker, John. *Striking a Balance: Environment and Natural Resources Policy in the Nixon-Ford Years.* Washington, D.C.: American Enterprise Institute for Public Policy Research, 1976.

Whitman, Alice. "Transportation in Territorial Florida." *Florida Historical Quarterly* 17 (July 1938): 25–53.

Willis, R. F. *Archaeological and Historical Survey of the Southeastern Portion of the Ocala National Forest, Marion and Lake Counties, Florida.* Tallahassee: Bureau of Historic Sites and Properties, 1977.

"Work on Canal across Florida to Start Immediately." *Engineering News-Record*, September 12, 1935, 376–77.

Wright, Albert Hazen. *The Atlantic-Gulf or Florida Ship Canal.* Ithaca: A. H. Wright, 1937.

Yannacone, Victor, Bernard Cohen, and Steven Davison. *Environmental Rights and Remedies.* Rochester, N.Y.: Lawyers Co-Operative, 1972.

Youngberg, Gilbert A. "The Atlantic-Gulf Ship Canal across Florida, An Historical Summary: An address Presented by Gilbert A. Youngberg at the Twentieth Annual Meeting of the Florida Engineering Society at Daytona Beach, Florida, April 3, 1936." Florida State Archives, Department of State, Division of Library and Information Services, Bureau of Archives and Records Management, Tallahassee, Florida.

———. "The Gulf-Atlantic Ship Canal across Florida." *Florida Engineering Society Bulletin*, no. 6 (August 1935): 1–20.

———. "The Proposed Gulf-Atlantic Ship Canal across Florida." *Florida Engineering and Construction* 7, no. 11 (December 1930): 135–36, 143, 146–48.

———. "Waterway Improvement at Government Expense." *Southeastern Waterways*, May 1936, 12, 18.

Yulee, David Levy. "Connexion of the Atlantic and the Gulf of Mexico, The Florida Railroad." *DeBow's Review*, April 1856, 492–514.

Index

163–67, 169, 171–72, 176–77, 195–96, 211, 213, 221–25, 229–31, 246; politics of, 113, 247–48, 250–52, 258–61, 262–65, 273, 278–82, 285, 293–99; promotion of, 157–58, 171–72, 183, 211–12, 220, 231–32, 238–39, 252–54, 256; and recreation, 192, 195, 220; and Rodman Reservoir, 210, 212–13, 284–85, 286–87, 221; route establishment, 65, 78; and saltwater intrusion, 113–15; and Vietnam War–era public works, 196–99, 205–6; World War II–period support for, 118–21, 125–27

Cross Florida Canal Navigation District, 138, 168–69

Crusher (crusher crawler), 184–87, 213, 309; opposition to, 195–96, 210, 224–25, 227

Daytona Beach, Fla., 88
Daytona Beach News-Journal, 7
DeBow's Review, 19, 47
DeLand Sun News, 101
Delano, Frederic, 106
Disney World, 2, 9, 179, 220
Disston, Hamilton, 29–30, 44, 46–47
Dixie (steamship), 73–74, 78, 81, 137
Dixie Highway (U.S. 441), 79, 82–83, 96
Dorsey, Hugh, 56–57
Dosh, R. N. "Bert," 80, 117, 130–31, 136, 138–41, 142, 157, 166, 206, 238, 239–40, 307
Dosh Lock, 144, 165, 257
Douglas, Marjory Stoneman, 146, 162, 300
Dunn, John, 34–35, 38
Dunnellon, Fla., 3, 34–38, 82, 90, 144, 281
Dunnellon Press, 152, 171–72, 278

Earle, Lewis, 295
Earth Day (1970), 256
Ehrlichman, John, 249, 263, 266–67, 274
Eisenhower, Dwight, 131, 134, 135, 202
Ellender, Allen, 134
Environmental Defense Fund (EDF), 225, 292; EDF/FDE lawsuit, 226–29, 261, 274–75, 282, 283; establishment of FDE, 214–20
Erie Canal, 2, 13
Eureka Dam, 166, 181, 257
Evans, Giles, 38, 140, 158, 169, 183, 190, 195, 254, 259, 275, 292, 294, 296; *Aiple* incident, comments on, 255–56; and canal critics, 166, 171, 238–39, 252–53, 256; on CFBC project budget cuts, 205; Jimmy Carter's policies, view of, 304; Nixon decision, response to, 269, 275; Rodman Reservoir, support for, 253–54, 284
Evergreen Cemetery (Gainesville, Fla.), 325, 327

Federal Emergency Relief Administration (FERA), 84

Fernandina, Fla., 20, 25, 29, 44, 47, 56, 59–60
Fisher, Holly, 174, 203
Fisher, Richard, 174, 181, 203, 236
Fletcher, Duncan, 53–56, 58, 106, 308; death, 99, 103; Florida Ship Canal funding, 70–74; Florida Ship Canal groundbreaking (1935), 81–82; legacy, 103, 104, 105; Senate fight for ship canal, 91–99
Florida Audubon Society, 155, 159, 285
Florida Bi-Partisans, 169
Florida Board of Conservation. 156, 167, 174, 192, 193, 203, 212
Florida Citrus Producers Trade Association, 92, 119
Florida Defenders of the Environment (FDE), 22, 31, 195, 199, 223, 224, 227, 231, 235, 241, 248, 250, 251, 252, 253, 256, 276, 277, 280, 282, 295, 296, 298, 302, 303, 311, 315, 316, 317, 323, 325; 1970 environmental impact statement, 205, 229, 243–47, 293; EDF/FDE lawsuit, role in, 219–20, 226, 228, 234, 261–62, 274–75, 283, 291; formation of, 214–15, 218–19; fundraising, 225; and Jetport controversy, 241–43; political activism, 259–61, 263–64, 293, 303, 320–21, 325–26; publicity, 203, 222–23, 229–30, 236–38, 240, 254–55, 257; and Nixon decision, 266–67, 271; Ocklawaha restoration, role in, 285, 300, 303–4, 317, 320, 325–26; and Rodman controversy, 284–89, 301, 310, 313, 317–21, 324
Florida Department of Environmental Protection (DEP), 195–96, 321, 322, 323
Florida Department of Natural Resources (DNR), 295, 319, 320
Florida Game and Fresh Water Fish Commission, 280–81, 285, 288
Florida Grower, 89
Florida Power Corporation, 36, 145
Florida Overland Railway Barge System (FORBS), 281–82
Florida Senate Appropriations Committee, 319
Florida Senate Rules Committee, 318–19
Florida Ship Canal, x, 1, 3, 5, 8, 52, 59, 62, 65, 72, 73–74, 77, 81, 103, 107, 108, 109, 110, 112, 113, 114, 116, 118, 120, 121, 122, 123, 129, 130, 146, 157, 183, 200, 297; Congressional debate on, 93–99; criticism of, 87–92; early consideration of, 16, 19, 20, 29, 44, 45, 54, 58; end of construction, 99; excavation and construction, 76–79, 82–83; funding for, 68–69, 92–93; groundbreaking (1935), 4, 80–82; promotion of, 62–63; route establishment, 19, 48, 58, 60, 62, 64, 144; and saltwater intrusion, 78, 87–89, 103; town of Santos in path of, 84–85; as work project, 63–65, 83–84, 86–87

Florida Ship Canal Authority, 64, 67, 68, 69, 79, 103, 107, 118, 125, 128, 130, 132, 133,134, 138, 183, 206

Florida Ship Canal Navigation District, 70, 79, 130

Florida State Canal Commission, 56–57, 62

Florida State Senate, 252–53, 295

Florida Times-Union, 8, 35, 39–40, 99, 269

Florida Water Conservation League, 92, 101

Florida Waterways Association, 231

Florida Waterways Committee, 149, 192

Florida and Tropic Sportsman, 247–48

Floridan Aquifer, 8, 90, 113–14, 145, 149

Flournoy, Alyson, 324–25

Ford, Gerald, 303

Fortson, Malcolm, 130, 138, 149

Friends of the Earth, 250

Frye, Earle, 280–81

Fullerton, Avery, 286–87

Funk, Ben, 229–30

Fuqua, Don, 208, 251

Gainesville, Fla., 44, 100, 147, 148, 150, 155, 157, 169, 172, 200, 203, 220, 223, 225, 267, 318, 320, 322, 324, 325, 326, 327

Gainesville Sun, 70, 158, 167, 170–71, 174–75, 255, 257, 325

Gallatin, Albert, 12

Gamble, Robert, 46–47

Gee and Jensen, 133

Gibbons, Sam, 188

Gillmore, Quincy, 45–48

Gleason, William, 29–30, 44, 45

Godfrey, Arthur, 271, 285

Godfrey, David, 317, 321, 325

Gordon, Ulysses, 172–73

Graham, Bob, 307, 309, 312, 324

Grant, Bill, 327

Grant, James, 11–12

Graves, Ernest, 300–302

Great Depression, 3, 4, 59; and CFBC, 63–67

Green, R. A. "Lex," 57, 67, 69, 72, 73, 80, 98, 105, 107, 115, 119,121, 122, 125, 128

Gregg, F. Browne, 184–87, 195, 196–97

Gulf Atlantic Ship Canal. *See* Florida Ship Canal

Gulf Intracoastal Canal Association (GICA), 50

Hahn, Theodore, 146

Harper's, 30

Hart, Hubbard, 32, 56, 246; and Civil War shipping, 24–25; death of, 38; early career, 23–25; Hart Line, 26–30, 38–43; lumber operations, 26

Hassett, Joseph, 217

Hawkins, Paula, 309–10

Hazouri, Tommy, 306

Henderson, Warren, 248

Hendricks, Joe, 107, 119, 122–27, 128

Herlong, Sydney, 128, 134, 139, 143, 154, 157, 159, 164, 180–81, 191, 206

Hiawatha (steamboat), 40–42

Hickel, Walter, 242, 259, 262, 263; CFBC construction moratorium, support of, 249–52

High, Robert King, 180–82

Hills, George, 71, 99

Hills and Youngberg, 59, 71

Hodge, F. W. "Wally," 149, 150, 155, 158, 169

Hodges, Randolph, 135, 156, 174, 203–4, 205, 275, 280, 294

Holland, Spessard, 128, 129, 134, 141, 200, 251–52, 258, 259

Hopkins, Harry, 69, 70, 72, 84, 97

House Committee on Appropriations, 94–95, 125, 129, 139, 141, 208, 306

House Committee on Rivers and Harbors, 105, 107, 108, 110, 126

House Committee on Roads and Canals, 14, 17

Howe, James, 88–89, 91

Huntley-Brinkley Report, 211–12

Ickes, Harold, 68–69, 90, 92, 112, 119, 123, 126–28

Inglis, Fla., 36, 61, 144–45, 152, 154, 157, 189, *194*, 269, 278; marina, 317; pool, 280

Inglis, John, 34–38

Inglis Dam, 36, 189

Inglis Lock, 145, 152, 167–68, 184, 189, 240, 254, 311, 316; completion of, 239–40

Inland Waterways Commission, 51, 55

Interstate Inland Waterway League, 50

Izaak Walton League, 270, 285–86

Jacksonville, Fla., x, 3, 7–8, 9, 18, 21, 35, 40, 53–57, 59–60, 70–71, 76, 80, 89, 91, 99, 105, 107, 115, 116, 127, 130, 133, 153, 154, 166, 171, 209, 238, 266–67, 269, 271, 282, 291, 294, 297, 308; as eastern terminus of Florida Ship Canal, 62–65; port improvements, 47–48

Jacksonville Chamber of Commerce, 1, 59, 98, 114, 116, 119, 132, 280, 289

Jacksonville Journal, 119, 125, 128, 130, 289

Jacksonville Port Authority, 283

Jacksonville Seafarer, 134, 207, 220, 276, 282

Johnsen, Harvey, 283–85, 287, 290–92, 296, 301

Johnson, Andy, 308

Johnson, Lady Bird, 170

Johnson, Lyndon, 5, 141, 172, 197, 199, 205, 206, 209, 234; CFBC groundbreaking speech (1964), 142–43

Jones, Charles, 45

Kaufman, Jack, 286
Kaufmann, George, 228–29, 261
Kennedy, John F., 135–36, 139–41, 197, 234, 272
Kent, Frank, 112, 116
King, Jim, 326
Kirk, Claude, 180, 229, 258–59, 262, 263–64, 278
Kirkpatrick Dam. *See* Rodman Dam
Kirkpatrick, George, 317–25; death of, 326–27
Kirwan, Mike, 201, 208
Koisch, Frank, 302
Knotts, A. F., 61–62, 65, 80

Labor Day Hurricane (1935), 73, 74
Lake George, 161, 257, 276–77
Lake Rousseau (Withlacoochee Backwaters), 36
Lanier, Sidney, 30–32, 40, 58; language and vision, FDE use of, 31, 148, 163–64, 227, 236–37, 246, 324, 325, 328
Leopold, Aldo, 162, 242
Leopold, Luna, 242–43
Lewis, Gerald, 247–48, 279, 294
Linville, George, 292, 296, 300
Little, Joe, 324, 325
Livingston, Joe, 199
Lovett, Robert, 130
Lowry, Sumter, 71, 81, 100–101
Lucas, Edward, 38–40
Lucas, Jack, 280, 288–89

MacKay, Kenneth "Buddy," 308, 310, 325
Macomb, Alexander, 15
Maiden, Jane, 196
Mann, Frank, 306
Mansfield, Joseph, 105, 108, 109, 113, 118, 121–22, 126–27
Marion County, 36–37, 50, 83–84, 86, 119, 124, 154, 307, 328; Citizens for Conservation, establishment of, 169
Marion County Board of Trade, 36–37, 50
Marjorie Harris Carr Cross Florida Greenway, 2, 316, 325, 327, 328–29
Markham, Edward, 78, 101, 103
Marshall, Art, 162, 242–43
Martinez, Bob, 313
Matthews, Donald Ray "Billy," 128, 134, 152, 155, 171
McCree, William, 138, 139, 140, 143, 154–55, 157, 172, 238, 251, 252, 268, 269
McIntyre, Marvin, 71
Metamora (steamboat), 38–40
Miami, Fla., 5, 7, 90, 108, 109, 116, 134, 167, 180, 250, 296
Miami Beach Chamber of Commerce, 108, 113, 122
Miami Herald, 6, 105, 108, 117, 119, 140, 271, 306, 314

Miami Jetport (Everglades Jetport), 234, 241–43, 249, 250, 262, 266, 287
Micanopy, Fla., 2, 202, 223, 228, 267
Miller, Dale, 275–76
Miller, James Nathan, 236–40
"Missing Link," 77, 133, 188, 190, 203–5
Mississippi to Atlantic Inland Waterway Association, 53–55
Morrison, Helen, 204
Morrison, Ken, 169, 204, 240–41

National Audubon Society, 215, 216, 217–18, 231, 232, 253–54
National Environmental Policy Act (NEPA), 235, 273, 274, 292; and EDF suit, 261–62
National Gulf-Atlantic Ship Canal Association, 62–65, 81, 107
National Resources Committee, 106
National Rivers and Harbors Congress, 110, 206
New Deal, x, 2, 4, 5, 65, 69, 71, 78, 85, 90, 93, 94, 97, 100, 104, 105, 106, 110, 112, 113, 116, 117, 141, 197, 198, 272
Newsweek, 174, 272
New York, 2, 19, 26, 29, 30, 35, 39, 43, 73–74, 78, 86, 106, 118, 120, 121, 170, 174, 200, 205, 214, 215, 217, 219, 226, 228–29, 268
New York Times, 99, 101, 102, 122, 156, 158, 266, 272
Nixon, Richard, x, 6, 9, 214, 217–18, 240, 241, 242, 243, 248, 249, 250, 258, 260, 262, 287, 294, 304; 1970 State of the Union address, 235–36, 249, 250; decision to halt CFBC, 262–72, 274–80, 282–85; and EDF/FDE lawsuit, 291–92; environmental policy, 234–36, 302–3; legality of CFBC decision challenged, 291–93; Tennessee-Tombigbee Waterway proposal, response to, 272–74
Norris, Frank, 132

Ocala, Fla., 3, 34, 35, 58, 64, 90, 114, 115, 118, 130–31, 135, 154, 166, 239, 254, 294, 312, 321; and CFBC, 175, 304, 307; CLAC meeting (1992), 316; and Florida Ship Canal, 78–87, 100; Silver Springs, 42
Ocala/Marion County Chamber of Commerce, 86, 119, 124, 155, 189, 191, 192, 206
Ocala National Forest, 276, 281
Ocala Star-Banner, 119, 136, 156, 172, 206, 228, 291, 304, 328
Ocklawaha Navigation Company, 38
Ocklawaha River, x, 1, 5, 6, 8, 9, 21, 34, 36, 137, 150, 167, 172, 183, 193, 196, 204, 210, 231, 253, 257, 258, 259, 272, 276, 277, 281–82, 294, 295; canal construction and, 144–45, 207; as canal route, 3, 18–19, 56, 57–58, 64, 72, 76,

Roosevelt, Franklin D., x, 2, 4, 5, 77–81, 85, 88, 89, 100, 108, 112–13, 116, 120–21, 128,131; authorization of barge canal, 124–27; Claude Pepper and, 104–6; Florida Ship Canal groundbreaking (1935), 80–81; funding ship canal, vacillation on, 92–93, 96–97; New Deal and ship canal, 64–74; 80–81

Roosevelt, Theodore, 51, 55, 182, 199, 248

Route 13-B, 68, 69, 85, 120, 144; description of, 3, 65

St. Johns Lock, 144, 145, 168, 184, 189, 207, 288; construction completion ceremony, 208–9

St. Johns River, x, 3, 8, 15, 18, 19, 22, 25, 44–45, 76, 137, 165, 257, 281, 296; dredging of, 47–48, 76

St. Johns River Water Management District, 317

St. Lawrence Seaway, 116, 120

St. Marys, Ga., 17, 44, 46, 47, 51, 55, 56; canal and, 57, 61

St. Marys River, 15, 45, 57

St. Petersburg Independent, 255, 257, 293

St. Petersburg Times, 83, 167, 188, 189, 258, 297–98, 316

Sanford, Fla., 90, 91, 113, 117, 122, 138

Sanford-Titusville canal, 138

Santos, Fla., 84–86; bridge stanchions, 96–97, 100, 107

Save Rodman Reservoir, Inc., 323, 327

Sayler, Henry, 293, 300

Scribner's, 30

Senate Committee on Appropriations, 126

Senate Committee on Commerce, 106, 108, 123

Shaw, Clay, 306, 310

Sheppard, Morris, 65, 67, 109

Shevin, Robert, 279, 281, 285–86, 294, 295, 299–300

Sholtz, David, 67, 96–96, 104

Sierra Club, 159, 216, 217–18, 232

Sikes, Bob, 134, 139, 142, 293, 295, 304

Silver Springs, 3, 22, 34–35, 76, 87, 144, 161, 165, 168, 210, 211, 245, 257, 269, 297; as steamboat vacation destination, 23–24, 26, 38, 42–43

Silver Springs Lock. *See* Dosh Lock

Simpson, Charles Torrey, 162–63

Slezas, Romas, 174

Smathers, George, 128, 134, 141, 143, 167, 272

Smith, Jim, 305

Smith, Martin Luther, 18–19, 20

Smith, Rod, 326

Somervell, Brehan, 68, 78–80, 84, 86, 91, 97

Southeastern Waterways, 100

Sports Illustrated, 214–16

SS Gulfamerica, 118

Staats, Elmer, 159

Steamboats, 12, 16, 23, 26–28, 38–43, 44

Stearns, Cliff, 312

Stone, Richard, 281–82, 294, 304

Stowe, Harriet Beecher, 26–27, 31, 246

Suez Canal, 106, 137, 188

Summerall, Charles, 62, 64, 101, 114, 122

Sunbelt, ix, 178–80

State University of New York (SUNY), Stony Brook, 215

Tabb, R. P., 157, 192

Talbot, Lee, 277–78

Tallahassee, Fla., 1, 43, 46, 135, 139, 147, 157, 164–67, 171, 178, 179, 182, 203, 214, 222–23, 223–24, 253, 265, 280, 285, 287, 294, 304, 312; cabinet meeting, December 1976, 297–300; CFBC deauthorization, political issues surrounding, 295, 297, 306; CLAC meeting, December 1992, 319–20; Rodman disposition, government disputes over, 319–20

Tallahassee Showdown (1966): actual meeting, 174–76; preparation for, 172–73; push for, 166–67, 170–71; ramifications of, 176–77, 178, 192, 203, 223–24, 257, 276–77, 281–82

Tampa, Fla., 5, 7, 18, 24, 56, 61, 80, 81; and "Missing Link," 188, 204; opposition to canal in, 73, 84, 122–23

Tampa Times, 61, 91

Tampa Tribune, 83, 117–18, 124, 127, 255, 256, 261

Taylor, Ed, 327

Tennessee-Tombigee Waterway, "Tenn-Tom," 272–74

Tennessee Valley Authority, 77, 272

Texas, 5, 50, 51, 55, 62, 65, 105, 109, 118, 120–21, 131–32, 133, 139, 189, 205, 275

Thompson, R. H., 38–39, 42

Time, 71, 93, 104, 113, 125–26, 131–32, 233, 234, 272

Timmerman, George, 87

Train, Russell, 234, 248–52, 274, 276; and CFBC, 262, 264–65, 272–73

Trammel, Park, 104

Truman, Harry, 128–31, 135

The Truth about Taxes, 116

Tydings, Millard, 112

University of Florida, 100, 147, 149, 200, 211, 218, 223, 224, 241, 246, 269, 286, 324

U.S. Army Corps of Engineers, x, 1, 2, 3, 9–10, 36–37, 42, 59, 61, 68, 138, 204, 273, 274, 299, 300, 302, 307, 323; late 1930s, 101, 102–3; 1940s, 114, 120–21, 127, 130; 1950s, 132–33; 19th-century surveys, 15–19, 44–48; 20th-century surveys, 55–58, 60;

U.S. Army Corps of Engineers—*continued*
benefit-cost ratios, 65, 68, 132–33, 145–46,
218, 230, 246–47; and CFBC, 142, 208–9; and
CFBC alternate route, 165–66, 257, 263, 276;
and CFBC construction, 143–45, 150, 157,
166–69, 182, 184–90, 207–9, 220, 253, 263;
and CFBC deauthorization, 311–12, 313–14,
316; on CFBC as national defense, 130, 135,
190; CFBC promotion, 153–54, 157–58, 182,
195, 199, 221, 239, 253–56, 288; on CFBC
recreational benefits, 192–95, 231, 245–46;
and CFBC restudy, 292–93, 296, 301–2; and
Crusher, 184–86; and FDE, 202–3, 219–24,
225, 236–38, 243–44; and Florida Ship Canal,
64, 68–69, 76, 97; and Florida Ship Canal
construction, 76–79, 82–84, 86, 97; and EDF/
FDE lawsuit, 217, 226–28, 230, 243, 261–62,
274–75, 279, 282–83; Nixon decision and,
251, 268–69, 286–87; opposition to canal,
responses to, 91, 146–47, 151, 152, 159–61,
166–67, 174, 176–77, 196, 230, 237, 257–58,
262, 293, 297; and Rodman Reservoir,
210–11, 213, 239, 284, 288–90, 301, 312; and
Rodman Reservoir drawdown, 288–91; and
Route 13-B, 65; saltwater intrusion issue
and, 87–90, 101, 145; on Vietnam War as
impediment to canal progress, 190, 206
U.S. Bureau of the Budget, 146
U.S. Bureau of Sports Fisheries and Wildlife,
250–51
U.S. Department of Commerce, 59
U.S. Department of Interior, 145, 288, 301
U.S. Department of War, 80, 93
U.S. Forest Service, 150, 162, 285, 288, 301
U.S. Geological Survey, 89, 101
U.S. Shipping Board, 59

Vandenberg, Arthur, 5, 102, 103, 105, 106, 128,
200, 272; Florida Ship Canal, opposition
to, 93–98; Floridians' support of, 108–9;
and World War II–period support for barge
canal, 112–14, 119, 121, 123;
Vernon, Robert, 146, 149
Veysey, Victor, 302
Vietnam War, 7, 190, 196–99, 205–6, 228, 232,
233, 260, 303

Vogt, Albertus, 34

Wall Street Journal, 62, 64
Wallace, George C., 273
Wallace, Henry, 124
Walls, Josiah, 44–45, 54
Walt Disney World, 2, 9, 220
Washington, D.C., 3, 12, 14, 16, 17, 43, 53, 55,
56, 62–64, 67, 71, 73, 78, 80,91, 99, 101, 104,
105–6, 107, 117, 130, 134, 139–40, 176, 180,
206, 226, 235, 237, 250, 261, 265, 269, 274, 276,
282, 283, 285, 288, 304, 309, 311
Washington Post, 96, 125, 131, 266, 272
Water Resources Congress (WRC), 275–76
Watt, James, 307
Webb, B. C., 82
Wetherell, Virginia, 322
Whitaker, John, 234, 248, 272, 276, 285; and
CFBC, 250, 262–67, 271
White, Joseph, 16
Wilcox, Mark, 104
Wilkie, Wendell, 116
Winter Park Garden Club, 92
Wisdom, Donald, 297, 299
Withlacoochee River, 3, 18–19, 21, 28, 29, 34, 44,
45, 46, 56, 58, 60, 64, 137, 144, 316,328; and
CFBC, 144–45, 151, 189, 255, 280; and Florida
Ship Canal, 65, 72, 76, 82; and phosphate
boom, 34–38
Works Progress Administration (WPA), 69–70,
78, 82, 84, 86, 100
World War II, 78, 114, 137, 139, 255; and CFBC,
117–28
Wurster, Charles, 215–16, 218, 226, 228

Yankeetown, Fla., x, 3, 61–62, 65, 80, 107, 268
Yannacone, Victor, 231, 261; EDF, departure
from, 228–29; EDF/FDE lawsuit, 226–28;
FDE, contact with, 216–20; legal philosophy,
214–16
Young, Bill, 295
Youngberg, Gilbert, 59–62, 64, 65, 71, 72, 100,
308
Yulee, David, 20, 21

Steven Noll is a senior lecturer at the University of Florida's Department of History. His research interests include environmental history, Florida history, and the history of disabilities. He is the author of *Feeble-Minded in Our Midst: Institutions for the Mentally Retarded in the South, 1900–1940* (1995) and editor of *Mental Retardation in America: A Historical Reader* (2005).

David Tegeder is associate professor of history at Santa Fe College, Gainesville, Florida. His research interests include the history of Southern race and labor relations and environmental history.

The Seminole Wars: The Nation's Longest Indian Conflict, by John and Mary Lou Missall (2004)

The Mosquito Wars: A History of Mosquito Control in Florida, by Gordon Patterson (2004)

Seasons of Real Florida, by Jeff Klinkenberg (2004)

Land of Sunshine, State of Dreams: A Social History of Modern Florida, by Gary Mormino (2005, first paperback edition 2008)

Paradise Lost? The Environmental History of Florida, edited by Jack E. Davis and Raymond Arsenault (2005, first paperback edition, 2005)

Frolicking Bears, Wet Vultures, and Other Oddities: A New York City Journalist in Nineteenth-Century Florida, edited by Jerald T. Milanich (2005)

Waters Less Traveled: Exploring Florida's Big Bend Coast, by Doug Alderson (2005)

Saving South Beach, by M. Barron Stofik (2005)

Losing It All to Sprawl: How Progress Ate My Cracker Landscape, by Bill Belleville (2006)

Voices of the Apalachicola, compiled and edited by Faith Eidse (2006, first paperback edition, 2007)

Floridian of His Century: The Courage of Governor LeRoy Collins, by Martin A. Dyckman (2006)

America's Fortress: A History of Fort Jefferson, Dry Tortugas, Florida, by Thomas Reid (2006)

Weeki Wachee, City of Mermaids: A History of One of Florida's Oldest Roadside Attractions, by Lu Vickers and Sara Dionne (2007)

City of Intrigue, Nest of Revolution: A Documentary History of Key West in the Nineteenth Century, by Consuelo E. Stebbins (2007)

The New Deal in South Florida: Design, Policy, and Community Building, 1933–1940, edited by John A. Stuart and John F. Stack Jr. (2008)

Pilgrim in the Land of Alligators: More Stories about Real Florida, by Jeff Klinkenberg (2008)

A Most Disorderly Court: Scandal and Reform in the Florida Judiciary, by Martin A. Dyckman (2008)

A Journey into Florida Railroad History, by Gregg M. Turner (2008)

Sandspurs: Notes from a Coastal Columnist, by Mark Lane (2008)

Paving Paradise: Florida's Vanishing Wetlands and the Failure of No Net Loss, by Craig Pittman and Matthew Waite (2008)

Embry-Riddle at War: Civilian and Military Aviation Training During World War II, by Stephen G. Craft (2009)

The Columbia Restaurant: Celebrating A Century of History, Culture, and Cuisine, by Andrew T. Huse, with recipes and memories from Richard Gonzmart and the Columbia Restaurant family (2009)

Ditch of Dreams: The Cross Florida Barge Canal and the Struggle for Florida's Future, by Steven Noll and David Tegeder (2009)